REGIONAL FLOOD FREQUENCY ANALYSIS

REGIONAL FLOOD FREQUENCY ANALYSIS

Proceedings of the International Symposium on Flood Frequency and Risk Analyses, 14–17 May 1986, Louisiana State University, Baton Rouge, U.S.A.

Edited by

VIJAY P. SINGH

Department of Civil Engineering, Louisiana State University, Baton Rouge, U.S.A.

D. REIDEL PUBLISHING COMPANY

A MEMBER OF THE KLUWER ACADEMIC PUBLISHERS GROUP

DORDRECHT / BOSTON / LANCASTER / TOKYO

Library of Congress Cataloging in Publication Data

CIP

International Symposium on Flood Frequency and Risk Analyses (1986: Louisiana State University, Baton Rouge)
Regional flood frequency analysis.

Includes indexes.
1. Floods—Congresses. I. Singh, V. P. (Vijay P.) II. Title.
GB1399.I58 1986c 551.48'9 87-20631
ISBN 90-277-2575-6
ISBN 90-277-2576-4 (set)

Published by D. Reidel Publishing Company,
P.O. Box 17, 3300 AA Dordrecht, Holland.

Sold and distributed in the U.S.A. and Canada
by Kluwer Academic Publishers,
101 Philip Drive, Assinippi Park, Norwell, MA 02061, U.S.A.

In all other countries, sold and distributed
by Kluwer Academic Publishers Group,
P.O. Box 322, 3300 AH Dordrecht, Holland.

Printed in The Netherlands

PREFACE

Floods constitute a persistent and serious problem throughout the United States and many other parts of the world. They are responsible for losses amounting to billions of dollars and scores of deaths annually. Virtually all parts of the nation--coastal, mountainous and rural--are affected by them. Two aspects of the problem of flooding that have long been topics of scientific inquiry are flood frequency and risk analyses. Many new, even improved, techniques have recently been developed for performing these analyses. Nevertheless, actual experience points out that the frequency of say a 100-year flood, in lieu of being encountered on the average once in one hundred years, may be as little as once in 25 years. It is therefore appropriate to pause and ask where we are, where we are going and where we ought to be going with regard to the technology of flood frequency and risk analyses. One way to address these questions is to provide a forum where people from all quarters of the world can assemble, discuss and share their experience and expertise pertaining to flood frequency and risk analyses. This is what constituted the motivation for organizing the International Symposium on Flood Frequency and Risk Analyses held May 14-17, 1986, at Louisiana State University, Baton Rouge, Louisiana.

The objectives of this symposium were therefore (1) to assess the current state of the art of flood frequency and risk analyses, (2) to demonstrate applicability of flood frequency and risk models, (3) to establish complementary aspects of seemingly different models, (4) to enhance interdisciplinary interaction, (5) to discuss practice of flood frequency and risk analyses technology by federal agencies in the U.S., (6) to discuss research needs in frequency and risk analyses, and (7) to determine directions for further research.

We received an overwhelming response to our call for papers. It was indeed a difficult task to select amongst the many excellent papers that were submitted, and we regret that we could not include all of them. The sole criterion for selection of a paper was its merit in relation to Symposium objectives. The subject matter of the Symposium was divided into 20 major topics encompassing virtually all facets of flood frequency and risk analyses. Each topic comprised of an invited state-of-the-art paper and a number of contributed papers. These contributions blended naturally to evolve a synthesized body of knowledge on that topic. Extended abstracts of all the invited and contributed papers were assembled in a pre-Symposium proceedings volume. This helped stimulate discussion and exchange of ideas during the Symposium.

The papers presented at the Symposium were refereed in a manner similar to that employed for publishing a journal article. As a result, many papers did not pass the review and were therefore eliminated from inclusion in the final proceedings. The papers contained in this book, REGIONAL FLOOD FREQUENCY ANALYSIS, represent one part of the Symposium contributions. The other parts are embodied in three separate books, FLOOD HYDROLOGY, HYDROLOGIC FREQUENCY MODELING, and APPLICATION OF FREQUENCY AND RISK IN WATER RESOURCES, which are being published simultaneously. Arrangement of these books under four different titles was a natural consequence of the diversity of technical material discussed in the papers. These books can be treated almost independently, although some overlap does exist between them.

This book contains five sections encompassing major aspects of regional flood frequency analysis. Each section starts usually with an invited state-of-the-art paper, followed by contributed papers. Beginning with an assessment of regional flood frequency analysis, the papers go on to discuss regional frequency for ungaged watersheds, comparison and selection of regional frequency methods, and paleohydrologic flood analysis.

The book will of of interest to researchers as well as those engaged in practice of Civil Engineering, Agricultural Engineering, Hydrology, Water Resources, Earth Resources, Forestry and Environmental Sciences. The graduate students as well as those wishing to conduct research in flood hydrology will find this book to be of particular value.

I wish to take this opportunity to express my sincere appreciation to all the members of the Organizing and Advisory Committees, and the Louisiana State University administration for their generous and timely help in the organization of the Symposium. A lack of space does not allow me to list all of them by name here. Numerous other people contributed to the Symposium in one way or another. The authors, including the invited keynote speakers, contributed to the Symposium technically and this book is a result of their efforts. The session chairmen administered the sessions in a positive and professional manner. The referees took time out from their busy schedules and reviewed the papers. Graduate students assisted in smooth conduct of the sessions. I owe my sincere gratitude to all of these individuals.

If the success of a Symposium is measured in terms of the quality of participants and presentations then most people would agree that this was a very successful Symposium. A very large number of internationally well-known people, who have long been recognized for their contributions and have long been at the forefront of hydrologic research, came to participate in the Symposium. More than 35 countries, covering the five continents and most of the countries of the world active in hydrologic research, were represented. It is hoped that long and productive personal associations will develop as a result of this Symposium.

March 1987
Baton Rouge, Louisiana

Vijay P. Singh
Symposium Director

ACKNOWLEDGMENTS

The International Symposium on Flood Frequency and Risk Analyses was sponsored and co-sponsored by a number of organizations. The sponsors provided financial support without which the Symposium might not have come to fruition. Their financial support is gratefully acknowledged. The co-sponsors extended their help in announcing the Symposium through their journals, transactions, newsletters or magazines. This publicity helped with attendance at the Symposium, and is gratefully acknowledged. The following is a list of Symposium sponsors and co-sponsors.

SYMPOSIUM SPONSORS

Louisiana State University
- Department of Civil Engineering
- Louisiana Water Resources Research Institute

National Science Foundation
U.S. Army Research Office
U.S. Geological Survey, Louisiana District, Baton Rouge
Woodward-Clyde Consultants, Baton Rouge

SYMPOSIUM CO-SPONSORS

American Geographical Society
American Geophysical Union
American Meteorological Society
American Statistical Association
American Society of Agricultural Engineers
American Society of Civil Engineers
American Water Resources Association
Association of American Geographers
Association of State Floodplain Managers
International Association for Hydraulic Research
International Association of Hydrological Sciences
International Association of Theoretical and Applied Limnology
International Commission on Irrigation and Drainage
International Geographical Union
International Water Resources Association
National Wildlife Federation
North American Lake Management Society
Pan American Institute of Geography
Society for Risk Analysis
Soil Conservation Society of America

TABLE OF CONTENTS

REGIONAL ANALYSIS OF FLOOD FREQUENCIES

Gary D. Tasker
U.S. Geological Survey
430 National Center
Reston, Virginia 22092

ABSTRACT. Recent advances in the index-flood method and regional regression method of regional flood frequency analysis are briefly summarized. Wallis and others have developed a new index-flood method which uses probability- weighted-moments and have shown it to be a viable means of improving at-site flood frequency estimates. Stedinger and others have developed and tested a means of accounting for unequal record length and cross correlation in a regional regression. They have shown that these methods can improve estimates of flood frequency at ungaged sites.

1. INTRODUCTION

Hydrologists are interested in more accurate flood frequency estimation because it is an economically important problem, and it is a very difficult problem. It is difficult because one often needs to estimate the magnitude and frequency of rare flood events with relatively short systematic records. It is important to accurately estimate magitude and frequency of relatively rare flood events because these events are the ones that cause greatest economic losses and loss of life. This has led hydrologists to search for flood records outside the at-site systematic records to improve their estimates. For example, Thomson and others (1964) obtain prerecord flood information from documents, newspaper accounts, and interviews with long-time residents, and Costa (1978), Baker (1983), and Sigafoos (1964) obtain prerecord flood information from examining sediment deposits and botanical evidence. These and other investigators attempt to improve at-site flood frequency estimates by adding "new" data to lengthen the existing systematic record at a site.

Another method of obtaining "new" data to aid in estimating flood frequency at a site is to use observed data at "similar" sites within a hydrologic region. This method of improving flood frequency estimates is called regional analysis of flood frequency and is the subject of the

V. P. Singh (ed.), Regional Flood Frequency Analysis, 1–9.

remainder of this paper. Regional analysis can be used to make flood frequency estimates at sites that have no flow records (ungaged sites) or can be used to improve estimates at sites with relatively short records. Two methods of regional analysis will be considered: (1) a regional regression that relates flood characteristics to physiographic and climatic characteristics through a least squares regression relationship, and (2) the index-flood method that uses a regional flood frequency curve that is normalized by an index flood (usually the mean annual flood). The recent advances in both of these methods will be discussed along with the applicability, advantages, and disadvantages of each. The final section discusses the areas where new research in regional analysis is most needed.

2. REGIONAL REGRESSION

One of the most used and useful methods of regional analysis of flood frequency is regional regression of flood characteristics on basin characteristics. Early studies used graphical methods of regression (Kinnison and Colby, 1944, and Potter, 1957). Benson's 1962 study used ordinary least squares regression and was the first of many such studies within the U.S. Geological Survey. McCuen and others (1977) provide an annotated bibliography of many of these studies. The Hydrology Committee of the U.S. Water Resources Council (1981) tested several methods of estimating peak flows at ungaged sites and found that the regional regression methods compared favorably with more complex and difficult to use watershed models.

Recently, Stedinger and Tasker (1985) proposed a new method of estimating the parameters of regional-regression models that takes into account the length of record available at gaged sites and the between-site cross correlation among concurrent flows. This new method uses a generalized least squares (GLS) estimator of the regression model's parameters. By contrast, hydrologists often use ordinary least squares (OLS) estimators which are appropriate and statistically efficient if the flow records are equally reliable (of equal length), the natural variability of flow at each site is the same, and concurrent flows at any pair of stations are independent. Stedinger and Tasker (1985) showed in a Monte Carlo simulation that the proposed GLS method (1) provided estimates of regression parameters with smaller mean square errors than the OLS estimates, (2) provided relatively unbiased estimates of the variance of the regression parameters compared to OLS, and (3) provided a more accurate estimate of the regression model error than OLS.

2.1. OLS-GLS Regression

The problem addressed here is how to estimate the parameters of a linear model specifying the value of the T-year flow event (such as the 100-year peak) as a function of various drainage basin characteristics (such as the drainage area, channel slope, and percent forest and lakes). The data available at each of N sites in the region are summarized by the vectors of basin characteristics, $\underline{x}_i$, and a streamflow record of n_i values

$$\left\{ y_{i,1},\ y_{i,2}, \ldots y_{i,n} \right\}$$

available for each site i. These may be the annual maximum flow or a transform thereof so that the y's are normally distributed. The normality assumption is adopted to facilitate this explanation. Other distributions, such as the log-Pearson III, could be employed with appropriate adjustments in the explanation that follows. The normal assumption worked well in Monte Carlo experiments reported by Stedinger and Tasker (1986) even when the actual distribution was Pearson III.

$$\text{Let } \hat{\underline{Y}} = (\bar{y}_1 + z_p s_1, \ \ldots, \bar{y}_N + z_p s_N)^T \tag{1}$$

be the vector of at site estimates of the (1/p)-year flow, where $\bar{y}_i$ and s_i are the usual sample estimates of mean and standard deviation of the flows at site i and z_p is the standard normal deviate with exceedance probability p. Also, let $\underline{Y} = (\mu_1 + z_p\sigma_1, \ \ldots, \ \mu_N + z_p\sigma_N)^T$ be the vector of true (1/p)-year flows, where μ_i and σ_i are the true mean and standard deviation of the $y_{i,t}$ at site i.

The ordinary least squares regional regression model can be written

$$\hat{\underline{Y}} = \underline{X}\underline{\beta} + \underline{e} \tag{2}$$

where $\underline{X}$ is an (Nxk) matrix of basin characteristics augmented by a column of ones, $\underline{\beta}$ is a (kxl) vector of regression coefficients, and $\underline{e}$ is an (N x 1) vector of random errors with expected values of zero and a covariance matrix which is assumed to be of the form $I_N\sigma^2$; here I_N is an N-dimensional identity matrix, and σ^2 is a constant. The OLS estimate of β is

$$\hat{\beta}_{OLS} = (\underline{X}^T\underline{X})^{-1}\underline{X}^T\hat{\underline{Y}} \tag{3}$$

The estimators' sampling covariance matrix, given the stated assumptions, is

$$\text{Var } (\hat{\underline{\beta}}_{OLS}) = \sigma^2 \ (\hat{\underline{X}}^T\hat{\underline{X}})^{-1}. \tag{4}$$

The OLS model is appropriate when all estimates of T-year events have equal variance and concurrent flows at different sites are uncorrelated. In general, these two conditions are not met because gaged records are of different length and concurrent annual floods are often cross correlated.

A more appropriate model for use in hydrologic regression is the generalized least-squares (GLS) model. In the GLS model, the assumptions of equal variance of the T-year events and zero cross-correlation for concurrent flows are relaxed. The covariance matrix for the errors in the GLS model are organized in a matrix Σ whose elements are

$$\Sigma_{ij} = \begin{cases} \gamma^2 + n_i^{-1}\,\sigma_i^2\,(1+0.5z_p^2)\ , & i=j \\ \rho_{ij}n_i^{-1}n_j^{-1}m_{ij}\ (1+0.5\rho_{ij}z_p^2)\ , & i\neq j \end{cases} \tag{5}$$

where ρ_{ij} is the estimated cross correlation between flood flows at sites i and j, m_{ij} is the number of concurrent years of record, and γ^2 is the model error variance. The model error variance is a measure of the precision of the true regression model and is defined by

$$\gamma^2 = E[(\underline{Y} - \underline{X\beta})^2] \tag{6}$$

The value γ^2 is assumed to be independent of $\underline{x}_i$, where x_i is the vector of basin characteristics at site i. The GLS estimate of β is

$$\underline{\hat{\beta}}_{GLS} = (\underline{X}^T\Sigma^{-1}\,\underline{X})^{-1}\,\underline{X}^T\Sigma^{-1}\underline{\hat{Y}} \tag{7}$$

This estimator has a sampling covariance matrix, given the assumptions, equal to

$$\mathrm{Var}(\underline{\hat{\beta}_{GLS}}) = (\underline{X}^T\Sigma^{-1}X)^{-1}. \tag{8}$$

The greatest barrier to using GLS methods has been that the value of the covariance matrix Σ is unknown because it depends on the population values of γ^2, σ_i, and ρ_{ij}. Stedinger and Tasker (1985) proposed use of a reasonable estmator of Σ. Their Monte Carlo simulations showed that use of that estimator of Σ in equation (7) led to an improved estimator of $\underline{\beta}$ (in terms of mean square error) and more accurate estimates of the model variance model γ^2, than did use of competing ordinary and weighted least squares procedures.

The method used to estimate Σ will be briefly described; Stedinger and Tasker (1985) provide a more detailed description. If Σ were known, then

$$E\left\{(\hat{Y} - \underline{X\hat{\beta}}_{GLS})^T\,[\Sigma(\gamma^2, \rho_{ij},\sigma_i)]^{-1}\,(\hat{Y} - \underline{X\hat{\beta}}_{GLS})\right\} = N - k \tag{9}$$

where Σ is written as $\Sigma(\gamma^2, \rho_{ij}, \sigma_i)$ to emphasize its dependence on γ^2, ρ_{ij}, and σ_i. The problem is to obtain estimates of γ^2, ρ_{ij}, and σ_i to enter into (5) to obtain an estimate of Σ.

Reliably estimating the cross correlation between flows at each pair of stations, ρ_{ij}, when only short concurrent records are available is a difficult task. It is best done subjectively using good hydrologic judgment. The goal of estimating the ρ_{ij}'s is to capture the essential underlying cross correlation structure among the streamflows within a region. In subjectively assigning values for ρ_{ij}, one should consider the distance between the two stations, and whether the stations are affected by the same rainfall events. One suggestion for estimating the ρ's is to classify each pair (i,j) of stations as likely to have a high, medium, or low cross correlation based on the sample cross correlation for the pair and hydrologic judgment. In classifying pairs of stations, one could consider the proximity of the stations, their drainage areas, any tributary relation between them, the type of storms that cause peak flows in each basin, and other factors.

At first, one may consider estimating the σ_i s in equation 9 by their usual sample estimate s_i. However, this leads to poor estimates of β (Stedinger and Tasker, 1985). A better approach is to first regress the s_i against basin characteristics and then use the regression estimate of $E(\sigma_i|x_i)$. Finally, γ^2 can be estimated by iteratively solving equation (9) to obtain an estimate, $\hat{\gamma}^2$, of γ^2 for the specified values of ρ_{ij} and $E[\sigma_i|x_i]$.

Some refinements in the procedure actually result in an algorithm which is slightly more involved than the procedure outlined here.

3. COMBINING AT-SITE AND REGIONAL REGRESSION ESTIMATES

The regional regression method has proved useful for estimating flood frequencies of ungaged sites. It can also be combined with an at-site estimate at a gaged site to "improve" the site-specific estimate. Several investigators including Shane and Garver (1970), Vicens, Rodriquez-Iturbe, and Schaake (1975), and Kucera (1982) have addressed the issue of how best to combine at-site and regional regression estimates. The Water Resources Council (1976, 1977, 1982) recommends combining these estimates by weighting inversely with their variances. It is the present policy of the U.S. Geological Survey to provide two flood frequency estimates at sites which have both gaged record and regional regression estimate: (1) an estimate based on the site-specific data and (2) an estimate made by combining the site-specific estimate with the regional regression estimate (U.S. Geological Survey Surface Water Branch Technical Memorandum No. 76.07).

If one assumes a relative quadratic loss (quadratic loss for estimators in log space), then the empirical Bayes estimator is approximately

$$y(\text{combined})=w*y(\text{at-site}) + (1-w)*y(\text{regression})$$

where y(.) is an estimator of a flow characteristic (such as the log of the 100-year peak) using the indicated method of estimation, and w is a weight given by:

$$w=v_p / (v_s+v_p)$$

where v_p is the variance of prediction of the regression model at the site in question and v_s is the sampling variance for y(at-site). If the GLS regression model of Stedinger and Tasker (1985) is used then v_p is easily calculated as

$$v_p = \hat{\gamma} + x_o(X^T \hat{\Sigma}^{-1}X)^{-1}x_o^T$$

where x_o is a vector of basin characteristics at site o. If one assumes annual peaks are distributed as a Log-Pearson III, v_s can be calculated as in Bobee (1975) or Hardison (1969). Note that Hardison (1971) and Stedinger and Tasker (1986) show that, on average, v_p is less than the squared standard error of prediction based on observed residuals for hydrologic regression models.

4. INDEX FLOOD

Another method of incorporating regional information in a flood frequency analysis is the index-flood method (Dalrymple, 1960). In the index-flood method observed annual peaks are first standardized by dividing each by their sample mean, then all the standardized observations within a region are used to estimate a dimensionless average frequency curve. To obtain an estimate of the T-year peak at a specific site the standardized T-year event is multiplied by the at-site mean annual peak. Thus, the index-flood method is able to combine at-site information (the sample mean annual peak) with regional information (the standardized T-year peak).

Wallis (1980) and Greis and Wood (1981, 1983) introduced the use of probability weighted moments (PWM) (Greenwood, Landwehr, Matalas, and Wallis, 1979) to the index flood method. This innovative technique calculates the PWM's at each site in a region from the standardized annual peaks, then the weighted regional average PWM's are used to compute the dimensionless average frequency curve. Several Monte Carlo studies (Hosking, Wallis, and Wood, 1985, Wallis and Wood, 1985, Lettenmaier, 1985, and Lettenmaier and Potter, 1985) have shown that the use of the index-flood method can improve the average mean square error of flood frequency estimates at the gaged sites in a region when compared to other methods such as the at-site WRC Bulletin 17B method. The index-flood method is very efficient when the region is homogeneous except for differences in mean annual peaks. The referenced Monte Carlo studies show that it also works well when there are mild deviations from this kind of homgeneity within a region. The advantage of the index-flood method are that it is simple, it uses only flow data, and it incorporates both at-site information in the form of the at-site sample mean and regional information in the form of the average frequency curve. The disadvantage of the method is that one must accept that the regional average frequency curve applies to all sites within the region. The index-flood method can also be used at ungaged sites if one can estimate the mean annual flood at the site by some means such as a regional regression equation.

5. CONCLUDING REMARKS

Both the index-flood method and the GLS regression method assume that one can define a "reasonably homogeneous" hydrologic region. To the index-flood method a homogeneous region is one in which the standardized frequency curves are the same for every site in the region. To the regression method a homogeneous region is one in which quantiles of the frequency curves are linearly related to some known basin characteristics. In either case one needs to define a region. The usual procedure is to use subjective methods to define geographical boundaries. Some investigators, including DeCoursey (1973), Mosley (1981), Tasker (1982), and Gottschalk (1985), have used multivariate analysis techniques such as cluster analysis, discriminant (classification) analysis, and principal components analysis. More work needs to be done in the area of defining "homogeneous" regions. Monte Carlo studies by Wallis and Wood (1985) indicate that while regional analysis improves the mean square error on average, some specific sites in a region may be adversely affected by regionalization. If one is interested in a specific site, is there some means of deciding whether one should use a regionaliation method at that site and, if so, which method whould be used? This is another problem that needs to be addressed in future studies.

6. REFERENCES

Baker, V.R., 1983, 'Paleoflood hydrologic techniques for the extension of streamflow records.' Transportation Research Record 922, Transportation Research Board, Washington, D.C., 18-23

Benson, M.A., 1962, 'Evolution of methods for evaluating the occurrence of Floods.' U.S. Geological Survey Water-Supply Paper 1580-A, 30p.

Bobee, B., 1973, 'Sample error of T-year event computed by fitting a Pearson Type 3 distribution.' *Water Resources Research*, Vol. **9**, No. 5, pp 1264-1270.

Costa, J.E., 1978, 'Holocene stratigraphy in flood-frequency analysis.' *Water Resources Research*, Vol. **14**, pp 626-632.

Dalrymple, T., 1960, Flood-frequency analysis. U.S. Geological Survey Water-Supply Paper 1543-A, 80 p.

DeCoursey, D.G., 1973, 'Objective regionalization of peak flow rates' in: *Floods and Droughts*, E.F. Koelzer, V.A. Koelzer, and K. Mahmood (editors) Proceedings of the Second International Symposium in Hydrology, Fort Collins Colorado, Water Resources Publications, Fort Collins, Colorado, pp 395-405.

Gottschalk, L., 1985, 'Hydrological regionalization of Sweden.' *Hydrological Sciences Journal*, Vol. **30**, No. 1,3, pp 65-83.

Greenwood, J.A., Landwehr, J.M., Matalas, N.C., Wallis, J.R., 1979, 'Probability weighted moments: Definitions and relation to parameters of several distributions expressable in inverse form.' *Water Resources Research*, Vol. **15**, No. 5, pp 1049-1054.

Greis, N.P. and E.F. Wood, 1981, 'Regional flood frequency estimtion and network design.' *Water Resources Research*, Vol. **17**, No. 4, pp 1167-1177.

Greis, N.P. and Wood, E.F., 1983, 'Correction to regional flood frequency estimation and network design.' *Water Resources Research*, Vol. **19**, No. 2, pp 589-590.

Hardison, C.H., 1969, 'Accuracy of streamflow characteristics.' *U.S. Geolocical Survey Prof. Paper 650-D*, pp D210-D214.

Hardison, C.H., 1971, 'Prediction error of regression estimates at ungaged sites.' *U.S. Geological Survey Prof. Paper 750-C*, pp C228-C236.

Hosking, J.R.M., Wallis, J.R., and Wood, E.F., 1985, 'An appraisal of the regional flood frequency procedure in the UK Flood Studies Report.' *Hydrological Sciences Journal*, Vol. **30**, No. 1,3, pp 85-109.

Hydrology committee, 1981, Estimating peak flow frequencies for natural ungaged watersheds: A proposed nationwide test. U.S. Water Resources Council, Washington, D.C., 346 p.

Kinnison, H.B., and Colby, B.R., 1944, 'Flood formulas based on drainage basin characteristics.' *Proceedings American Society of Civil Engineers*, Vol. **70**, No. 3, pp 267-294.

Kuczera, G., 1982, 'Combining site-specific and regional information: An empirical Bayes approach.' *Water Resources Research*, Vol. **18**, No. 2, pp 306-314.

Lettenmaier, D.P., 1985, 'Regionalization in flood-frequency analysis: Is it the answer?' *Proceedings U.S.-China Bilateral Symposium on the Analysis of extraordinary Flood Events*, Nanjing, China, October 1985.

Lettenmaier, D.P., and Potter, K.W., 1985, 'Testing flood frequency estimation methods using a regional flood generation model.' *Water Resources Research*, Vol. **21**, No. 12, pp 1903-1914.

McCuen, R.H., Rawls, W.J., Fisher, G.T., and Powell, R.L., 1977 'Flood flow frequency for ungaged water sheds: A leterature evaluation.' *Agricultural Research Service* ARS-NE-86, 135 p.

Mosley, P., 1981, 'Delimination of New Zewland hydrologic regions.' *Journal Of Hydrology*, Vol. **49**, pp 173-192.

Potter, W.D., 1957, Peak rates of runoff in the New England hill and lowland area. U.S. Department of Commerce, Washington, D.C.

Sigafoos, R.S., 1964, Botanical evidence of floods and flood plain development. U.S. Geological Survey Prof. Paper 485-A, 35 p.

Shane, R.M., and Garver, 1970, 'Statistical decision theory techniques for the revision of mean flood flow regression estimates.' *Water Resources Research*, Vol. **6** No. 6, pp 1649-1654.

Stedinger, J.R., and Tasker, G.D., 1985, 'Regional hydrological analysis 1: Ordinary, weighted, and generalized least squares compared.' *Water Resources Research*, Vol. **21**, No. 9. pp 1421-1432.

Stedinger, J.R., and Tasker, G.D., 1986, 'Regional hydrological analysis 2: Model error estimators, estimation of sigma, and Log-Pearson type 3 distributions.' *Water Resources Research*, Vol. 22, no. 10, pp 1487-1499.

Tasker, G.D., 1982, 'Comparing methods of hydrologic regionalization.' *Water Resources Bulletin*, Vol. **18**, No. pp 965-970.

Thomson, M.T., Gannon, W.B., Thomas, M.P., and Hayes, G.S., 1964, Historical floods in New England, U.S. Geological Survey Water-Supply Paper 1779-M, 105 p

Vicens, G.J., Rodriguez-Iturbe, I., and Schaake, J.C., Jr. 1975, 'A Bayesian framework for the use of regional information in hydrology.' *Water Resources Research*, Vol. **11**, No. 3, pp 405-414.

Wallis, J.R., 1980, Risk and uncertainities in the evaluation of flood events for the design of hydrologic structures presented at Seminar on Extreme Hydrological Events: Floods and Droughts, Erice, Italy, March 22-25, 1980.

Wallis, J.R., and Wood, E.F., 1985, 'Relative accuracy of Log Pearson III procedures.' *Journal of Hydraulic Engineering*, Vol. **111**, No. 7.

Water Resources Council, 1976, Guidelines for determining flood flow frequency. Water Resources Council Bulletin 17, Washington, D.C.

Water Resources Council, 1977, Guidelines for determining flood flow frequency. Water Resources Council Bulletin 17A (revised), Washington, D.C.

Water Resources Council, 1982, Guidelines for determining flood flow frequency. Water Resources Council Bulletin 178 (revised), Washington, D.C.

REGIONAL FLOOD DISCHARGE ANALYSIS SOUTHWEST REGION OF THE KINGDOM OF SAUDI ARABIA

Ali Unal Sorman and Mohamed J. Abdulrazzak
Department of Hydrology and Water Resources Management
Faculty of Meteorology, Environment and Arid Land Agriculture
King Abdulaziz University, P.O. Box 9034
Jeddah 21413. Kingdom of Saudi Arabia

ABSTRACT. This study deals with the regional analysis of flood characteristics for the wadis of the southwestern region of Saudi Arabia which drains towards the Red Sea. Nine gauged wadis are selected and their flood discharge parameters are determined. Pearson type III and Extreme type I distributions are found to be the best representative probability distributions. The flood magnitudes with various recurrence intervals are estimated using both distribution functions. The ratios of estimated peak floods for various recurrence are computed to derive the regional frequency values for each of the subregions. Further, the 2 year annual flood discharges are statistically correlated with hydrologic, morphometric and soil data parameters so that the maximum flood discharges can be estimated for ungauged wadis with similar climatic and hydrologic conditions. The area, mean annual rainfall, infiltration area, relative relief, and mean effective hydraulic conductivity values are found to be the most significant parameters affecting the annual 2 year flood discharges.

1. INTRODUCTION

Saudi Arabia is located in a semi-arid region of the world. The Kingdom, therefore, is generally characterized by low and sporadic rainfall of short duration and high intensity, with the exception of the southwestern part. This region usually experiences seasonal rainfall that can be as high as 600 mm and the sharp reliefs of the terrain induce extreme flood magnitudes.

Efficient utilization of the surface water generated, as well as reduction of frequent flood damage, depends heavily on flood estimation and forecasting. Flood discharge estimation using the frequency analysis method can assist in the efficient design of hydraulic structures and can result in averting flood related disasters.

In this study, the regional flood discharge analysis of the western wadis in the southwest zone of the Kingdom is presented. The study area contains about 18 main wadis covering an area of approximately 30×10^3 km^2 and receives about 10×10^9m^3 of rainfall annually (Ministry

V. P. Singh (ed.), Regional Flood Frequency Analysis, 11–25.

of Agriculture and Water - MAW, 1963∿1985). All of the wadis drain towards the Red Sea, and the amount of surface runoff generated in this area is estimated to comprise 40% of the water resources for the entire country.

2. BACKGROUND

The southwest zone of the Kingdom has two distinct areas; the western and eastern slopes, which are separated by the escarpment ridge of the Asir mountains, Figure (1). The western zone encompasses a coastal trip between the Red Sea and the mountain range. The upstream sections of the western slope wadis have very steep topography near the escarpment and, consequently, the runoff is carried by incised wadis of limited area. The surface materials in the mountains are largely impermeable, thus, groundwater resources at the upstream sections are negligible. As the drainage system enters the foothills, the topography flattens, and the alluvial wadi beds widen. The shallow alluvial deposits begin to hold significant amounts of groundwater. Figure (1a) illustrates the topographic sections of the drainage system in the western zone, and Figure (1b) shows the isohyetal map of annual precipitation for the area. This map indicates the geographic variability over the western and eastern slopes of the zone, as well as showing the location of the 18 wadis.

Most of the surface runoff stations are located at the foothill side of the wadis. Most of the runoff records vary from 1967 to the present, as presented in MAW (1963∿1985) publications, with the exception of few stations which were in operation since 1953. The common length of record is selected as 1970-83. In some wadis, it is not possible to provide an accurate estimate of the natural surface runoff because some water is diverted for agricultural use, or the wadi runoff is measured in only one of the branches.

According to the hydrological similarities of the flow regimen, the main wadis in the study area are grouped into three sub-regions: Region I contains the Jizan plain, Region II comprises the Asir mountains and the Al Birk plateau, and Region III covers the Tihamah Al Sham plain.

Region I, the Jizan plain, is located between the Yemen frontier in the south and wadi Baysh in the north. There are eight main wadis, namely: Liyyah, Khulab, Khums, Jizan, Damad, Sabya, Shahdan, and Baysh. Some of the wadis have records of varying lengths, however only four had records of sufficient length for the purpose of this study. The drainage areas of the wadis cover approximately 11000 km^2 and the area represents the Kingdom's most active hydrological regimen. The mean annual precipitation and surface runoff is estimated to be 374 mm and 251 MCM respectively (Development Analysis Associates-DAA, 1980; Italconsult, 1969; MAW, 1963∿1985; Sogreah, 1968). The relatively large storage capacity of the alluvial adds to the area's potential for future water resources development.

Region II, the Asir mountains and Al Birk plateau, contains three major wadis: Itwad in the south, Hali in the middle, and Yiba in the

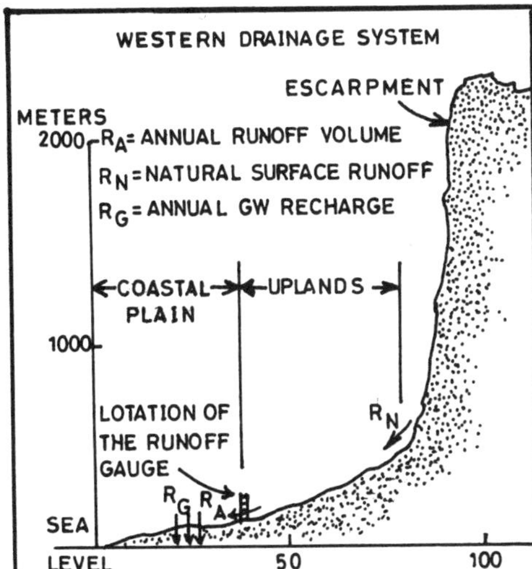

Topography of the Southwest Zone

FIGURE (1a) TOPOGRAPHIC SECTION OF THE ZONE.

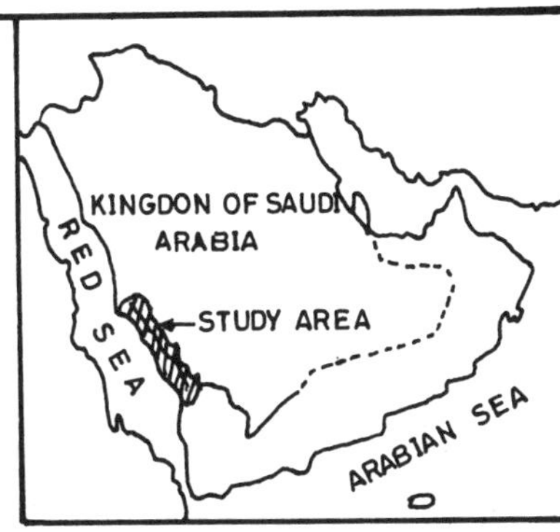

LOCATION MAP OF THE STUDY AREA

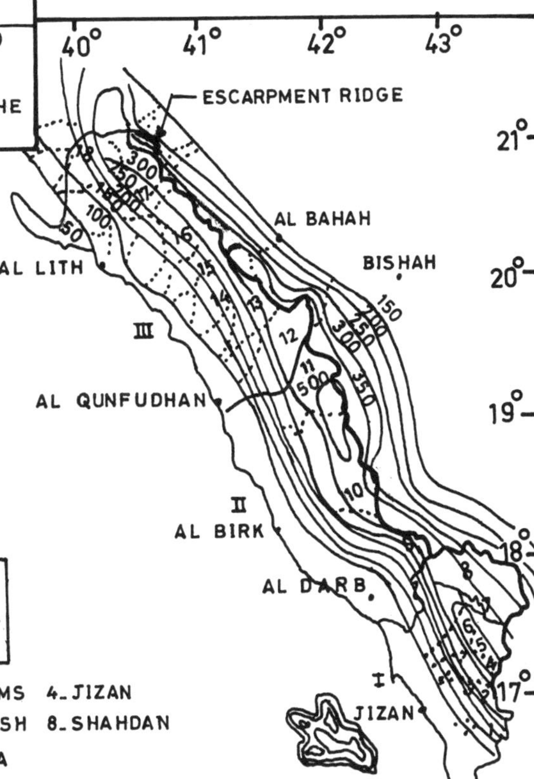

LEGEND:

- INTERNATIONAL BOUNDARY.
- WATERSHED REGION BOUNDARY.
- ISOHYET.

REGION I { 1. LIYYAH 2. KHULAB 3. KHUMS 4. JIZAN 5. DAMAD 6. SABYA 7. BAYSH 8. SHAHDAN

REGION II { 9. ITWAD 10. HALI 11. YIBA

REGION III { 12. QANUNAH 13. AHSIBAH 14. DAWQAH 15. S. YAMANIYAH 16. S. SHAMANIYAH 17. AL. LITH 18. SADIYAH

FIGURE (1b) ISOHYETAL MAP AND WATERSHED BOUNDARIES OF MAIN WADIS, AFTER, DAA (1980)

Table I - A quantitative description of the main wadis

Wadis	Area (Km^2)	Location N	Location E	Precipitation ($10^6 m^3$)	Runoff station No.	Observed mean annual discharge (MCM)	Operation	Location N	Location E
Region I (Jizan Plain)									
1- Liyyah	1200	16°34'	42°55'	296.3	SA 425		(1984-)	16°41'	43°05'
2- Khulab	900	16°38'	42°44'	403.7	SA 421	37.15	(1969-)	16°43'	43°01'
3- Khums	500	16°47'	42°44'	129.6	SA 420		(1967-68)		
4- Jizan	1200	16°57'	42°33'	564.0	SA 418	60.53	(1953-55,58-)		
5- Damad	1000	16°59'	42°33'	462.2	SA 417	41.29	(1959-)	17°09'	42°53'
6- Sabya	660	17°03'	42°27'	275.4	-				
7- Shahdan	730	17°11'	42°34'	232.4	-				
8- Baysh	4700	17°06'	42°25'	1713.2	SA 415	72.47	(1959-65,68-)	17°35'	42°37'
Sub total:	10980 Km^2			$4076.8x10^6$					
Region II (Al Birk Plateau)									
9- Itwad	1350	17°34'	42°09'	410.3	SA 414	3.05	(1967-74,78-81)	17°48'	42°21'
10- Hali	4500	18°36'	41°18'	1729.8	SA 411	85.67	(1967-)	17°46'	41°35'
11- Yiba	2700	18°51'	41°13'	1072.7	SA 401	21.70	(1969-)	19°17'	41°27'
Sub total:	8550 Km^2			$3212.8x10^6$					
Region III (Tihamat Al Sham Plain)									
12- Qanunah	1962	19°08'	41°05'	680.6	J-413		(1983-)	19°42'	41°41'
13- Ahsibah	1196	19°18'	40°58'	390.0	J-412		(1983-)	19°36'	41°19'
14- Al Duqah	970	19°34'	40°50'	211.2	J-404	8.24	(1968-)	19°45'	41°02'
15- S.Yamaniyah	996	19°42'	40°48'	246.6	J-414		(1983-)	20°04'	40°50'
16- S.Shamiyah	1413	19°48'	40°39'	358.9	-				
17- Al Lith	3013	20°09'	40°16'	643.9	J-409		(1971-73)	20°08'	40°11'
18- Sadiyah	1420	20°35'	39°40'	300.8	J-407	1.15	(1970-77)	20°44'	39°55'
Sub total:	10930 Km^2			$2832.6x10^6 m^3$					
Grand total:	30370 Km^2			$10121.2x10^6 m^3$					

Table II - Mean annual runoff discharges

	Gauged basin	Area A(Km)	Observed mean annual: Short term mean	Observed mean annual: 1970-83 mean	Observed mean annual: Long term mean	Available net surface run-off[(3)] $E(R_A)$ $10^6 m^3$	Annual surface run-off[(4)] $R_A(50\%)$ $10^6 m^3$
I	Khulab	900	38.79	37.15		26.8	24.1
	Jizan	1200	-	60.53	54.76	33.7	30.0
	Damad	1000	-	41.29	32.52	23.8	21.5
	Baysh	4700	-	72.47	68.16	102.5	93.0
II	Itwad	1350	[(1)] 3.05	[(2)] 24.30	-	24.3	21.8
	Hali	4500	93.00	85.67	-	84.1	76.0
	Yiba	2700	25.66	21.70	-	51.2	46.0
III	Al Duqah	970	8.20	8.24	-	10.3	9.3
	Sadiyah	1420	0.67	1.15	-	14.6	12.9

NOTES (1) The gauging station may be measuring runoff of only one of wadi branches, where wadi divides into several channels separated by sand banks as reported in DAA (1980).

(2) Extended mean annual $\bar{Q}$, using runoff stations in wadi Baysh (SA 415) and wadi Hali (SA 411).

(3) $E(R_A) = m_\nu\ E(B)\ A\ P\ (h>h^*)$, where

$E(R_A)$ = expected annual surface runoff in (MCM)

m_ν = mean number of storms per year

$E(B)$ = expected value of runoff index (mean of storm runoff per unit area (mm)

A = catchment size (Km^2)

$P(h>h^*)$ = probability of obtaining a storm of depth $h>h^*$ (which produces runoff)

(4) When ν and B are uncorrelated approximation of the above equation for annual surface runoff is $R_A = \nu\ B\ A\ P\ (h\ h^*)$ where $R_A(50\%)$ is the median value of the annual runoff volume being exceeded within 2 year return period

Call ν, B as a new random variable (z) and $A\ P\ (h>h^*) = D$ then the CDF of z is $CDF\ (z<a) = CDF\ (R_A<a.D)$.

The parameter a corresponds to the different values that the variable z can take.

north. The total estimated drainage area is approximately 8550 km^2. The upstream catchments are located in the Asir mountains which receive significant amounts of rainfall, particularly the Itwad catchment. The mean annual precipitation depth is estimated as 376 mm and the mean annual surface runoff is 160 MCM (DAA, 1980; Italconsult, 1969; MAW, 1963∿1985; Sogreah, 1968). Surface water is found to be relatively abundant, however groundwater availability is limited by shallow alluvial deposits. The aquifers are confined to the wadi beds and the alluvial fans.

Region III, the Tihamat Al Sham plain, includes seven main wadis which extend from wadi Qanunah in the south, to wadi Sadiyah in the north near the city of Jeddah. Rainfall in this region is lower than in the other two regions, and as a result, water resources are scarcer. The mean annual runoff is estimated to be 137 MCM (DAA, 1980; Italconsult, 1969; MAW, 1963∿1985; Sogreah, 1968). The mean precipitation is estimated to be 259 mm. Groundwater availability in the upper portion of the region is confined to the narrow channels of the wadis which are carved into the underlying impermeable formations. The alluvial aquifers increase in thickness in the downstream direction, so the capacity for water storage is increased, but groundwater utilization is limited by salt water intrusion.

A quantitative description of the main wadis of the study area is presented in Table I. The observed mean annual runoff records for each selected gauged basin is presented in Table II, and values are grouped under three categories: short term, long term, or standard, covering the period from 1970-1983 (MAW, 1963∿1985). Additionally, the findings from the most recent study (DAA, 1980) which were based on only a few years of observed runoff records are presented for reference. This background information will be used at a later stage in comparing the findings of this study with others, particularly the Eagleson's model presented in (DAA, 1980; Eagleson, 1978) for gauged and ungauged wadis.

3. DETERMINATION OF MEAN ANNUAL DISCHARGE

3.1 Collection and Extension of Hydrologic Data

In order to carry out complete and thorough analysis it was necessary to establish records of equal length for all stations. Review of the runoff records of the wadis to be modelled revealed than the common length for most of them was 14 years (1970-1983). However, the runoff stations at some of the wadis such as Itwad, Hali, Yiba, and Sadiyah had short term records that had to be extended using long term records for stations on the same wadi or nearby basins. It was found that the correlation coefficient varied from 0.70 to 0.80. For modelling purposes, watershed parameters, either measured or collected, are presented in Table III.

3.2 Multivariate Regression Analysis

Based on the data available and the length of record, several multi-

Table III - Characteristics of the main wadis in the western slope

		Mean annual prec. $Px10^6m^3$ (2,4,9)	Hydraulic cond. K(m/hr) (2,4,9)	Infilt. area $A_I(Km^2)$ (2)	Drain area $A(Km^2)$	Main stream length L(Km)	Relief ΔH(m)	Derived parameters: Soil $K*A_I(10^6)$	Derived parameters: Topog. S=ΔH/L
1	Liyyah	296.3	*0.168∿ 0.108	14.7	1200	39.50			
2	Khulab	403.7	0.108	6.2	900	63.00	301	0.6696	0.0048
3	Khums	129.6	0.072	16.7	500	14.64			
4	Jizan	564.0	0.020	25.0	1200	72.00	980	5.0000	0.0128
5	Damad	462.2	0.360	20.0	1000	71.70	1075	7.2000	0.0150
6	Sabya	275.4	0.360	12.9	660	51.14	2100		
7	Shahdan	232.4	1.800	7.2	730				
8	Baysh	1713.2	1.800	45.0	4700	117.10	2100	81.0000	0.0180
1	Itwad	410.3	7.200	5.8	1350	58.60	2050	41.7600	0.0350
2	Hali	1729.8	1.440	12.9	4500	117.10	2522	18.5800	0.0215
3	Yiba	1072.7	0.720∿ *0.600	10.6	2700	99.50	1520	7.6300	0.153
1	Qanunah	680.6	0.720∿ *0.600	2.5	1962	106.73			
2	Ahsibah	390.0	"	8.0	1196	70.23			
3	Duqah	211.2	0.720	1.4	970	65.78	320	1.0100	0.0049
4	S.Yamaniyah	246.6	0.720	2.0	956	60.00			
5	S.Shamiyah	358.9	0.720	8.0	1413	76.20			
6	Al Lith	643.9	25.200∿ *12.200	9.2	3013	98.00			
7	Sadiyah	300.8	7.200	6.2	1420	73.20	2130	44.6400	0.0283

$A_I(Km^2)$ values are total infiltration areas over the wadi beds and numbers in paranthesis refer to references

*K, conductivity values determined by newly operated observation wells reported in DAM(1985). Rest of the conductivity values are obtained from previous studies done by Sogreah (1968) etc.

variate regression models are developed and tested after collection and transformation of the variables for the nine gauged wadis: Jizan, Damad, Baysh, Itwad, Hali, Yiba, Al Duqah, and Sadiyah. The selected transformed variables, one of which is presented as dependent ($\bar{Q}$), and the other five variables are independent (AP, K, KA_I, L, and $\Delta H/L$). The formulated models are grouped under three categories depending upon the number of the independent variables used as m = 1, 2, or 3. Each model of more than one parameter includes also several parameters combinations out of five as a, b, or c. As a result of this study six models are selected for presentation in Table IV, which have either single or multiple parameters. The regression parameters, (A, B_1, B_2, B_3), standard error of estimate for residuals (Se) and the correlation coefficient (R) are determined and presented in Table IV. If the model selection is based on criteria of the high correlation coefficient, with least estimate of standard error of residual, the result would be the selection of either model (2a) or (3c), depending on the number of available parameters.

$$\bar{Q} = 0.0108\ (AP)^{1.204}\ (K)^{-0.459} \quad \ldots \quad \text{Model (2a)}$$

$$\bar{Q} = 0.020\ (AP)^{1.086}\ (K)^{-0.60}\ (S)^{0.39} \quad \ldots \quad \text{Model (3c)}$$

In the given models, the volume of annual precipitation (AP) plays a predominant role, with an exponential of 1.204 and 1.086. On the other hand, the hydraulic conductivity (K) inversely affects the annual discharge with the power of -0.46 to -0.60. Relative relief plays a tertiary role in the formulation of the model which may vary from 0.005 in flat areas to 0.035 in some steep wadis. Actually, the estimates in standard error of residuals vary due to chance, and due to unexplained differences in basin characteristics. In general, the size of catchment, mean annual precipitation, main channel slope and percent of area cover, as well as soil type are found in the literature (Riggs, 1973) as main model parameters producing a standard error very close to the practical minimum.

The predicted mean annual discharges, using some of the previously formulated models for the gauged wadis, are presented in Table V. The sums of the predicted mean annual values range from 306.6 MCM to 335.2 MCM. The total sum of the long term observed records for the same wadis provided 331.9 MCM (MAW, 1963∿1985). Although for some of the basins, the observed records do deviate from the predicted ones, the deviation from some of the developed models is small enough to enable their use.

In predicting the mean annual discharges of the ungauged wadis, model (2a) is used due to the fewer number of parameters and the least standard error of residuals. The predicted values for the ungauged wadis are tabulated in Table V along with the estimated annual runoff values for 50% probability of exceedance determined by other investigators (DAA, 1980).

Table IV - Formulation of models

No. of para-meters		Model parameters	A Cons	B1	B2	B3	Se	R	Model	Model #
m=1		$Y=\bar{Q}\rightarrow(AP)$	-4.31	1.17	-	-	22.05	0.71	$\bar{Q}=0.0134(AP)^{1.175}$	(1)
m=2	a	$Y=\bar{Q}\rightarrow(AP,K)$	-4.53	1.20	-0.45	-	18.92	0.78	$\bar{Q}=0.0108(AP)^{1.204}(K)^{-0.459}$	(2a)
	b	$Y=\bar{Q}\rightarrow(AP, KA_I)$	-6.00	1.57	-0.38	-	26.54	0.61	$\bar{Q}=0.0025(AP)^{1.571}(KA_I)^{-0.3778}$	(2b)
m=3	a	$Y=\bar{Q}\rightarrow(AP, KA_I, \Delta H/L)$	-6.00	1.60	-0.52	0.36	32.80	0.44	$\bar{Q}=0.0025(AP)^{1.60}(KA_I)^{-0.518}(S)^{0.357}$	(3a)
	b	$Y=\bar{Q}\rightarrow(AP, K, L)$	41.71	2.80	-0.37	-.500	21.51	0.81	$\bar{Q}=1.30(AP)^{2.81}(K)^{-0.375}(L)^{-5.0}$	(3b)
	c	$Y=\bar{Q}\rightarrow(AP, K, \Delta H/L)$	-3.93	1.09	-0.60	0.39	20.19	0.83	$\bar{Q}=0.02(AP)^{1.086}(K)^{-0.60}(S)^{0.39}$	(3c)

Table V - Comparison of observed ($\bar{Q}_{ob}$) and

A) Predicted $(\hat{Q})_p$ for selected wadis, and

B) Estimation of annual surface water for western part of southwestern region of Saudi Arabia

Gauged wadis	Observed $\bar{Q}x10^6m^3$	Predicted $\hat{Q}x10^6m^3$			
Parameters		Model(1) (AP)	Model(2a) (AP),(K)	Model(3b) (AP,(K),(L)	Model 3(c) (AP),(K),(S)
SA 421	37.15	15.49	41.26	59.50	38.20
SA 418	60.53	22.96	46.52	62.05	55.65
SA 417	41.29	18.17	27.97	29.08	33.52
SA 415	72.47	81.45	64.65	53.62	56.77
SA 414	24.30	15.80	6.12	18.56	6.77
SA 411	85.67	85.71	72.53	59.98	70.32
SA 402	21.70	48.86	56.09	46.06	55.59
J-404	8.24	7.24	7.92	3.83	6.07
J-407	1.15	10.96	4.21	2.54	4.44
Period (1970-83)	$352.50x10^6$	306.64	327.27	335.22	327.33
Σlong term	$331.90x10^6$				
Un-gauged	$R_A(50\%)$*				
Liyyah	20.9		23.17		
Khums	8.8		12.65		
Sabya	15.2		14.98		
Shahdan	13.3		5.83		
Qanunah	29.0		32.43		
Ahsibah	16.8		16.58		
S.Yama-niya	10.6		9.54		
S.Sham-aya	15.0		15.00		
Lith	28.5		8.25		
Sub Total	158.1		143.71		
TOTAL	$492.7x10^6$		$470.98x10^6$		

(*) See footnote #4 in Table II for estimated runoff with 50% probability of exceeding $R_A(50\%)$ determined by Eagleson's Model.

4. REGIONAL FREQUENCY STUDIES

There are several procedures of regional frequency analysis, but most of these depend on information transfer depending on length of record and number of recording runoff stations. The methods discussed in the literature concerning streamflow data usually follow the extreme type 1 probability function. Nevertheless, the index flood method, described by Dalrymple (1960) is the one most commonly used by U.S. Geological Survey.

Alternative regional analysis procedures proposed by other investigators (Kite, 1977; Riggs, 1973) use some other probability functions such as log-normal or Pearson type III. Whichever method is used, the statistical parameters of the distribution functions must be determined.

4.1 Determination of Statistical Parameters

The frequency curves for flood discharge at the gauged wadis are obtained by fitting the theoretical frequency distribution to data and then the deviations are described by the computed parameters of the theoretical distribution. For that purpose the statistical parameters for the normal and transformed data are computed through computer packages and the results are presented in Table VI. Furthermore, these parameters require two adjustments due to length of record and outliers (highest observed records). In this study, ten years is considered to to be the minimum record length. No extension of records is intended due to low cross correlations between the flood records at short and long-term stations. Secondly, an adjustment is required in the determination of the parameters due to handling of outliers with or without considering them. For region I and III in the study area, outliers are kept in the data series, because the extreme flood records were within the product of the second largest flood with the square root of the ratio between the 2nd and the 8th largest extreme values. But in Region II, for recording stations SA 401, SA 411, and SA 414 they are disregarded in the second computer run to get more uniformly distributed sample parameters.

4.2 Selection of Probability Density Functions

Six probability distribution functions, namely, truncated normal (TN), log-normal distribution with two and three parameters (LN2, LN3), type I extreme (Type 1) and Pearson and log Pearson type III (PT3, LPT3) are fitted using adapted computer programs packages. The sum of squares of the differences between computed values and observed discharges at specific return periods are determined. The selection of the best fitted distributed function is based on the criteria of the standard error or estimate (SE). The results are presented in Table VII and the distributions are ranked by order of least standard error in each of the three sub-regions, in Region I, PT3, LN2 and Type 1; in Region II, LPT3, PT3 and LN2; and in Region III, Type 1, LN3 and PT3. The standard errors are computed using the same computer packages for estimation of peak discharges for various return periods by

Table VI – Statistical parameters of the hydrologic variables

Region	Station #	Basin	Area $A(Km^2)$	Disch. $\bar{Q}$	Std.Dev. S_Q	$SQ/\bar{Q}$ C_V	Skew. g	Disch. $\ln\bar{Q}$	Std.Dev. $\ln S_y$	Skew. V	Record n
I	SA 421	W.Khulab	900	500	372	0.744	1.96	6.015	0.632	0.568	14
	SA 418	Jizan	1200	663	838	1.264	1.95	5.67	1.41	–0.044	25
	SA 417	Damad	1000	628	521	0.83	1.60	6.17	0.746	0.438	25
	SA 415	Baysh	4713	670	568	0.848	1.348	6.161	0.903	–0.2196	15
II	SA 414	Itwad	1350	401	701	1.75	2.72	5.064	1.333	0.92	10
		"		188.8	216	1.143	2.01	4.766	1.00	0.613	9
	SA 411	Hali	4576	1070	1587	1.48	2.60	6.29	1.144	0.616	15
		"		718	847	1.18	2.56	6.12	0.966	0.352	14
	SA 404	Hali	119	37.3	47.1	1.26	2.58	3.14	0.967	0.789	9
	SA 402	Yiba	2722	486	451	0.929	0.766	5.65	1.33	–0.83	6
	SA 401	Yiba	784	225	256	1.138	2.24	4.819	1.242	–0.43	14
		"		166	135	0.814	0.423	4.66	1.132	–0.728	13
III	J 404	Dogah	970	194.3	166.7	0.86	1.43	4.6116	1.602	–1.49	16
	J 407	Sadiyah	1420	151.7	238.3	1.57	1.50	1.7146	4.60	–0.75	7

Table VII – Comparison of frequent distribution based on the standard errors (SE)

Region	Runoff Station	TN	LN2	TYP I	LN3	PT3	LPT3	Regional characteristics $\bar{C}_v$	$\bar{g}$
I	SA 421	184	96(2)	118(3)	127	92(1)	140		
	SA 418	442	354	274	282(3)	122(1)	406		
	SA 417	230	109(2)	141	151	89(1)	133(3)	0.92	1.715
	SA 415	221	107(2)	124(3)	1114	92(1)	183		
II	SA 414	520	350(2)	404	437	385(3)	175(1)		
	SA 411		590(3)	803	855	513(2)	265(1)	1.38	2.535
	SA 402	156	285	130(1)	157(2)	160(2)	450		
	SA 401	144	76(1)	104(3)	107	95(2)	157		
III	J 404	66	347	47(1)	49(2)	51(3)	1052	1.215	1.63∿1.75
	J 407	140	139(3)	106(1)	123(2)	106(1)	2916		

additional subroutines following the maximum likelihood estimation procedure.

4.3 Regional Flood Frequency Analysis

Selection of parameters and estimation of peak discharges for regionalization process require averaging the individual frequency curves at each station in the region. This was accomplished after expressing the flood magnitudes at selected recurrence intervals of Tr = 100, 50, 20, 10, 5, and 2 years, as well as the ratios of the index flood which has 2 year return period. This method is satisfactorily applied in many parts of the world and is also fairly easy to perform due to its simplicity (USWRC, 1976). The results can be transferred to ungauged areas because only a few parameters need to be measured.

From the previous analysis, it was decided that there are basically two common probability distributions which provide the least error of estimate, namely PT3 and TYP1. The flood discharges of the indicated return period, are determined for each theoretical fitting function. Also, the ratios to the 2 year return period flood discharges are computed in order to illustrate their variation in dimensionless form of the regional frequency values for the study zones.

The estimated flood discharges for different return periods of region I and III are presented in Table VIIIa. However, for Region II, two sets of data are given for fitting, one with outliers and the other without outliers, in order to illustrate the influence, Table VIIIb. However for the basins, Itwad, Hali, and Yiba, in Region II, the parameters and the estimated flood discharges are significantly affected by the outliers.

To summarize the findings of the regional frequency analysis, two different ratios Q_{100}/Q_2 and Q_{10}/Q_2 are presented, as sampled data for the two theoretical probability distribution functions, namely PT3 and TYP1 and shown below.

Region:	(I)		(III)		(II)	
Ratios	PT3	TYP1	TP3	TYP1	PT3	TYP1
Q_{100}/Q_2	4.6∿8.9	4.5∿7.0	4.8∿9.9	5.0∿10.0	8.4∿23.2 *(3.0∿ 9.75)	6.6∿11.0
Q_{10}/Q_2	2.5∿4.3	2.5∿3.7	2.7∿4.8	2.8∿ 5.0	4.0∿ 9.2 *(1.4∿ 4.3)	3.85∿5.5

*not including outliers

From the regionalized dimensionless frequency ratios presented above, it can be concluded that:

1- The dimensionless values of the regional frequency ratios do not differ much for each probability distribution in geographical regions of zone I and III. But in region II where statistical parameters of the regional frequency distribution are high, it does differ for PT3 and TYP1 distributions.

2- The largest flood magnitudes (outliers) affect considerably the dimensionless frequency ratios as tabulated in paranthesis above. These

Table VIIIa- Estimated flood discharges and ratio to the mean annual flood (Region I and III)

Station #		Tr=100	50	20	10	5	2	Q100/Q2	Q50/Q2	Q20/Q2	Q10/Q2	Q5/Q2
SA 421	PT3	1833	1579	1232	985	730	388	4.72	4.07	3.18	2.54	1.88
Khulab	TYP1	1990	1735	1397	1133	860	446	4.46	3.89	3.13	2.54	1.93
SA 418	PT3	3661	3090	2311	1457	1179	411.6	8.89	7.50	5.62	4.27	2.86
Jizan	TYP1	3787	3250	2536	1983	1407	536.8	7.06	6.05	4.72	3.69	2.62
SA 417	PT3	2394	2076	1642	1320	980	495	4.83	4.19	3.32	2.67	1.98
Damad	TYP1	2570	2237	1792	1449	1090	550	4.68	4.07	3.26	2.64	1.98
SA 415	PT3	2510	2196	1761	1430	1076	546.7	4.59	4.02	3.22	2.615	1.97
Baysh	TYP1	2945	2556	2039	1637	1219	588.0	5.00	4.35	3.47	2.78	2.07
J 404	TYP1	862	748	596	478	355	170	5.07	4.4	3.50	2.81	2.09
W.Duqah	PT3	741	647	516	417	311	156	4.75	4.15	3.31	2.67	1.99
J 407	TYP1	1200	1033	795	604	409	120	10.00	8.60	6.62	5.03	3.40
W.Sadiyah	PT3	948	805	614	466	315	96	9.88	8.4	6.40	4.85	3.28

Table VIIIb- Estimated flood discharges and ratio to the mean annual flood (Region II)

Station #		Tr=100	50	20	10	5	2	Q100/Q2	Q50/Q2	Q20/Q2	Q10/Q2	Q5/Q2
SA 414*	PT3	3162	2572	1810	1257	730.5	136	23.2	18.9	13.3	9.24	5.37
Itwad	TYP1	3431	2916	2228	1697	1143	306	11.20	9.53	7.28	5.54	3.74
SA 411*	PT3	7242	5944	4236	3035	1861	486	14.9	12.2	8.72	6.24	3.83
Hali	TYP1	7426	6340	4895	3772	2605	842	8.82	7.53	5.81	4.47	3.09
SA 404*	PT3	220	182	132	96	61	20	10.9	9.1	6.6	4.8	3.05
Hali	TYP1	240	206	160	124	87	30	7.8	6.9	5.3	4.0	2.9
SA 401*	PT3	1178	988	739	552	368	140	8.40	7.06	5.28	3.94	2.63
Yiba	TYP1	1250	1075	842	660	473	188	6.64	5.72	4.48	3.81	2.52
SA 414**	PT3	968	818	620	469	320	122	7.9	6.7	5.1	3.84	2.62
SA 411**	PT3	3955	3311	2421	1770	1141	410	9.75	8.1	5.9	4.32	2.78
SA 401**	PT3	521	473	403	344	276	1565	3.33	5.7	4.35	3.33	2.32

*with outliers
**without outliers not including first highest value

ratios may increase about 200 percent.

3- The median values for each range of distribution can now be used to determine the magnitude of a peak flood discharge after multiplying the ratio by the estimated two year flood discharge. The mean annual 2-year flood is determined by the multivariate regression models presented in this study, using size of the basin, annual mean precipitation and also the hydraulic conductivity values.

ACKNOWLEDGEMENT

The authors would like to express their appreciation to the Department of Water Resources, Ministry of Agriculture and Water for providing hydrological data, without which this study would not have been possible. The encouragement and support provided by the Department of Hydrology and Water Resources Management, Faculty of Meteorology, Environment & Arid Land Agriculture, King Abdulaziz University, is gratefully acknowledged.

REFERENCES

1. Dalrymple, T. 1960. Flood Frequency Analysis, U.S. Geological Survey Water Supply Paper, 153-A.
2. Development Analysis Associates (DAA), Inc. 1980. The Annual Water Balance in Saudi Arabia. Prepared for the Saline Water Balance Conversion Corporation, Kingdom of Saudi Arabia.
3. Eagleson, P. 1978. Climate, Soil and Vegetation. Water Resources Research Journal, Vol. 14, No. 5.
4. Italconsult, 1969. Water and Agricultural Development Surveys. Prepared for the Ministry of Agriculture and Water, Kingdom of Saudi Arabia.
5. Kite, G.W. 1977. Frequency and Risk Analysis in Hydrology. Water Resources Publications, U.S.A.
6. Ministry of Agriculture and Water, Hydrology Division. 1985. Hydrological Publications No. 97, 98, 100, 101, 104.
7. Riggs, H.C. 1973. Regional Analyses of Streamflow Characteristics. Techniques of Water Resources Investigations of the U.S. Geological Survey. Chp. B3.
8. Saudi Arabian Dames and Moore. 1985. Representative Basins Study for Wadis. Prepared for the Ministry of Agriculture and Water, Kingdom of Saudi Arabia.
9. Sogreah. 1968. Water and Agricultural Development Studies. Prepared for the Ministry of Agriculture and Water, Kingdom of Saudi Arabia.
10. United States Water Resources Council. 1976. Guidelines for Determining Flood Flow Frequency. Bulletin No. 1708. The Hydrology Committee.

REGIONAL FLOOD FREQUENCY ANALYSIS FOR SOME SELECTED BASINS IN SAUDI ARABIA

ABDULAZIZ S. AL-TURBAK AND ALI AKHTAR QURAISHI
Department of Civil Engineering
King Saud University
P.O. Box 800, Riyadh 11421
Saudi Arabia

ABSTRACT. Annual peak flow rates from thirteen runoff gaging stations in Saudi Arabia are considered in this study. These stations are located in two different regions of the south-western part of the Kingdom. Seven stations are within Wadi Bishah drainage area while six are in Tihamah-South region. The first step in the analysis was to calculate the 5, 10, 25, 50 and 100 years floods for the 13 stations using both the Extreme-Value Type I (EV1) and the Log-Pearson Type III (LP3) distributions. Results showed EV1 to give better estimates than LP3. Therefore, the EV1 values were the only ones used in the rest of the work. The obtained results (EV1) were then used to check for homogeneity for all the stations within each area. A regional flood frequency analysis was then performed. Multiple regression equations relating peak flow rates in each area to four watershed and precipitation variables were established. These independent variables are: drainage area, mean annual precipitation, main channel length and average main channel slope. Using SPSS computer program, regression coefficients for all the equations were obtained and checked. The resulting equations will be important tools for design as well as other studies related to surface water in these two developing regions of the Kingdom.

1. INTRODUCTION

The lack of adequate runoff information on many wadis in Saudi Arabia is one of the main problems associated with design of water control structures and in the evaluation of surface water potentials in many parts of the country. Frequently, engineers in Saudi Arabia are called upon to estimate peak discharges at a certain site where there are no runoff data. Another problem is that some of the existing gaging stations (a total of 75) have been in operation only for a few years.

Flood frequency data at individual stations have limited transfer capability. Therefore, estimates of flood characteristics of ungaged basins should be based on regional analysis done on data from gaged basins. Regional analysis techniques provide means of combining records from several gages. This provides the two advantages of reducing stan-

V. P. Singh (ed.), Regional Flood Frequency Analysis, 27–34.

dard errors of estimates at gaged sites and enabling estimates to be obtained for ungaged locations.

The south-western part of Saudi Arabia has the most reliable surface water potential in the whole country. It is also the area with the highest and most dependable precipitation. The primary objective of this work was to carry out a regional flood frequency analysis for two large areas; both located in the south-western part of the country. These are Bishah and Tihamah-South areas. Flood data from a total of thirteen stations located in these two regions were analyzed. The EV1 results for 5, 10, 25, 50 and 100 years return periods were used in developing multiple regression equations for the two areas.

2. THE STUDY AREAS

Seven of the thirteen stations included in this analysis are all for subbasins within Wadi Bishah catchment area. Wadi Bishah basin is one of the most important areas for development in Saudi Arabia. The Ministry of Agriculture and Water (MAW) is preparing plans for development projects within that area. MacLaren report (1979) indicates that Bishah basin contains a considerable percentage of the overall agricultural production in the south-west.

Wadi Bishah basin lies between the two latitudes 18° 00' and 20° 30' N and the two longitudes 41° 50' and 43° 10' E. The drainage area of Wadi Bishah at Sada (Station B-406) is approximately 16920 km^2. The western and southern parts of Wadi Bishah have both steppe and mountain features. This include the basins: Wadi Ashran (Station A-402), Wadi Hani (Station A-404) and Wadi Tindahah (Station A-405). Wadi Tabalah (Station B-405) starts in the mountains and drains to semi-desert areas. Wadi Bishah itself, north of Ben Hashbal (Station A-403) can be characterized as being semi-desert, with relatively flat undeveloped drainage areas, sparse vegetation and areas of deep sediment deposits. Previous investigations have shown that the downstream wadi bed has a high capability to absorb flows, most notably in the reach from Heifa (Station B-402) to Sada (Station B-406).

The other area taken in this analysis is Tihamah-South. A total of six stations from this region were included. The wadis in this area start in the high mountains of Asir and travel from east towards the west. After the high escarpments, these wadis traverse the Tihamah gently sloping plain until they reach the Red Sea. The catchments studied all lie between the latitudes 16° 30' and 20° 05' N and the longitudes 41° 05' and 43° 30' E. Tihamah-south area is also an important region for development in the Kingdom. Some of the wadis in this region have the highest surface water potentials in the whole country.

3. REGIONAL FLOOD FREQUENCY ANALYSIS

3.1 Basic Data

The basic raw data of peak flow rates in m^3/s, date of occurrence, the period of records, the coordinates, elevations and drainage areas were obtained for all the stations used in the analysis. The Hydrology Division of the MAW supplied these basic data. Table I show the number, drainage area, coordinates, elevation and years of records for all the selected stations.

3.2 Flood Frequency Analysis

Annual peak discharges for the stations were then used to estimate the 5, 10, 25, 50, and 100 years floods for each station. Both the Extreme-Value Type I and Log-Pearson Type III distributions were used. From the results, it seemed that the EV1 is more suitable distribution than the LP3 in these areas. The expected floods for the shown return periods at the stations using the EV1 distributions are shown in Table II. In further analysis only the results of EV1 distributions has been used.

3.3 Homogeneity Test

The homogeneity test was performed for the stations in each basin by Index-Flood Method as suggested by Dalrymple (1960) and elaborated by Kite (1977). The results showed that all the stations within each area are homogeneous.

3.4 Regional Regression Analysis

Frequently, hydrologists in Saudi Arabia are called upon to estimate runoff where there are no runoff stations or data at all. Therefore, estimates of flood characteristics of ungaged basins should be based on regional analysis done on data from gaged basins. Regional analysis techniques provide a means of combining records from several gages. The regional analysis also reduces the errors from individual stations and it is possible to estimate flood magnitudes at sites where no observations have been taken by using the regional analysis results.

The regional equations used in this study are non linear multiple regression equation that relate peak discharges of 5, 10, 25, 50 and 100 years recurrences intervals to (i) the drainage area (ii) the basin mean annual precipitation, (iii) the main channel length and (iv) the average main channel slope.

The regression equation has the general form

$$Q_t = a(DA)^b \quad (AP)^c \quad (CL)^d \quad (CS)^e$$

where

Q_t is the peak discharge (m^3/sec) for a recurrence interval of t- years,
DA is the drainage area (km^2),
AP is the mean annual precipitation (mm),
CL is the main channel length (km),
CS is the average main channel slope (m/m),
a, b, c, d, and e are regression coefficients.

TABLE I NUMBER, DRAINAGE AREA, COORDINATES, ELEVATION AND YEARS OF RECORDS FOR ALL SELECTED STATIONS

Station No.	Drainage Area (Km^2)	Latitude	Longitude	Elevation (m)	Years of Record
(a) Bishah Area					
A-402	80	18° 17' 00"	42° 28' 00"	2150	18
A-403	2285	18° 28' 00"	42° 42' 00"	1800	15
A-404	146	18° 25' 00"	42° 31' 00"	2145	19
A-405	440	18° 20' 00"	42° 52' 00"	1880	18
B-402	11350	19° 52' 00"	42° 32' 00"	1280	17
B-405	1090	20° 02' 00"	42° 16' 00"	1290	13
B-406	16920	20° 09' 00"	42° 42' 00"	1090	18
(b) Tihama-South Area					
J-404	970	19° 45' 00"	41° 02' 00"	80	17
SA-401	784	19° 16' 55"	41° 48' 45"	420	14
SA-411	4576	18° 46' 30"	41° 35' 00"	298	16
SA-415	4713	17° 34" 40"	42° 37' 00"	200	16
SA-417	1000	17° 09' 00"	42° 53' 00"	130	26
SA-421	900	16° 43' 00"	43° 01' 00"	99	15

In the regression analysis for each station Q_t is taken from flood frequency analysis by EV1 (Table II), DA from the basic data supplied by MAW (Table I). The basin mean annual precipitation, AP, was calculated from the rainfall recording stations in and around the basin, CL was estimated from maps available at MAW and CS was obtained by dividing elevation difference of the two ends of the main channel (as estimated from contour maps and available records) by the channel length CL. The values of AP, CL and CS for each stations are shown in Table III.

The flood frequency results (Table II) along with values of the four independent variables: drainage area, basin mean annual precipitation, main channel length and average main channel slope were used in the multiple regression analysis. The computer program SPSS which is available at the computer center of King Saud University was used to obtain the regression coefficients. Table IV shows summaries of the regression coefficients obtained from the two areas of the study.

3.5 Discussion

The standardized regression coefficients were obtained as part of the SPSS computer program output. These coefficients give at least a rough estimate of the relative contribution of each of the independent variables used in the multiple regression equations. From the values of these coefficients, it was found that the basin mean annual precipitation was very important in determining peak discharges for any return period in both regions of the study. The main channel length was found to have almost the same relative importance in both regions. However, the values of the standardized regression coefficients for the main channel length in both areas were found less than those for the basin mean annual precipitation. The drainage area was found to be less significant in Wadi Bishah area than in Tihamah-South. For the average main channel slope, the opposite was true.

The regression coefficients shown in Table IV were used in the equations to check how well the resulting equations can be used to estimate peak flow rates for the given return periods. The obtained flow rates were then compared to the values given in Table II. It was found that the regression equations with the coefficients shown in Table IV give reasonable results (less than 8% difference between predicted values and values of the EV1 distribution) if the drainage area is 1000 km^2 or less. The predictions seem to worsen as the drainage area increases.

4. CONCLUSIONS

In this work, annual peak discharge records from thirteen basins located in two important regions of Saudi Arabia were analyzed. Flood frequency analysis for the individual stations was first carried out using both the Extreme-Value Type I (EV1) and the Log-Pearson Type III (LP3) distributions. The results were the flood magnitudes for each of the stations for 5, 10, 25, 50 and 100 years return periods. The EVI disbution was found to be more suitable than LP3 in predicting peak discharges. Using the EV1 values, all stations within each area were

TABLE II FLOOD FREQUENCY ANALYSIS FOR SELECTED STATIONS

Station No.	Magnitude of flood, in cubic meters per second, for the indicated return period in years				
	5	10	25	50	100
A-402	148	206	279	334	388
A-403	710	987	1338	1598	1856
A-404	90	126	171	205	238
A-405	220	285	368	429	489
B-402	1187	1581	2078	2447	2814
B-405	214	301	410	491	572
B-406	643	816	1035	1198	1359
J-404	318	414	535	624	713
SA-401	409	559	749	889	1029
SA-411	2147	3049	4189	5034	5873
SA-415	1042	1374	1794	2105	2413
SA-417	939	1219	1573	1835	2092
SA-421	743	962	1237	1442	1645

TABLE III BASIN MEAN ANNUAL PRECIPITATION (mm), MAIN CHANNEL LENGTH (km) AND AVERAGE MAIN CHANNEL SLOPE (m/m) FOR ALL SELECTED STATIONS

Station No.	AP (mm)	CL (km)	CS (m/m)
A-402	347.20	12.5	0.03
A-403	243.94	75.5	0.00596
A-404	202.05	18.5	0.015
A-405	215.97	34.0	0.00647
B-402	225.49	243.0	0.00399
B-405	184.23	94.3	0.00793
B-406	225.23	279.0	0.00415
J-404	222.16	55.1	0.0058
SA-401	299.48	28.5	0.00724
SA-411	351.10	113.0	0.0175
SA-415	340.50	104.1	0.01633
SA-417	438.87	45.8	0.0146
SA-421	372.125	39.0	0.0075

TABLE IV SUMMARY OF REGRESSION COEFFICIENTS

(a) Bishah Area

t-years	a	b	c	d	e
5	9.4421×10^{-7}	0.03886661	2.5284840	0.2692622	-0.9335391
10	2.4297×10^{-6}	0.03077384	2.456960	0.2932729	-0.8721999
25	4.6542×10^{-6}	0.02250668	2.418189	0.3112359	-0.8366683
50	6.9373×10^{-6}	0.02185488	2.393050	0.3163070	-0.8138542
100	9.0099×10^{-6}	0.01727890	2.3815509	0.3265585	-0.8001028

(b) Tihamah-South Area

t-years	a	b	c	d	e
5	1.7794×10^{-5}	0.2965655	2.089295	0.5460759	-0.2356713
10	5.7480×10^{-5}	0.3599283	1.976838	0.4419608	-0.1744814
25	1.5335×10^{-4}	0.3630956	1.889393	0.4060093	-0.1259114
50	2.3837×10^{-4}	0.3263985	1.853680	0.4287483	-0.1060902
100	3.4124×10^{-4}	0.2971848	1.823983	0.4470053	-0.09204098

then checked for homogeneity. A regional flood frequency analysis was then performed for the two areas. Multiple regression equations were used to predict peak discharges at any site in the two regions for the specified return periods. The independent variables used in the regression were drainage area, mean annual precipitation, average main channel slope and main channel length. The resulting equations were checked and found to give reasonable results if the drainage area is about 1000 km^2 or less.

5. ACKNOWLEDGEMENTS

The authors would like to thank both Mr. Sa'ad Swat and Mr. Abdallah Al-Mousa, from the Hydrology Division, Water Resources Development Department, MAW for their help in obtaining the necessary data for the study.

6. REFERENCES

Dalrymple, T. 1960, 'Flood-Frequency Analyses', U.S. Geological Survey, Water Supply Paper 1543-A.

Kite, G.W. 1977, 'Frequency and Risk Analysis in Hydrology', Water Resources Publications, Fort Collins, Colorado.

MacLaren International Limited, 1970, 'Water and Agricultural Development Studies, Arabian Shield-South', (Report submitted to Ministry of Agriculture and Water, Kingdom of Saudi Arabia).

HIERARCHICAL APPROACH FOR REGIONAL FLOOD FREQUENCY ANALYSIS

Mauro Fiorentino
Dipartimento di Difesa del Suolo, Università della Calabria.
C/da S. Antonello, 87040 Montalto Uff. (Cosenza), Italy.

Salvatore Gabriele
Consiglio Nazionale delle Ricerche-IRPI.
Via Verdi 1, 87030 Rende (Cosenza), Italy.

Fabio Rossi
Dipartimento di Idraulica, Gestione delle Risorse
Idriche ed Ingegneria Ambientale.
Via Claudio 21, 80125 Napoli, Italy.

Pasquale Versace
Dipartimento di Difesa del Suolo, Università della Calabria, and CNR-IRPI. Cosenza, Italy.

ABSTRACT. In this paper a hierarchical approach for regional flood frequency analysis is proposed. It is characterized by three investigation levels: i) identification of homogeneous regions wherein the skewness coefficient is assumed not to vary from site to site; ii) identification of homogeneous sub-regions wherein the coefficient of variation is assumed to be constant too; iii) use of regression models to relate a location parameter of the flood distribution to climatic and physiographic factors.

The hierarchical approach is herein presented with the use of the two-component extreme value (TCEV) distribution. The validity of the presented approach is demonstrated using twenty-eight Italian annual flood series (AFS). In particular, it is shown that the sample skewness coefficient exhibits an observed variability comparable with the theoretical sampling variability, the latter derived by computer simulation experiments from a regionalized TCEV distribution considered as representative of the entire region in question. Instead, for the same AFS's, the observed variance of the coefficient of variation is nearly twice the one derived from the sampling experiments.

V. P. Singh (ed.), Regional Flood Frequency Analysis, 35–49.

1. INTRODUCTION

Regionalization techniques are used in flood frequency analysis mainly to estimate T-year flood flow Q at ungauged river basin sites and also to improve site-specific estimates based on limited single site data. In an implicit or explicit manner, regional analysis is based on a flood distribution which describes how Q varies with the return period T. It is also based on a regionalization model which associates annual flood characteristics with physiographic and climatic causative factors by a number of relationships equal to the number of flood distribution parameters.

The analysis can be greatly simplified since in large climatic areas the moment-ratios of the annual flood distribution - e.g. the coefficients of variation and skewness - appear to be non-significantly variable and practically independent of the physiographic factors, except perhaps the basin area. This empirical evidence has suggested the use of regional flood frequency distribution representing the average ratios of the floods at various return periods to a scaling index flood, e.g. the mean annual flood. To obtain estimates of the T-year flood at ungauged sites in the region, one can use this regionalized distribution, usually referred to as growth curve, together with regression models, the latter being used for estimation of the index flood. The approach referred to as the index flood method was first introduced in the U.S.A. by the Geological Survey (Dalrymple, 1960; Riggs,1973).

Furthermore a procedure for identifying homogeneous regions is required. A realistic regionalization model was proposed by Matalas and Gilroy (1968), who adopted the classical analysis of the variance into two components representing different sources of random variation. Regionalization was then considered to extend lengths of the historical sequences and led to a reduction in time sampling errors, though at the price of introducing space disturbance errors. In addition, the reduction of the time sampling variance as the number of stations increases is limited by the correlation among concurrent flood flows.

The transfer of regional information is then affected by three factors, viz. time sampling variance, space disturbance variance, and interstation correlation. However, these play a different role as the order of the moment increases from the mean to the coefficient of skewness. It is known that the sampling variability increases with the order of moments, while in this paper the ratio of space disturbance variance to the sampling variance will be shown to decrease. Furthermore, Stedinger (1983) showed that the influence of the interstation correlation decreases passing from the mean to the skewness.

Such a varying regional behavior of the statistical parameters suggested to Fiorentino et al. (1985) a modified version of the index flood approach. Instead of using two levels of analysis, the first one dealing with ratio of the T-year flood to the scaling value and the second with the location parameter, regional analysis can be hierarchically broken down into three distinct levels, one each for the following statistical parameters: coefficient of skewness, coefficient of variation, and location value. A homogeneous region with respect to the coefficient of skewness is firstly identified. Within this region more homogeneous sub-regions with respect to the coefficient of variation can be discriminated to reduce large space disturbance errors. This approach, herein referred to as hierarchical approach, optimizes the transfer of regional information.

In this paper a regionalization estimation procedure is presented. It is based on the two-component extreme value (TCEV) distribution, on the maximum likelihood (ML) estimation method, and on two different averaging procedures for regional estimation. However the hierarchical approach is suitable for application with different distributions and different estimation procedures.

2. REGIONALIZATION MODELS

A conceptual framework to transfer regional information was articulated by Matalas and Gilroy (1968). Suppose that there are k gauged sites in a region, each with n years of record. Let α be a parameter or a moment-ratio of the annual flood distribution. The simplest regionalization model assumes that in a given region α is not significantly affected by climatic and physiographic factors. However, this does not mean that α is identical at any site. Since presumably there are many unaccounted for causative factors at work, it is reasonable to assume for the basin j $(j=1,2,\ldots,k)$

$$\alpha_j = \alpha_0 + \delta_j \tag{1}$$

where α_0 is a regional parameter equal to the mean of the α's and δ_j represents a random space disturbance error with mean $E(\delta)=0$ and variance $VAR(\delta)=\omega^2$, the latter being equal to that of the α_j's. At station j let us suppose that an unbiased estimate $\hat{\alpha}_j$ of α_j can be obtained from data and is given as

$$\hat{\alpha}_j = \alpha_j + \eta_j \tag{2}$$

where the random variable η_j is a time sampling error having $E(\eta_j)=0$ and $VAR(\eta_j)=\sigma_j^2$ which is equal to that of the $\hat{\alpha}_j$'s. Combining (1) and (2) we get

$$\hat{\alpha}_j = \alpha_0 + \delta_j + \eta_j \tag{3}$$

Assuming that the variance $\sigma_j^2 = \sigma^2$, i.e. it is the same for all sites, and that δ_j and η_j are independent, the mean and variance of the $\hat{\alpha}_j$'s over both the time and space are

$$E[\hat{\alpha}_j] = \alpha_0 \quad ; \quad VAR[\hat{\alpha}_j] = \sigma^2 + \omega^2 = V \tag{4}$$

The simplest regionalization model then contains three parameters : $\alpha_0, \omega^2, \sigma^2$. Estimates of these parameters can be obtained from the $\hat{\alpha}_j$'s. Particularly an estimate of α_0 is obtained by

$$\hat{\alpha}_o = \sum_{j=1}^{k} \hat{\alpha}_j / k \tag{5}$$

while estimates of ω^2 can be derived by

$$\omega^2 = \hat{V} - \sigma^2 \tag{6}$$

where

$$\hat{V} = \sum_{j=1}^{k} (\hat{\alpha}_j - \alpha_o)^2 / (k-1) \tag{7}$$

and the sampling variance σ^2 can be evaluated by analitical computation or computer simulation experiments.

The spatial mean $\hat{\alpha}_o$ of the $\hat{\alpha}_j$'s can be used as a regional estimator of α_j at every site j. The mean square error (MSE) of $\hat{\alpha}_o$, i.e. the variance of $\hat{\alpha}_o$ about α_j, was given by Matalas and Gilroy (1968)

$$MSE[\hat{\alpha}_o] = E[(\hat{\alpha}_o - \alpha_j)^2] = \sigma^2 / k_e + \omega^2 (1 \pm 1/k) \tag{8}$$

where the sign plus or minus differentiates the ungauged from the gauged site j. In (8) k_e is the equivalent number of independent stations and can be put in the form:

$$k_e = k / [1 + (k-1) \bar{\rho}_\alpha] \tag{9}$$

where k is the number of stations and $\bar{\rho}_\alpha$ is the average interstation correlation coefficient of the $\hat{\alpha}_j$'s. Matalas and Benson (1961), Matalas and Gilroy (1968), and Stedinger (1983) showed that $\bar{\rho}_\alpha$ is more or less equal to $\bar{\rho}, \bar{\rho}^2$ and $\bar{\rho}^3$, α being equal to the mean, variance, and skewness respectively, and where $\bar{\rho}$ is the average interstation correlation of concurrent annual floods.

The regionalization leads to a space-time trade-off that is regulated by the relation (8). Increasing the number of sites, one obtains a reduction of time sampling variability which is partially counterbalanced by the space disturbance errors and is limited by interstation correlation. A criterion of minimizing MSE of the regionalized estimator can be used for identifying homogeneous regions together with cluster analysis techniques.

When the space disturbance variance ω^2 is large, it can be reduced by dividing the region into more homogeneous sub-regions or introducing some climatic or physiographic factors which are able to account, at least partially, for ω^2. In the latter case the simplest regionalization model of equation (1) becomes a regression model, and the MSE of the regionalized estimator is given by a regionalized expression of the relation (8) (Matalas and Gilroy, 1968).

In this paper it will be shown, using Italian data, that the skewness coefficient exhibits a smaller space disturbance variance than the coefficient of variation. Thereupon, and also because it is less affected by correlation, more stations can be used to transfer information for estimating skewness or distribution parameters depending on skewness.

3. HIERARCHICAL REGIONALIZATION APPROACH USING THE TCEV DISTRIBUTION

3.1. Flood Distribution.

The TCEV distribution estimation (Rossi et al., 1984) constitutes a reasonable basis for a hierarchical-type regionalization model. This distribution originates from a phenomenological model of the flood process by considering the annual flood to be the maximum of a Poisson-distributed number K of independent and identically distributed random variables Z's, also independent of K (Versace et al., 1982). The distribution of Z variate was empirically assumed to be a mixture of two exponential components. Its probability density function (PDF) is

$$f_X(x) = \exp(-\Lambda_1 e^{-x/\theta_1} - \Lambda_2 e^{-x/\theta_2})\left(\frac{\Lambda_1}{\theta_1} e^{-x/\theta_1} + \frac{\Lambda_2}{\theta_2} e^{-x/\theta_2}\right) \quad ; \; x>0$$

$$= \exp(-\Lambda_1 - \Lambda_2) \quad ; \; x=0 \qquad (10)$$

Λ_1 ($>\Lambda_2$) and Λ_2 (>0) being the mean annual number of independent floods coming from the first and second component respectively. Since the finite probability that the value is exactly zero is negligible, PDF (10) is indistinguishable in practice from a continuous function also allowing for negative values, except perhaps in arid or semiarid basin sites, where small values of $\Lambda = \Lambda_1 + \Lambda_2$ are likely to be found.

In order to develop the regionalization model it is useful to introduce the reduced variate

$$Y = \frac{X}{\theta_1} - \ln\Lambda_1 \qquad (11)$$

having PDF which for x>0 is

$$f_Y(y) = \exp(-e^{-y} - \Lambda_* e^{-y/\theta_*})\left(e^{-y} + \frac{\Lambda_*}{\theta_*} e^{-y/\theta_*}\right) \qquad (12)$$

where

$$\theta_* = \theta_2/\theta_1$$
$$\Lambda_* = \Lambda_2/\Lambda_1^{1/\theta_*} \qquad (13)$$

Equation (10) can then be written as

$$f_X(x) = \exp(-\Lambda_1 e^{-x/\theta_1} - \Lambda_1^{1/\theta_*}\Lambda_* e^{-x/\theta_1\theta_*})$$

$$\left(\frac{\Lambda_1}{\theta_1} e^{-x/\theta_1} - \frac{\Lambda_1^{1/\theta_*}\Lambda_*}{\theta_1\theta_*} e^{-x/\theta_1\theta_*}\right) \quad ; \qquad x>0 \qquad (14)$$

The coefficients of skewness $\gamma_1 = \mu_3/\mu_2$^(3/2) of TCEV distribution depends on θ_* and Λ_* while its coefficient of variation $\gamma = \mu_2^{1/2}/\mu$ depends on θ_*, Λ_* and Λ_1, μ_2 and μ_3 being the second and third central moment respectively. The mean μ depends on all the four parameters of the distribution. Expressions of the moments were given by Beran et al. (1986).

3.2. Regionalization Model.

Using the TCEV distribution, and according to the considerations outlined in the Introduction, a hierarchical regionalization approach can be developed in the following manner (Fiorentino et al., 1985).

At the first level an extensive region, wherein the theoretical coefficient of skewness γ_1 is assumed to be constant, is selected. Within this region dimensionless parameters θ_* and Λ_* are also constant and the distribution of Y variate is therefore identical at any site.

The second level assumes that homogeneous sub-regions with respect to γ exist within the region where γ_1 is constant. In each sub-region, parameters θ_*, Λ_*, and Λ_1 are assumed not to change from site to site, and therefore the ratio of X variate to the index flood is supposed to be identically distributed regardless of the site.

Finally at the third level the hypothesis that the index flood is correlated with some climatic and physiographic factors of the basin is made.

3.3. Estimation Procedure.

Estimation of parameters is strictly related to the regionalization model and differs from level to level.

Four parameters of the TCEV distribution can be estimated from a site-specific data set (it can be thought of as the zeroth level of regionalization) using ML estimation method (Rossi et al., 1984). However, estimates of θ_* and Λ_* obtained in this manner are too uncertain,especially in case very short samples are used. This prompts one to invoke the use of regionalization techniques, at least for estimation of θ_* and Λ_*, it also influences the choice of the regionalization procedure.

The method proposed by Fiorentino and Gabriele (1985), and currently used to estimate θ_* and Λ_*, is an iterative ML procedure, being essentially a station-year method which, as well known, is sensitive to the interstation correlation effects. Nevertheless this method was chosen because it does not invoke an explicit averaging between site-specific estimates of θ_* and Λ_*. Furthermore it can be noted that (1) estimation of θ_* and Λ_* implicitly involves estimation of moments with order larger than two, and (2) correlation between moments progressively decreases as the order increases.

More precisely, the first level of the regional estimation procedure consists of carrying out the following steps using k annual flood series (AFS) each with n_j (j=1,2,...,k) years of record:

a) for each series, assign tentative values to parameters θ_1 and Λ_1;

b) transform all the data by (11), and form a pooled sample with the so obtained y's. The length of this sample is

$$N = \sum_{j=1}^{k} n_j \;;$$

c) estimate θ_* and Λ_* using (12): θ_*, Λ_* : $\ln L_1 = \max$, where

$$L_1 = \prod_{i=1}^{N} f_Y(y_i \,;\, \theta_*,\, \Lambda_*)$$

d) for each series estimate θ_1 and Λ_1 using the ML method constrained to the values of θ_* and Λ_* derived at step c): θ_1, Λ_1: $\ln L_2 = \max$, where

$$L_2 = \prod_{i=1}^{n_j} f_X(x_i \,;\theta_1, \Lambda_1, | \theta_*, \Lambda_*)$$

and $f_X(x_j)$ is given by (14);

e) repeat steps b) to d) till estimates of θ_* and Λ_* no longer change.

In short, at the first level two regional estimates, of θ_* and Λ_*, and two site-specific estimates constrained to them for each series, are attained.

At the second level, regional estimation of Λ_1 is carried out by introducing the parameter CV1 which is obtained by the transformation

$$CV1 = 0.557/(\log\Lambda_1 + 0.251) \tag{15}$$

and whose estimator is less variable and biased than that of Λ_1 (Fiorentino and Gabriele, 1985). More precisely, average of the site-specific estimates of CV1 is used to give regional estimate of Λ_1 by inverting relationship (15).

At this level only the parameter θ_1 assumes different values from site to site in the sub-region investigated. Its estimates are obtained using the ML method and constraining the solution to the regional estimates of θ_*, Λ_*, and Λ_1:

θ_1 : $\ln L_3 = \max$, where

$$L_3 = \prod_{i=1}^{n_j} f_X(x_i \,;\, \theta_1 | \theta_*,\, \Lambda_*,\, \Lambda_1)$$

At the third level regression models are used to achieve index flood estimates at ungauged sites.

4. STATISTICAL PROPERTIES OF THE REGIONALIZED TCEV-ML ESTIMATOR FOR $\omega^2 = \bar{\rho} = 0$

The regional flood-frequency estimation procedure using the TCEV distribution was assessed in two previous papers (Fiorentino and Gabriele, 1985; Arnell and Gabriele, 1986) exploiting Monte Carlo experiments. The hypothesis that floods come from a homogeneous world was made throughout the simulation works. That is disturbance effects caused (1) by floods drawn from a population different from that of others and (2) by data errors were left out of account. Moreover spatial (interstation) correlation effects were not simulated in the experiments. Fiorentino and Gabriele (1985) gauged the accuracy in estimating parameters and quantiles with increasing exploitation of the regional estimation procedure levels. Synthetic series were generated from a TCEV distribution, with theoretical coefficients of skewness $\gamma_1=2.75$ and of variation $\gamma=0.59$, which was considered as a reasonable "parent" of the central and southern Italian annual floods.

The fitting of the TCEV distribution to a site-specific data set was first simulated. Then the performance of the regional estimator at both first and second level was assessed. Quantile estimators were shown to be little biased regardless of the amount of regional information utilized, though the bias increased with return period. Root mean square error was also used as a measure of performance. It was standardized to ensure compatibility between different experiments by RMSE = $[E(\hat{x}-x)^2]^{\frac{1}{2}} / x$, where $\hat{x}$ and x are the estimated and population value respectively of either at-site quantile or one of the parameters as appropriate. As regards the quantile estimates, RMSE values are shown in Table 1.It can be seen that, as expected in an idealistic situation of having homogeneous data, the RMSE drastically decreased everywhere as regional information was acquired. It is worth emphasizing the reasonable low value of the RMSE of estimates, even of upper quantiles, once the second level of the regionalization procedure is exploited. On the other hand, admittedly, quantile estimation based upon a single data set appears very uncertain.

	Regionalization level		
F	0	1	2
0.9	0.15	0.10	0.08
0.98	0.27	0.12	0.09
0.99	0.31	0.14	0.10
0.998	0.38	0.16	0.12
0.999	0.40	0.17	0.13

Table 1 - RMSE of quantile estimates.

Statistical properties of the regionalized TCEV-ML estimators were compared with those of other regionalization procedures by Arnell and Gabriele (1986), who investigated the robustness of competing estimators with respect to changes in parent distribution. Herein the real world of floods is assumed to have a TCEV distribution.

Three regional estimation algorithms were selected for comparison. The first two follow a regionalization procedure based on the probability-weighted moment (PWM) estimation method for the Wakeby (WAK) distribution and the generalized extreme value (GEV) distribution, which are described in Wallis (1980) and in Hosking et. al. (1984) respectively. The third one is an up-to-date computerized version of the graphical procedure which was employed in the 1975 U.K. Flood Studies Report (NERC, 1975). As quoted by the Authors this version slightly differs from that developed by Hosking et. al. (1985). It will be referred to as FSR.

Some of the results obtained by Arnell and Gabriele (1986) are herein shown.

	Quantiles		Growth factors	
F	WORLD-1	WORLD-2	WORLD-1	WORLD-2
GEV-PWM				
0.95	0.076	0.097	0.036	0.015
0.98	0.076	0.104	0.034	0.047
0.99	0.075	0.118	0.038	0.075
0.998	0.227	0.127	0.220	0.087
0.999	0.279	0.125	0.276	0.080
WAK-PWM				
0.95	0.076	0.098	0.035	0.017
0.98	0.087	0.101	0.053	0.035
0.99	0.082	0.109	0.045	0.057
0.998	0.171	0.122	0.161	0.074
0.999	0.214	0.132	0.206	0.085
FSR				
0.95	0.073	0.104	0.050	0.058
0.98	0.102	0.101	0.090	0.045
0.99	0.114	0.106	0.102	0.053
0.998	0.149	0.143	0.134	0.113
0.999	0.181	0.183	0.169	0.164
TCEV				
0.95	0.055	0.082	0.013	0.022
0.98	0.061	0.089	0.028	0.037
0.99	0.078	0.099	0.053	0.054
0.998	0.156	0.121	0.146	0.090
0.999	0.182	0.128	0.171	0.101

Table 2 - RMSE of quantile estimates (second level of the regionalized TCEV-ML procedure)

Two different TCEV distributions were taken into account as parent. The former was considered as representative of the U.K. flood experience (γ_1=3.5, γ=0.40) (WORLD-1) the latter was chosen to be the one which has been referred above (γ_1= 2.75, γ=0.59; WORLD-2).

Table 2 refers to the RMSE of quantile and growth factor estimates, growth factor being the ratio of quantile to the index flood. It can be inferred that in comparison to the other estimators, the TCEV-ML procedure performs better for at-site quantiles than for growth factor.

With regard to quantile estimators, TCEV procedure exploited the advantage that the population was a TCEV-type world, and in both the experiments, whereas it was outperformed by the competing estimators only in the growth factor estimation for WORLD-2. By the way, something needs to be said about small sample properties of ML estimators of θ_* and Λ_*. In fact θ_* reflects (RMSE=0.153), the good performance of quantile regional estimators, whereas the RMSE of Λ_* (=0.717) is undoubtedly unsatisfactory. As a consequence, this may suggest that rather better statistical properties of quantile and especially of growth factor regional estimators could be attained by enhancing the estimates of Λ_*.

5. IMPACT OF THE REGIONALIZATION MODEL WITH DATA

The use of regionalization techniques requires hydrologists to check the suitability of available data in order to ensure homogeneity between flood series.

A re-examination of the whole flood data base is currently being done in Italy. Since one of the aims of this paper is to show the different spatial heterogeneity of the skewness and variation coefficients, the local factors which can affect skewness and variability of the AFS's have been examined, particularly: i) lateral spilling immediately upstream the flood gauging station; ii) effect of the basin area; iii) ephemeral streams in semiarid areas.

Flood series recorded downstream of a river branch tends to be bounded at the upper end when a significant and systematic lateral spilling of high flows occurs at the upstream branch. This leads to outlying low values of both coefficients of variation and skewness. In some cases, however, it was impossible to get sufficient information to discard an AFS showing a suspicious upward curvature of the right-hand tail when plotted on extremal probability paper. Several of these series exhibiting an evident upper bound (a typical example is shown in Fig. 1) represent the basins having drainage area greater then 3000 Km^2 , which according to central and southern Italy geomorphology correspond to rivers self-formed mostly in alluvial flat plains.

The basin drainage area does not seem to affect the coefficients of variation CV and skewness G of the Italian AFS's, except for areas greater than 3000 Km^2 where lower coefficients of variation are generally found. It is difficult to recognize if this is due to the upstream spilling phenomenon or to climatic reasons, although with regard the latter hypothesis, there is the observation that the annual number Λ =Λ_1+Λ_1 of independent floods tends to increase with the area.

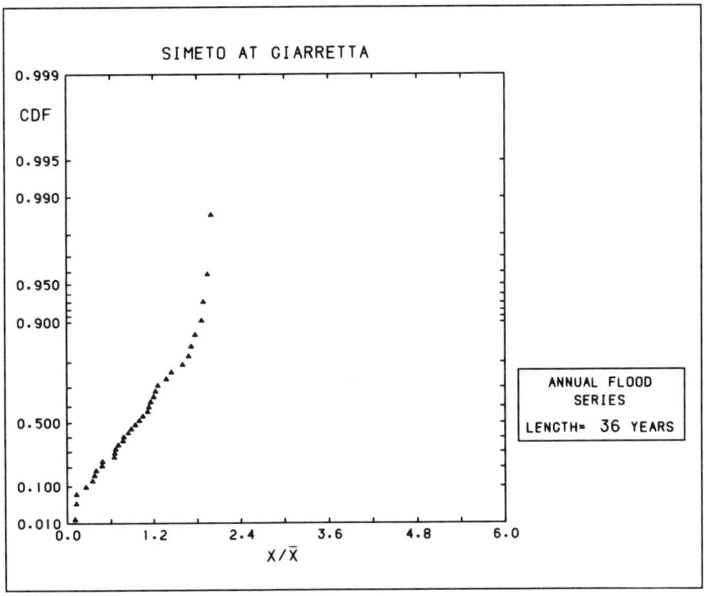

Fig. 1 - Observed CDF of a typical AFS whose largest values are affected by lateral spilling phenomena.

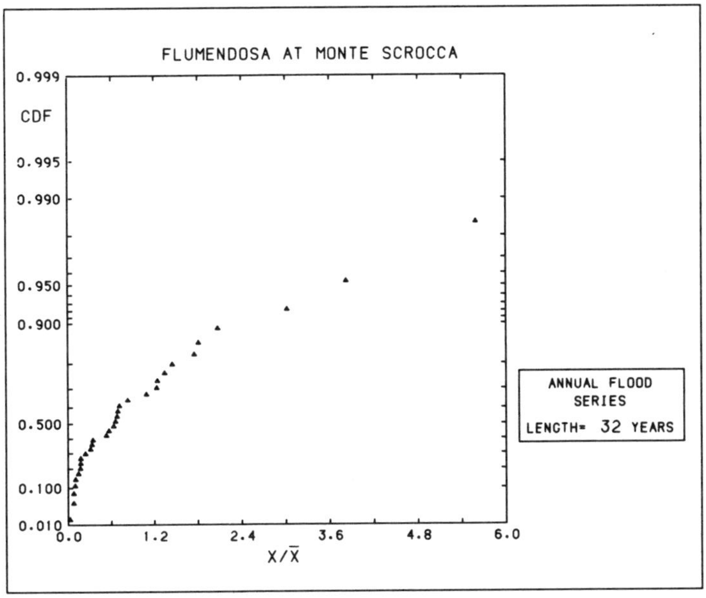

Fig. 2 - Observed CDF of a typical AFS recorded in semiarid regions.

Anyway, for the sake of simplicity, basins with area larger than 3000 Km^2 have not been considered so far.

Coefficient of variation tends to increase from humid to arid regions. In Italy, sub-regional median values of CV range from 0.40 to 0.67 except in semiarid regions, where it reaches unity. AFS's recorded in that region, because of the scanty average annual number of floods, lead to a non-negligible probability of having zero, or practically zero values of annual maximum flow. Clearly, this contrasts with the use of the proposed ML estimation procedure which assumes that the PDF is undistinguishable from a continuous function also allowing for negative values. These AFS's were not considered, either. A typical example of such AFS, recorded in the Sardinia Island, is shown in Fig. 2.

According to the observations outlined above, a data base for inferring the regionalization model was formed by selecting, amongst the AFS's recorded at stations operating in Italy on non- Alpine streams, twenty-eight AFS's having an average length nearly equal to 39, and with a total of 1091 station-year data. The AFS's having less than 35 years of record were not taken into account, as suggested by Arnell and Gabriele (1986) for the first level of the regional estimation procedure.

A sensitivity analysis demonstrated that the selected 28 AFS's were not significantly affected by small changes such as the removal of individual data sets.

TCEV-ML estimation of parameters was carried out using the procedures detailed above and assuming that the parameters θ_*, Λ_*, and Λ_1 have constant values regardless of the site in the entire region in question, θ_1 (and thus any location parameter) being variable from site to site. The following regional estimates of θ_*, Λ_*, and Λ_1 were attained: $\theta_*=2.654$, $\Lambda_*=0.350$, $\Lambda_1=12.15$.

Errors depending on neglecting interstation correlation effects are likely insignificant, for the 28 AFS's are quite scattered over the entire region and neither do cover a homogeneous time duration.

Theoretical sampling distribution of the sample coefficients of skewness and variation of the regionalized TCEV distribution with the attained regional values of θ_*, Λ_1 and Λ_1 were derived by exploiting Monte Carlo experiments.

Ten-thousand independent synthetic series were generated with their lengths distributed in the manner of the 28 AFS's. Coefficients of skewness G and variation CV of each series were calculated. Experiment-derived mean and variance of G and CV worked out to be the values listed in Table 3 where they are compared with the respective values calculated using the 28 AFS's.

One may note that the observed variance of G is almost completely accounted for by the sampling variability. This means that the contribution of the spatial disturbance is so little that there is no evidence to reject the hypothesis that the region wherein the 28 AFS's have been recorded is homogeneous with respect to the skewness.

This does not apply to the coefficient of variation CV. In fact its observed variance is almost twice the experiment-derived variance which reproduces the expected sampling variability. Moreover, since the ratio of the time sampling variance to the observed variance is only

0.0127/0.0225=0.564, a division of the entire region into homogeneous sub-regions with respect to CV should be advocated.

		Observed	Experiment-derived
CV	Mean	0.536	0.537
	Variance	0.0225	0.0127
G	Mean	1.431	1.587
	Variance	0.634	0.586

Table 3 - Mean and variance of the coefficients of variation CV and skewness G for 28 28 observed Italian annual flood series and 10,000 experiment-derived series.

In other words the 28 AFS's exhibit a variability of G, though very large,but comparable with the theoretical one, whereas for the same series the variability of CV is much larger than that expected. This constitutes the basis for suggesting the use of the hierarchical approach. In fact, it enables us to utilize the information contained in a large number of AFS's over an extensive region for estimating the parameters for which the information provided by a specific series is particularly scanty. This approach exploits a smaller number of AFS's only at the next levels, when estimation of parameters for which a specific AFS provides a relatively larger amount of information, is being considered. The hierarchical approach can then perform better than the classical index flood method, which,in order to preserve the simultaneous homogeneity with respect to all the parameters to be estimated, is forced to exploit a limited number of AFS's for estimation of high-order moment-ratios also.

6. CONCLUSION

The hierarchical approach for regional flood frequency estimation herein proposed arises from the observation that the hydrologic homogeneity with respect to the shape parameters of the annual flood distribution can be hypothesized in large regions, while as regards the coefficient of variation the homogeneity is evident only in smaller areas. Therefore, for estimation of the shape parameter, one can use a larger number of data sets than for estimating the coefficient of variation. As a consequence, the hierarchical approach exploits, better than the classical index flood method, the hydrologic information.

A regional estimator which uses: i) the two-component extreme value distribution (TCEV), ii) the maximum likelihood (ML) estimation method, iii) a station-year type regionalization algorithm, is shown to account for the observed distribution of the skewness coefficient of a number of annual flood series (AFS) recorded at stations operating in Italy on non-Alpine streams. For the same series the observed distribution of the coefficient of variation appears to be more variable than the theoretical sampling distribution. This suggests that the considered region should be broken down into different sub-regions to obtain valuable regional estimates of the dispersion parameter.

Computer simulation experiments showed that the regionalized TCEV-ML estimator exhibits good performance in estimating quantiles of the annual flood distribution, whereas the estimates of the parameters θ_* and, especially Λ_*, are relatively worse. It was shown that local disturbance factors can heavily affect the extreme values of floods, thus making AFS's exhibit anomalous shapes which are independent of the general characteristics of the basin. Some such shapes are frequently accounted for by the stream morphology, a typical example being constituted by those AFS's whose higher values are bounded because of the spilling phenomena occurring immediately upstream the gauging station. On the other hand it seems that these problems deserve more attention than those, largely discussed so far in the literature, devoted to ensure the efficiency and resistance of the estimators with respect to more or less unrealistic parent flood-distribution.

ACKNOWLEDGMENTS. This research has been supported by the Italian National Research Group for the Prevention of Hydro-Geological Disasters.

7. REFERENCES

Arnell, N.W., and S.Gabriele, 'Regional flood frequency analysis using the two-component extreme value distribution: an assessment using computer simulation experiments'. Proceedings of the Seminar on Combined Efficiency of Direct and Indirect Estimates for Point and Regional Flood Prediction, Perugia,Italy, 1986.

Beran, M.A., J.R.M. Hosking, and N.W. Arnell, Comment on 'Two-component extreme value distribution for flood frequency analysis' by Rossi et al. (1984). Water Resources Research, **22(2)**,263-266, 1986

Dalrymple, T., 'Flood-frequency analyses'. U.S. Geological Survey Water Supply Paper **1543-A**, 1960.

Fiorentino, M., and S.Gabriele, (in Italian), 'Distribuzione TCEV:metodi di stima dei parametri e proprietà statistiche degli stimatori'. Consiglio Nazionale delle Ricerche-IRPI, Geodata n. **25**, Cosenza, Italy, 1985.

Fiorentino, M., P.Versace, and F.Rossi, 'Regional flood frequency estimation using the two-component extreme value distribution'. Hydrological Sciences Journal, **30**,51-64,1985.

Hosking, J.R.M., J.R.Wallis, and E.F.Wood, 'Estimation of the generalized extreme value distribution by the method of probability weighted moments'. Institute of Hydrology Report **89**. Wallingford, U.K., 1984.

Hosking, J.R.M., J.R.Wallis, and E.F.Wood,'An appraisal of the regional flood frequency procedure in the U.K. Flood Studies Report'. Hydrological Sciences Journal, **30**, 85-89,1985.

Matalas, N.C., and M.A.Benson, 'Effects of interstation correlation regression analysis'. Journal of Geophysical Research, **66(10)**, 3285-3293, 1961.

Matalas, N.C., and E.J.Gilroy, 'Some comments on regionalization in hydrologic studies'. Water Resources Research, **4(6)**,1361-1369, 1968.

Natural Environment Research Council (NERC), Flood Studies Report. Five volumes,London, 1975,

Riggs, H.C., 'Regional analysis of streamflow characteristics'. U.S. Geological Survey Water Resources Invest. Tech. Book 4, 1973.

Rossi F., M.Fiorentino, and P.Versace,'Two-component extreme value distribution for flood frequency analysis'. Water Resources Research, **20(7)**, 847-856, 1984.

Stedinger, J.R., 'Estimating a regional flood frequency distribution'. Water Resources Research, **19(2)**, 503-510, 1983.

Versace P., M.Fiorentino, and F.Rossi, 'Analysis of flood series by stochastic models', in Time Series Methods in Hydrosciences, edited by A.H.El-Shaarawi and S.R.Ester by, 315-324, Elsevier, New York,1982.

Wallis, J.R., 'Risk and uncertainties in the evaluation of flood events for the design of hydrologic structures'. In Piene e Siccità, edited by A.Guggino, G.Rossi, and E.Todini, CLUP, Milan, 3-48, 1980

APPLICATION OF REGIONAL FLOOD FREQUENCY ANALYSIS TO PHILIPPINE RIVERS

Eluderio S. Salvo
Development of Environmental Systems, Inc.
16 Assets St. GSIS Village
Quezon City, Metro-Manila
Philippines

ABSTRACT. The Philippines is divided into 12 water resources regions. Two regions, one in the Visayas and one in Mindanao were analyzed. After annual peaks within a region were listed and missing data filled-in by correlation, preliminary frequency curves were prepared for each station using the log-Pearson Type III distribution. Using the preliminary frequency curves, the homogeneity test was applied on all rivers within a region. For each region, a regional flood frequency curve was derived using the ratio of discharge to the mean annual flood as ordinate and recurrence interval as abscissa. Another curve was prepared showing the relation of mean annual flood to the size of the area. An equation derived was $Qm = CA^{0.77}$.

INTRODUCTION

The Philippines is divided into 12 water resources regions. The subdivision is shown in Fig. 1. While the subdivision conforms more or less with the 12 political regions, the boundaries are so chosen that the areas falling under each region took into account the prevailing climatic patterns (Quimpo et al, 1983, p. 324).

The National Water Resources Council of the Philippines, the government agency that is responsible for conducting and integrating all activities related to water resources development and management, published "Philippine Water Resources Summary Data" (NWRC, 1980). All the river discharges used in the analysis were taken from this publication.

It has been found that 48 years of record were needed to define the 100-year flood within 25% and 115 years of record to define the same flood within 10% with a 95% probability in both cases (Benson, 1952). It was considered dangerous to extrapolate a 100-year flood using a 20-year record from a single station (Cole, 1965, p. 42). Even if the distribution can be linearized on probability paper, hydrologists are warned against extrapolation because the effect of sampling errors is magnified (Chow, 1964, p. 8-27).

In a hydrologically homogenous area, which need not be contiguous, the data from $\underline{n}$ stations covering the same number of years can be

V. P. Singh (ed.), Regional Flood Frequency Analysis, 51–66.

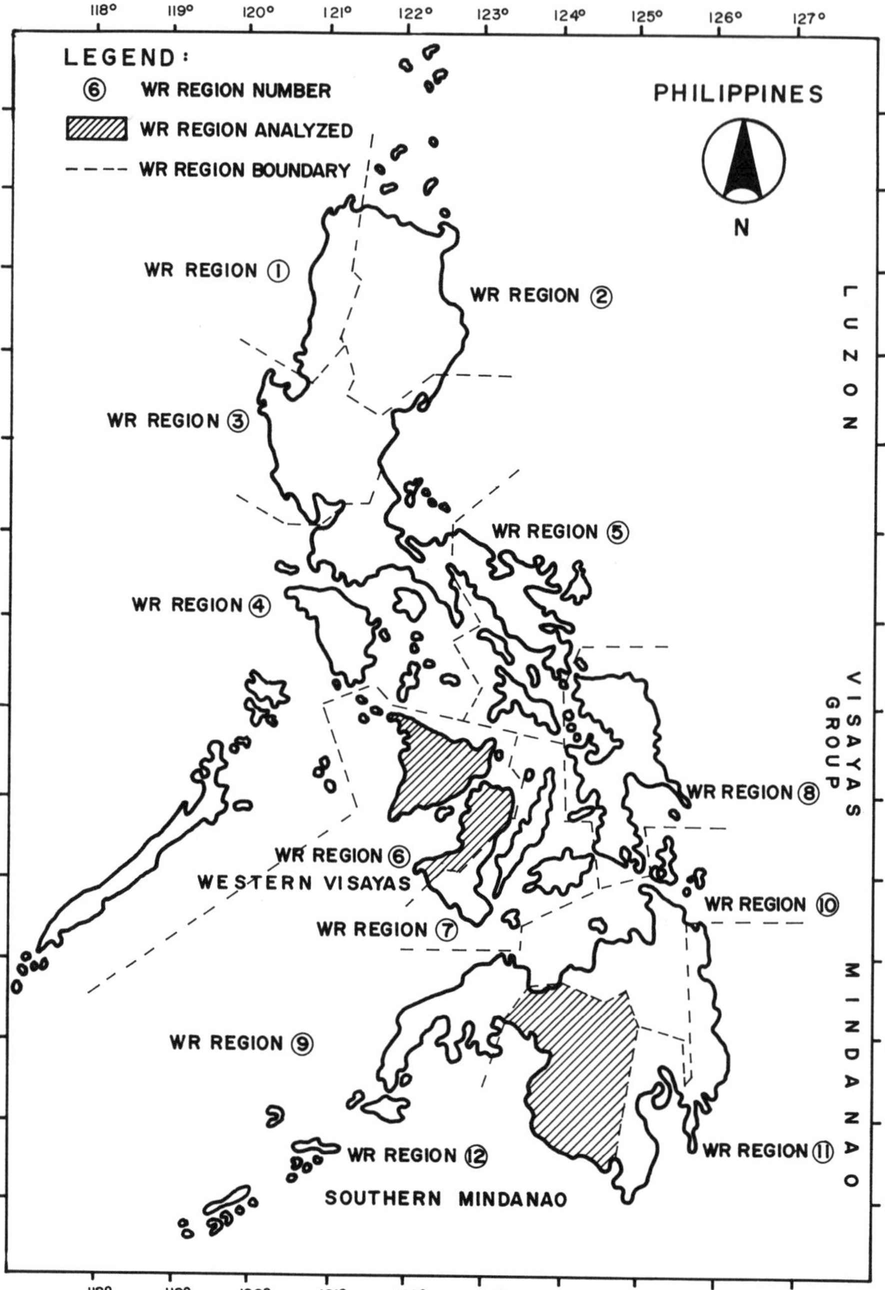

FIG. 1 THE TWELVE WATER RESOURCES REGION

combined to form n measurements of the same data and thus improve the statistical quality of the derived frequency distribution (Ward, 1978, p. 87). This is the principle behind regional flood frequency analysis and is applied in two water resources regions in the Philippines.

WATER RESOURCES REGIONS ANALYZED

Water Resources Region 6

This region, located in Western Visayas as shown in the cross-hatched portion of Fig. 1, lies between latitudes 9° 45' and 12° 00' and longitudes 121° 50' and 123° 35'. It is comprised of the island of Panay and a portion of the island of Negros, called Negros Occidental.

It has 38 gaging stations with watershed areas varying from a minimum of 12 sq km to a maximum of 2059 sq km. Out of the 38 stations, only 31 were used in the analysis because the 6 stations eliminated do not have discharge records but only gage heights while one has only four years of record with many months of missing data. The names of the gaging stations used and their corresponding areas are presented in columns (2) and (3) of Table I.

Water Resources Region 12

This region is located in the southern part of the island of Mindanao as shown also in the cross-hatched portion of Fig. 1. It lies between latitudes 6° 00' and 8° 15' and longitudes 123° 35' and 125° 25'.

It has 34 gaging stations having watershed areas varying from 9 sq km to a maximum of 19406 sq km. Out of the 33 stations, only 19 were used because 5 stations discarded do not have discharge records but only gage heights while 9 have less than five years of record. Similarly, the names and corresponding areas are shown in columns (2) and (3) of Table II.

Usable Records

A minimum of 5 years of record was used in the analysis. The usable records for region 6 were divided into two sets. One set is composed of 14 years of annual peak discharges for each station corresponding to the years 1957-1970. Where there were gaps in the record, the annual peaks were filled-in by correlation with other rivers within the region.

Another set of 21 years of annual peaks corresponding to the years 1950-1970 were prepared. Similarly, all gaps were filled-in by correlation with other rivers.

The same method of listing was done with the data from region 12. There were also two sets of annual peaks for each station corresponding to a short base period of 12 years, from 1959-1970 and a longer base period of 20 years, from 1951-1970.

In the determination of the missing data, rivers with incomplete records were correlated with the annual values for the same years with rivers with complete records. The missing values were estimated from the resulting curve which was fitted graphically by eye without resorting to mathematical computation. It was accepted that the scatter of the data did not warrant any mathematical correlation since the only purpose is to be able to allocate an order number in the annual peaks.

Determination of Recurrence Interval

Utilizing the set of annual peaks of each region, the recurrence interval T for the annual peaks were computed by the well known Weibull formula, T = (n+1)/m. The recurrence intervals for the filled-in data were not computed and they did not enter into the computation for the parameters of the frequency curves.

Log-Pearson Type III Distribution

The annual peak series were converted to logarithms and the mean, standard deviation and the skew coefficient were computed.

The value of flood peak, X, for any recurrence interval was computed from the formula

$$\log X = \overline{\log X} + K\,\sigma_{\log x} \tag{1}$$

where $\overline{\log X}$ and $\sigma_{\log x}$ are the mean and standard deviation respectively and K is taken from the table of values for skew coefficient (Linsley et al, 1982, pp. 361-363).

Preliminary Frequency Curves

For each station the annual peak values were plotted as ordinates and the recurrence intervals as abscissa on log-normal probability paper. The annual peak values for recurrence intervals of 2, 10 and 100 years were computed using equation (1) and also plotted. These three annual peak values, designated Q_2, Q_{10} and Q_{100} defined the preliminary curve.

Although the theoretical plot on log-normal probability paper should be a straight line, the meagerness of the data does not justify insistence on a straight line. For the smaller recurrence intervals, graphical fitting by eye was utilized.

As an example, the resulting preliminary frequency curve of Aklan river of region 6 is shown in Fig. 2.

The suggested method for low and high outliers test (Thomas, 1985) was not utilized because there is no generalized skew in the Philippines at present. It was accepted that graphical plotting in this case has granted greater weight to the lower and medium values rather than allowing greater influence by the high outliers.

Homogeneity Test

As previously mentioned, the data from several stations within a hydrologically homogenous region can be combined so as to increase the size of the sample and thus improve the statistical quality of the frequency curve. In order to define a homogenous region, a test has been developed by Langbein for regional frequency analysis practiced by the U.S. Geological Survey (Chow, 1964, pp. 8-37).

For each station within a region to be tested, the 10-year flood was determined from the preliminary frequency curve. For the same curve, the mean annual flood, defined as the flood having a recurrence interval of 2.33 years, was also determined. These 10-year floods expressed as ratios to mean annual floods were averaged to obtain the mean for the region.

For each station a new 10-year flood was obtained by multiplying the mean annual flood by the mean of the ratio. The computations are shown in Table I. The recurrence intervals for the new 10-year floods were plotted against the adjusted period of record as shown in Fig. 3 for region 6. The convention followed was that the adjusted period of record was equal to the actual record plus one-half the missing years.

The same procedure was done for the data of region 12 and the results are shown in Table II.

From Table I, col. (9), and Fig. 3, it can be seen that the data from Tangalan, Mambuasao, Paliuan, Sibajao and Binalbagan rivers had to be discarded because they plotted outside the test curve.

Similarly, from Table II, col. (9), the data from Mindanao, M'lang, Rio Grande de Mindanao, Pulangui, Carmen and Pulangui, Valencia had to be discarded because they also failed to pass the homogeneity test.

Confidence Limits

Confidence limits for the preliminary curves were computed using the approximate values presented by Stedinger (1983). In Table III, likewise in Table IV, are presented the computed values of the 90% confidence limits for the 10-year and 100-year events for stations having records that extended beyond the shorter base period. The flood flows used for the 10-year and 100-year values in the preliminary curves are also presented for comparison. As can be seen in Table IV all the floods flows lie within the confidence interval.

The control curves defining the confidence limits are incorporated in Fig. 2 for region 6.

Regional Frequency Curve

From each preliminary frequency curve, the ratio of discharge to the mean annual at chosen recurrence intervals were computed. All the preliminary frequency curves for each region were processed. The recurrence intervals chosen are for convenience of plotting only.

The median values of the ratios were plotted and a curve drawn. The resulting curve for the short base period was compared with the resulting curve for the longer base period. By giving more weight to the stations common to both base periods, the final regional frequency curve for each region was derived. The curves are shown in Fig. 4 for region 6 and Fig. 5 for region 12.

Mean Annual Flood vs. Area

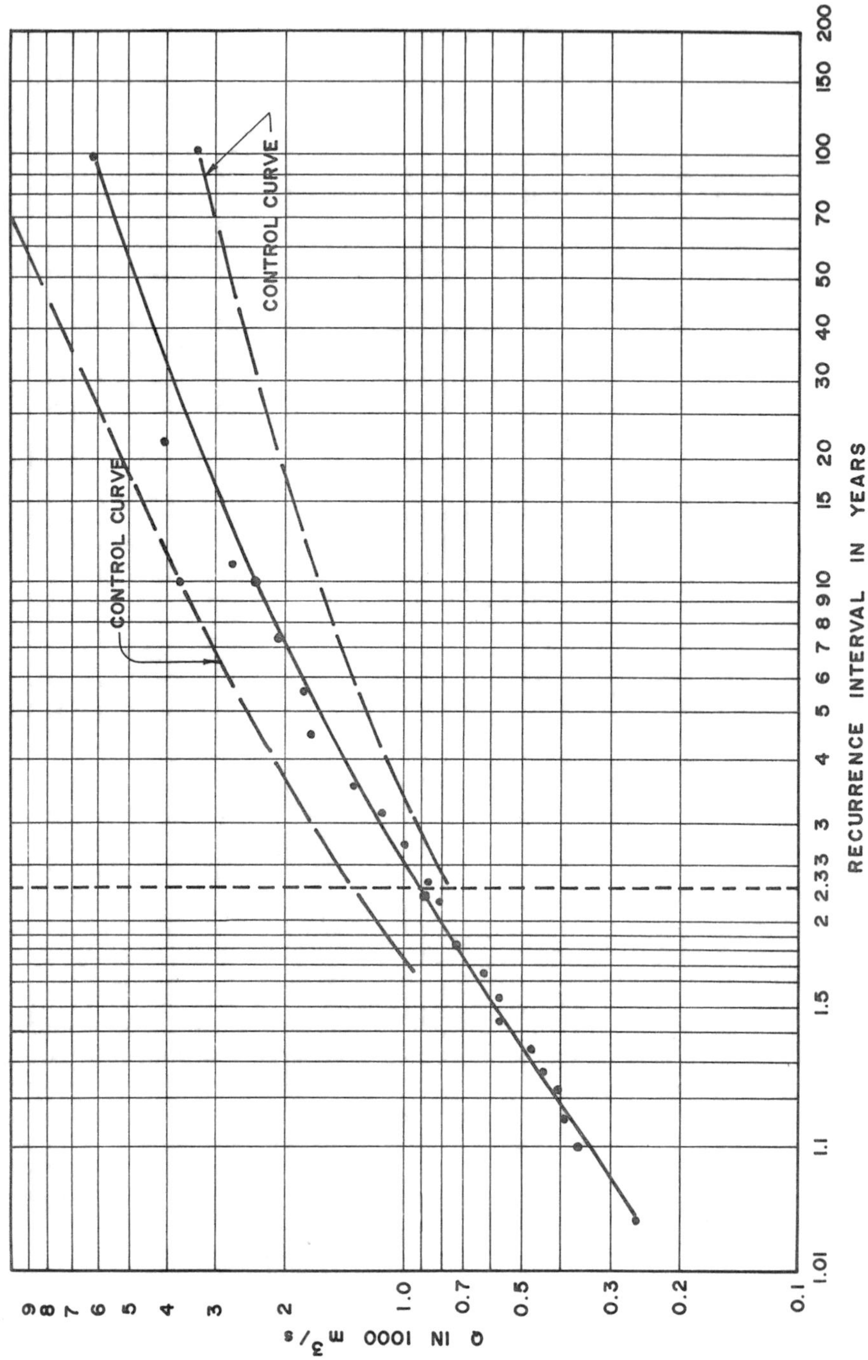

FIG. 2 PRELIMINARY FREQUENCY CURVE FOR AKLAN RIVER OF REGION 6

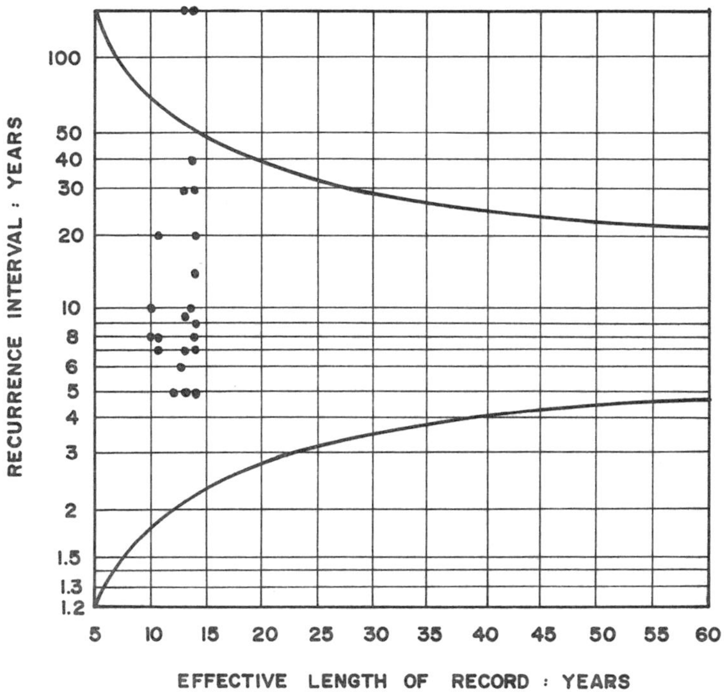

FIG. 3 HOMOGENEITY TEST FOR REGION 6

TABLE I DATA FOR HOMOGENEITY TEST, BASE PERIOD 1957-1970

[1] No./ Region	[2] River	[3] Water shed area (km^2)	[4] Mean annual flood $Q_{2.33}$ (m^3/s)	[5] 10-year flood Q_{10} (m^3/s)	[6] Ratio Q_{10}: $Q_{2.33}$	[7] $Q_{2.33}$ X ave. ratio (m^3/s)	[8] T for Q of col. (7) (yrs)	[9] Adjusted period of record (yrs)
01/6	Tangalan	38	320	450	1.41	617	150	14
02/6	Aklan	705	750	1750	2.33	1447	7	14
03/6	Bacong	54	50	100	2.00	96	10	13
04/6	Maayon	265	95	170	1.79	183	15	14
05/6	Mambuasao	307	370	510	1.38	714	150	14
06/6	Panay	880	850	1400	1.65	1640	20	14
07/6	Paliuan	176	900	1200	1.33	1737	150	14
08/6	Barotac	90	55	125	2.27	106	8	14
09/6	Sibalom, Pang.	635	540	920	1.70	1042	30	13
10/6	Sibalom, Omamb.	117	225	480	2.13	434	8	14
11/6	Inabasan	97	65	150	2.31	125	7	14
13/6	Jalaur, Dingle	1065	800	1500	1.88	1544	10	13.5
14/6	Jalaur, Pototan	1499	900	1500	1.67	1737	30	14
15/6	Suague	186	150	250	1.67	289	20	14
16/6	Ulian	247	230	480	2.09	443	8	14
17/6	Jalaur, Passi	534	950	1700	1.79	1833	15	14
18/6	Jalaur, Sim.	169	500	800	1.60	965	30	14
19/6	Jalaur, Alibu.	120	120	350	2.92	231	5	12
20/6	Sibajao	12	20	25	1.25	38	150	14
21/6	Malogo	129	400	1000	2.50	772	6	12.5
22/6	Bago	683	660	1400	2.12	1273	8	10.5
24/6	Binalbagan	350	260	360	1.38	501	150	13
25/6	Piguian	12	17	35	2.06	32	10	10
26/6	Imbang	33	95	250	2.63	183	5	14
31/6	Ilog, Camugao	1959	1000	1630	1.63	1930	20	10.5
32/6	Ilog, San Juan	1947	1100	2230	2.03	2123	7	10.5
34/6	Hilabangan, Dang.	431	350	770	2.20	675	7	14
35/6	Hilabangan, Tag.	392	100	250	2.50	193	5	13
36/6	Ilog, Pg. Orang.	1453	1300	2000	1.54	2509	40	14
37/6	Ilog, Dahile	1390	500	1000	2.00	965	9	14
38/6	Ilog, Inapoy	1245	410	845	2.06	791	8	10
	Average ratio				1.93			

TABLE II DATA FOR HOMOGENEITY TEST, BASE PERIOD 1959-1970

[1] No./Region	[2] River	[3] Water shed area (km^2)	[4] Mean annual flood $Q_{2.33}$ (m^3/s)	[5] 10-year flood Q_{10} (m^3/s)	[6] Ratio Q_{10}: $Q_{2.33}$	[7] $Q_{2.33}$ X ave. ratio (m^3/s)	[8] T for Q of col. (7) (yrs)	[9] Adjusted period of record (yrs)
01/12	Mandulog	576	210	350	1.67	437	25	10
02/12	Maigo	74	90	260	2.89	187	6	10
03/12	Maranding	345	100	185	1.85	208	14	11.5
06/12	Libungan	534	260	950	3.65	541	4	12
07/12	Mindanao	17744	1100	1500	1.36	2288	100	9.5
08/12	Dansalan	3749	400	600	1.50	832	80	9
09/12	Allah,Ampatuan	1496	220	410	1.86	458	15	12
10/12	Allah, Isulan	936	110	340	3.09	229	5	11.5
12/12	Banga	324	50	85	1.70	104	25	11.5
15/12	Marbel	290	13	40	3.07	27	6	12
16/12	Alip	380	260	430	1.65	541	70	10
17/12	Malasila	145	80	225	2.81	166	5	11.5
18/12	M'lang	164	110	180	1.64	229	100	11
20/12	Rio Grande	12999	900	1100	1.22	1872	150	12
21/12	Maridagao	1333	480	950	1.98	998	10	12
22/12	Kabakan	698	480	820	1.71	998	10	12
23/12	Pulangui, Car.	6752	980	1300	1.33	2038	150	12
29/12	Pulangui, Val.	2730	850	1200	1.41	1768	150	8.5
31/12	Suwaga	327	190	600	3.16	395	6	8
	Average ratio				2.08			

TABLE III 90% CONFIDENCE LIMITS FOR 10-YEAR EVENT

[1] No./ Region	[2] River	[3] Years of record	[4] $\overline{\log X}$	[5] $\sigma_{\log x}$	[6] 90% confidence limits for 10-year event		[7] Q_{10} of preliminary curve (m^3/s)
					lower (m^3/s)	upper (m^3/s)	
02/6	Aklan	19	2.95424	0.32902	1712	3940	2430
10/6	Sibalom	21	2.25573	0.26332	306	565	397
15/6	Suague	21	2.06296	0.23392	185	319	226
26/6	Imbang	21	1.95200	0.33998	178	392	250
03/12	Maranding	19	2.06375	0.23746	184	336	234
06/12	Libungan	20	2.53343	0.46519	856	2688	1200
10/12	Allah	16	2.09873	0.31097	225	538	320
15/12	Marbel	20	1.31714	0.51381	57	203	98

TABLE IV 90% CONFIDENCE LIMITS FOR 100-YEAR EVENT

[1] No./ Region	[2] River	[3] Years of record	[4] 90% confidence limits for 100-year event		[5] Q_{100} of preliminary curve (m^3/s)
			lower (m^3/s)	upper (m^3/s)	
02/6	Aklan	19	3355	11225	6347
10/6	Sibalom	21	520	1329	847
15/6	Suague	21	297	682	362
26/6	Imbang	21	352	1181	682
03/12	Maranding	19	299	716	420
06/12	Libungan	20	2224	11649	2320
10/12	Allah	16	422	1500	705
15/12	Marbel	20	164	1024	435

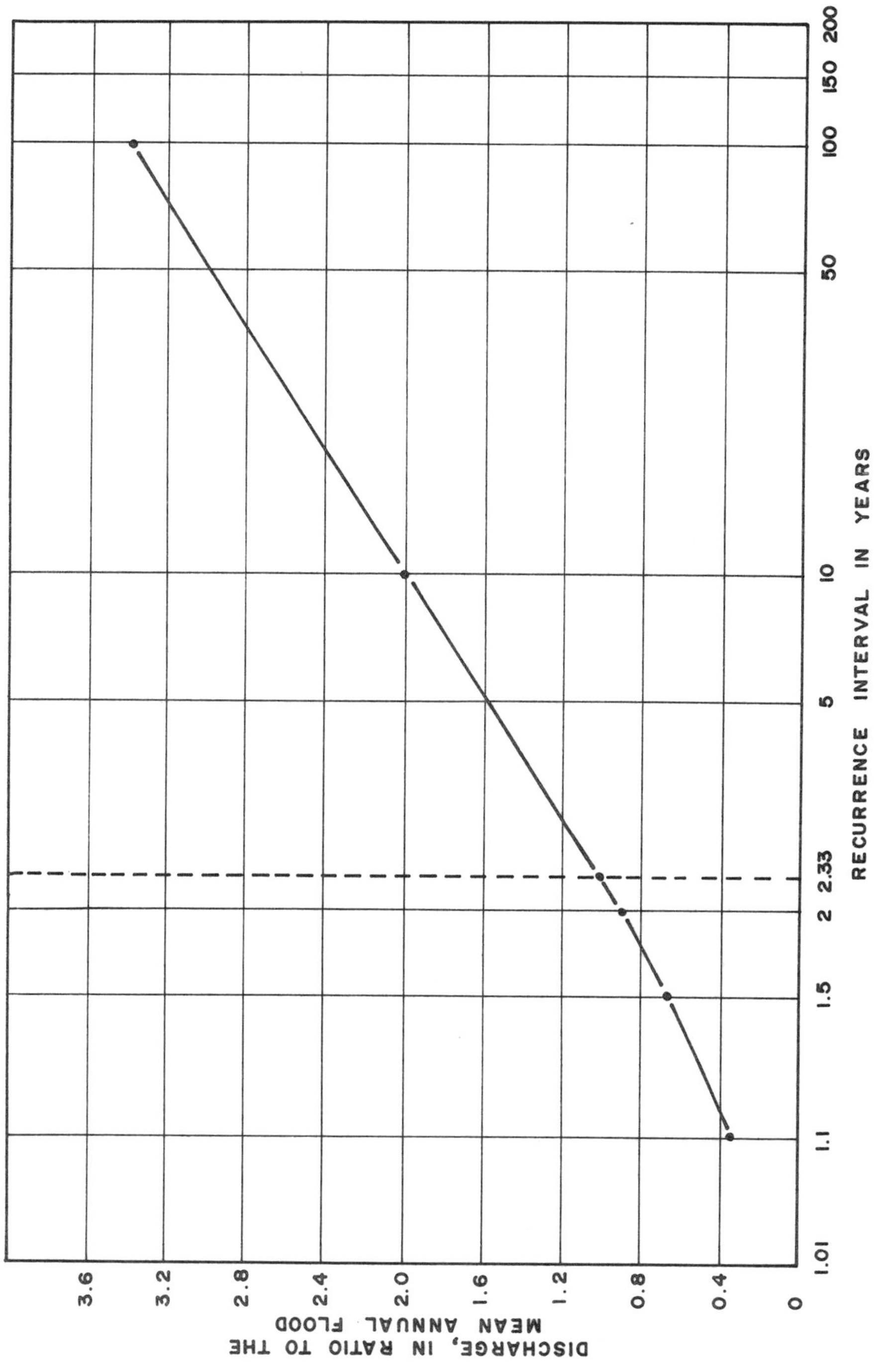

FIG. 4 REGIONAL FREQUENCY CURVE FOR REGION 6, 1950 - 1970

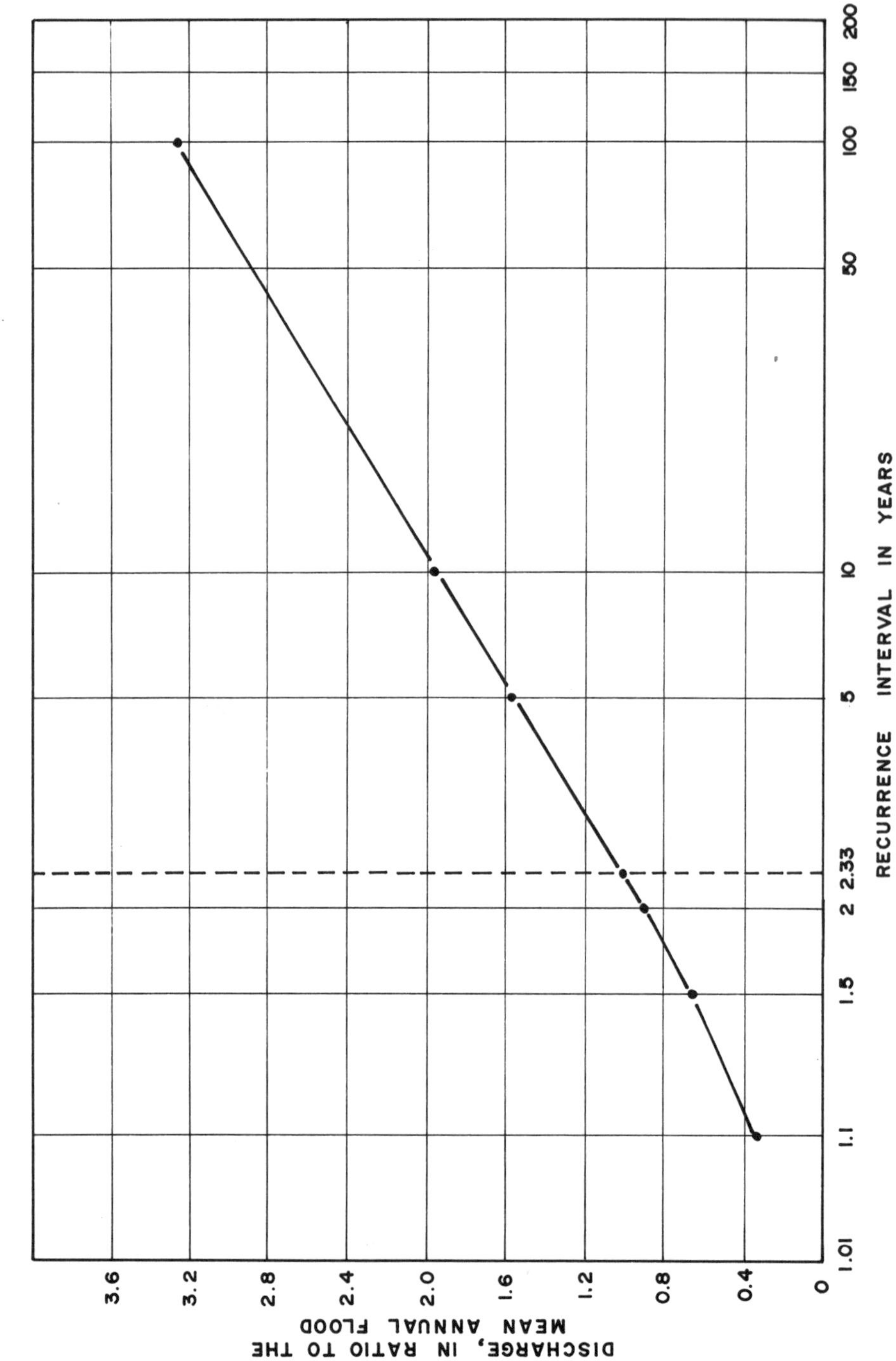

FIG. 5 REGIONAL FREQUENCY CURVE FOR REGION 12, 1951 - 1970

All the mean annual floods of all the stations for region 6 were plotted against their watershed area on logarithmic paper as shown in Fig. 6. By fitting a curve using graphical procedure (Yevdjevich, 1964, p. 8-46) it was found that the plotted points showed marked tendency to group along a family of straight lines that were parallel and defined by the equation

$$Q_m = CA^b \quad (2)$$

where:

Q_m = mean annual flood in m^3/s
A = watershed area in sq km
C and b are constants

It was found that the value of b could vary between b = 0.70 to b = 0.85. Taking b = 0.77, it can be seen that the mean annual flood vs. area are within the band C = 1 and C = 10.

The same was done for region 12 and shown in Fig. 7. Using the same value of b = 0.77, the mean annual floods vs. areas are within the band C = 1/2 and C = 4.

These two curves, regional frequency curve and mean annual flood vs. area curve, are the curves prescribed by the Survey method of analysis (Dalrymple, 1964, p. 25-6).

It is interesting to note that the mean annual floods of region 6 plot above the mean annual floods of region 12. This could be attributed to the fact that region 6 lies in the typhoon belt while region 12 does not.

Referring to Fig. 6, it can be also noted that the mean annual floods of the same river but different gaging stations, (designated 17, 18 and 19) do not plot along the same C line. This is also true for 36, 37 and 38. It can be inferred that slope should be considered as one of the parameters in the relation between mean annual flood and drainage area.

CONCLUSION

Regional flood frequency analysis can be applied without any doubt to increase the statistical quality of the regional frequency curve especially to a data-sparse country like the Philippines. In ungauged rivers within the region analyzed, the water resources engineer can use the Survey method of analysis.

Although there are approximately 200 stream gaging stations and 369 rain gaging stations all over the country, only 87 of the stream gaging stations and 86 rain gages have more than 20 years of record. It is therefore necessary that all the regions be analyzed and regional frequency curves derived. A generalized skew coefficient applicable to the country should be derived and utilized in order to minimize bias in estimating future frequencies (Beard, 1974, p. 25).

Many years have elapsed since the Philippine Water Resources

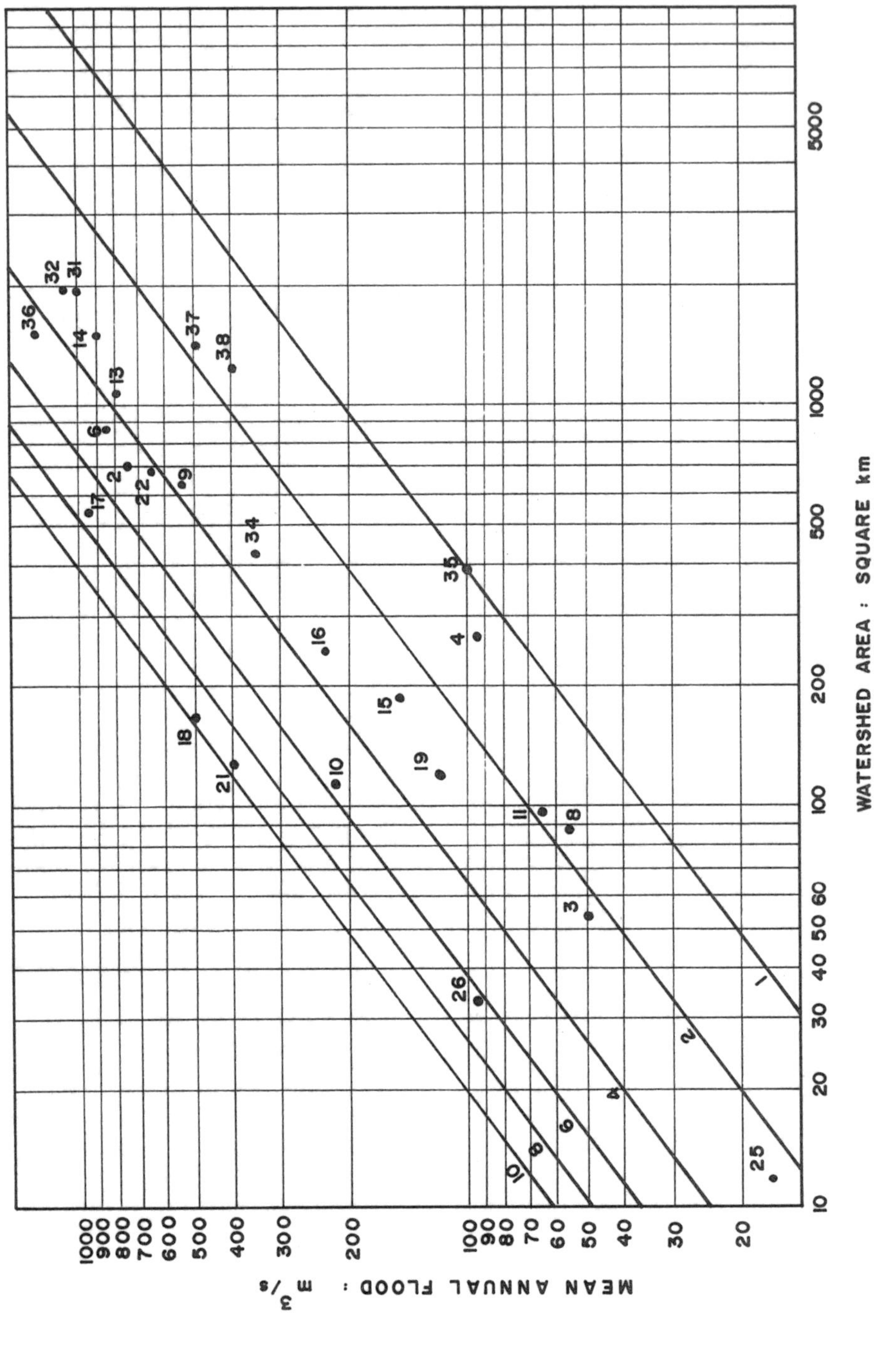

FIG. 6 VARIATION OF MEAN ANNUAL FLOOD WITH AREA FOR REGION 6.

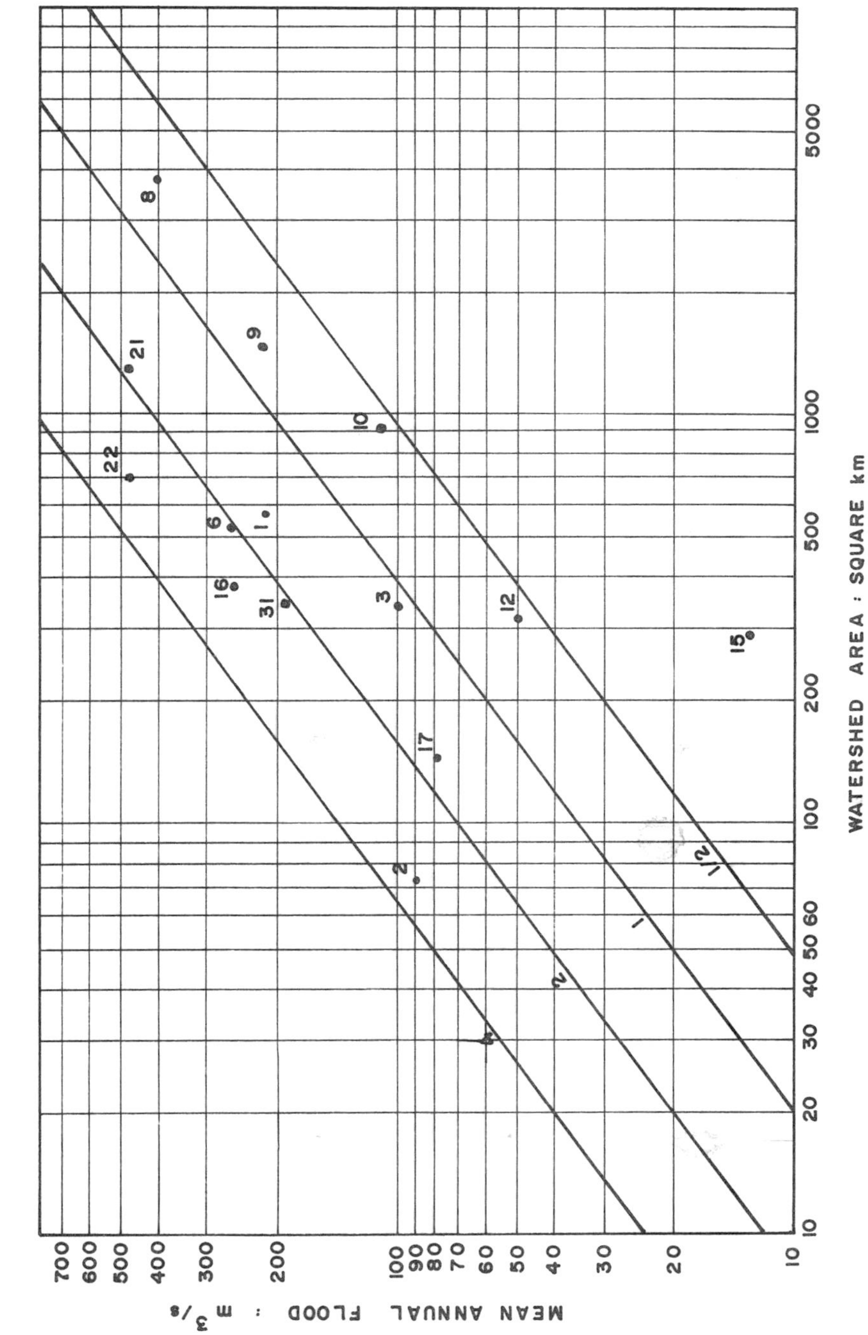

FIG. 7 VARIATION OF MEAN ANNUAL FLOOD WITH AREA FOR REGION 12.

Summary Data was published in 1980. It is hoped that this study will serve its purpose until such time as additional data that accumulated through the years will have been published.

ACKNOWLEDGEMENTS

The author wishes to acknowledge the considerable help provided by Engr. Rene Flordeliz and Arch. Nanette V. Salvo in preparing all the tables and illustrations in this paper.

REFERENCES

Beard, L.R. 1974. 'Flood Frequency Techniques' 59 pp., Office of Water Research and Technology and Water Resources Council, U.S. of America.

Benson, M.A. 1952. 'Characteristics of Frequency Curves Based on a Theoretical 1000-Year Record.' U.S. Geological Water Supply Paper 1543-A.

Chow, V.T. 1964. Part I. Frequency Analysis. in: V.T. Chow (Editor-in-Chief), Handbook of Applied Hydrology: A Compendium of Water Resources Technology, pp. 8-1 to 8-42.

Cole, G. 1965. 'An Application of the Regional Analysis of Flood Flows.' in: Institution of Civil Engineers, River Flood Hydrology Symposium, pp. 39-57, London, England.

Dalrymple, T. 1964. Part I. Flood Characteristics and Flow Determination. in: V.T. Chow (Editor-in-Chief) Handbook of Applied Hydrology: A Compendium of Water-resources Technology, pp. 25-1 to 25-33, McGraw Hill Book Co., New York.

Linsley, R.K., Kohler, M.A. and Pulhus, J.L.H. 1982. Hydrology for Engineers, McGraw Hill International Book Co., Tokyo, pp. 357-384.

Quimpo, J.R., Alejandrino, A.A. and McNally, T.A. 193. 'Regionalized Flow Duration for Philippines.' Journal of Water Resources Planning and Management, American Society of Civil Engineers, Vol. 109 No. 4, pp. 320-330.

Stedinger, J.R. 1983. 'Confidence Intervals for Design Events.' Journal of Hydraulic Engineering, American Society of Civil Engineers, Vol. 109 No. 7, pp. 13-27.

Thomas, W.O. Jr. 1985. 'A Uniform Technique for Flood Frequency Analysis.' Journal of Water Resources Planning and Management, American Society of Civil Engineers, Vol. 111 No. 3, pp. 321-337.

Ward, R. 1978. Floods, A Geographical Perspective, The McMillan Press Ltd., London, pp. 68-93.

Yevdjevich, V.M. 1984. Part II. Regression and Correlation Analysis. in: V.T. Chow (Editor-in-Chief), Handbook of Applied Hydrology: A Compendium of Water-resources Technology, pp. 8-43 to 8-67, McGraw Hill Book Co., New York.

National Water Resources Council, Republic of the Philippines. 1980. Water Resources Summary Data. Vol. 1: 'Streamflow and Lake or River Stage Ending December 31, 1970.' pp. 279-319 and pp. 423-451.

IMPROVING FLOOD QUANTILE ESTIMATION ON UNGAGED WATERSHEDS

Ellen P. Baldwin and Kenneth W. Potter
Department of Civil and Environmental Engineering
University of Wisconsin
Madison, Wisconsin 53706

ABSTRACT. For many situations regional regression is the best available method for estimating flood quantiles at ungaged sites. Quite often, however, the standard errors of existing regression equations are uncomfortably large. It is our belief that these errors can be reduced considerably. This paper reports on our preliminary attempts to do this for drainage basins in the Driftless Area of Wisconsin. Our overall strategy is to investigate gaged rivers for which quantiles estimated from the observed flood series differ significantly from those predicted by the available regression equations. This paper focuses on two such rivers - the Kickapoo and Pecatonica. The Kickapoo and Pecatonica Rivers are gaged at multiple locations. The variation in quantiles at these gages is considerably smaller than would be expected from the variation in drainage area. Hence the regional regression equations perform poorly, underestimating at the gages with the smallest areas and overestimating at the gages with the largest areas. We suspect that the timing of tributary contributions is responsible for the anomolous quantile relationships. This has been confirmed by time-area histograms, which we constructed for each gage. Furthermore, the peaks of the time-area histograms appear to be useful for estimating flood quantiles at sites upstream or downstream of existing gages. Further work is needed to explore the value of this approach.

1. INTRODUCTION

A number of methods have been used to estimate flood quantiles on ungaged drainage basins. Newton and Herrin (1982) demonstrate that, in terms of accuracy, reproducibility and efficiency, regional regression equations provide the most satisfactory method. Such equations are developed by regressing flood quantiles estimated at gaged basins against measured basin characteristics. In general, the basin characteristics used in regional regression equations are easily obtained from maps or other readily available references; few or no field measurements are required.

V. P. Singh (ed.), Regional Flood Frequency Analysis, 67–75.

In spite of the relative success of regional regression equations, they can have serious shortcomings. Standard errors of the predicted quantiles are sometimes uncomfortably high. Often the equations do not account for physical factors which experienced hydrologists know to be important. At other times they incorporate basin characteristics which have no physical relevance.

It is our belief that significant headway can be made in overcoming these shortcomings. This paper reports on our preliminary attempts to do this for drainage basins of the Driftless Area of Wisconsin. In particular we have explored the use of time-area histograms.

2. STUDY AREA

Conger (1981) developed sets of regression equations for each of five regions in Wisconsin, utilizing drainage basin and climatic characteristics to estimate the 2, 5, 10, 25, 50 and 100 year annual flood discharges. The standard errors of these estimates (defined as one half of the width of the 68 percent confidence intervals) range from 23 to 38 percent. The highest standard errors are for Conger's Area 1, where only two variables were found to be significant for predicting flood quantiles: drainage area and an annual precipitation factor. Since there is no apparent physical basis for the precipitation factor, it appears that it acts as a surrogate in the regression for another unidentified physical factor, which like precipitation increases to the southwest. Because of the relatively high standard errors for Area 1 and because of the anomolous correlation between flood quantiles and annual precipitation, we chose this area for study. In particular, we hoped to identify other variables which could be used to reduce the standard error of the regression equations.

Area 1 is located in the so-called Driftless Area of Wisconsin. This region, covering approximately the southwest one-sixth of the state, is the only part of Wisconsin which escaped Pleistocene glaciation. Hence it has relatively high relief and lacks the thick deposits of unconsolidated glacial material which characterize much of the state.

3. METHODS AND RESULTS

Our strategy for attempting to improve the regression equations for the Driftless Area is to investigate gaged rivers for which quantiles estimated from the observed flood series differ significantly from those predicted by the available regression equations. In the case of the Driftless Area, several such rivers exist. This paper focuses on the Kickapoo and Pecatonica Rivers.

The Kickapoo and Pecatonica Rivers are the only Driftless Area rivers for which there exist multiple long-term annual flood series (Figure 1). In the case of the Kickapoo River, three gages occur in series, at 689 km^2, 1598 km^2 and 1787 km^2. For the Pecatonica River there is a lowermost gage at 2678 km^2 and two upstream tributary gages, representing areas of 572 km^2 and 707 km^2. Flood quantiles estimated from the annual series at these gages differ significantly from those predicted by the regional regression equations. The latter underpredict the quantiles at the three smallest basins and overpredict the quantiles at the three largest basins, at all recurrence intervals. The reason for this pattern of errors is obvious - the flood quantiles at the largest basins are only marginally larger than those at the smallest basins. This is counter to the general pattern observed in the Driftless Area, whereby quantiles increase with about the 0.6 power of drainage area.

Obviously several factors affect the way in which quantiles increase with drainage area, including storm characteristics and the factors controlling the timing of tributary contributions to flood flows. Since we were dealing with a single region of reasonably homogeneous climate, we chose to focus on the latter factors. In particular we hypothesized that time-area histograms for the gages on the Kickapoo and Pecatonica Rivers would explain the anomolous quantile relationships, and furthermore might provide an improved method for quantile estimation.

A time-area histogram for a river is constructed by delineating isochrones, or lines of equal travel time to the location of interest, measuring the areas between successive isochrones, and plotting the results as a histogram. The peak of the time-area histogram is of particular interest in this study, as it crudely represents the fraction of total basin area which can be expected to contribute to peak flow at the gage, given a brief but uniform rainfall.

In order to compute travel times for our basins, we used a method suggested by Pilgrim (1977) which combines the Manning equation with an empirical geomorphic relationship between hydraulic radius and distance along the channel. The equation for travel time is given by

$$t_i = K\, l_i^{3/2} / (\Delta h_i^{1/2}\, L_i^{0.3})$$

where t_i = travel time along channel segment i;

l_i = length of channel segment i;

Δh_i = change in elevation over segment i;

L_i = length of channel above segment i;

K = proportionality coefficient accounting for channel roughness and the empirical relationship between hydraulic radius and channel length

The value of K, although unknown, can be assumed constant within a basin, and hence a time-area histogram can be constructed without knowledge of the absolute travel times involved.

We constructed time area histograms for the upper and lower Kickapoo gages (the middle gage is very close to the lower gage and hence was omitted), and the three gaged sites on the Pecatonica River. In doing so we chose an arbitrary travel time interval and measured subareas contributing flow to the gage within each interval. It should be noted that the absolute time represented by this interval is unknown, though we believe it to be of the order of two to four hours.

Figure 2 gives the time-area histograms for the upper and lower Kickapoo gages. It is apparent that very little of the additional drainage area between the two gages contributes simultaneously with the area above the upper gage. This is consistent with the very small increase in flood quantiles between the two gages.

Figure 3 gives the time-area histograms for the three gages on the Pecatonica River. Although the Pecatonica drainage basin is less elongated than that of the Kickapoo, it has been observed that flood peaks from its two main subbasins generally fail to coincide. This observation is supported by the time-area histogram of the largest Pecatonica basin.

Qualitatively, the time-area histograms explain the anomolous Kickapoo and Pecatonica quantiles. But are they of any quantitative value?

One crude test of the utility of time-area histograms for quantile estimation is to use the histogram peaks as predictors. For the case of estimating quantiles at a site upstream or downstream of an existing gage, the estimates can be made by multiplying the quantiles at the gage by the ratio of the histogram peak at the site to the histogram peak at the gage. The results of such a procedure for the 2 and 100 year quantiles are shown in Table 1 for the upstream and downstream Kickapoo gages and for the three gages on the Pecatonica River. Also shown are the results of using a ratio of total drainage areas raised to the 0.6 power and of using the regional regression equations. Note that the time-area approach is always better than the total area ratio, sometimes substantially. The time-area estimates perform significantly better than the regression estimates in three of the six cases, and approximately as well in two other cases. Note that it may be possible to improve the results by raising the histogram peak ratios to a power less than one. More data would be needed to determine the appropriate exponent.

4. CONCLUSIONS AND FUTURE WORK

The preliminary results reported here indicate that time-area histograms may be useful in estimating quantiles on ungaged watersheds.

Table I Percent errors in estimated quantiles using total ratios (A), time-area ratios (TA) and regression equation (Conger, 1981)

Gage No. (Drainage Area-km^2)		Percent Error $(A_2/A_1)^{0.6}$		TA_2/TA_1		Regression	
1. Site With Known Quantiles	2. Site of Estimate	Q_2	Q_{100}	Q_2	Q_{100}	Q_2	Q_{100}
5408000 (689)	5410500 (1787)	+74	+55	+6	-3	+108	+78
5410500 (1787)	5408000 (689)	-42	-36	-5	+3	-13	-26
5432500 (707)	5434500 (2678)	+57	+114	+33	+89	+48	+88
5434500 (2678)	5432500 (707)	-36	-53	-25	-49	+18	+22
5433000 (572)	5434500 (2678)	+18	+79	-8	+45	+48	+88
5434500 (2678)	5433000 (572)	-15	-44	+9	-31	-6	-29

Of particular promise is the use of the histogram peaks to estimate quantiles at a site upstream or downstream of an existing gage. However, this approach must be tested at other sites before any conclusions about its potential can be drawn. Such tests should include basins where quantiles increase more rapidly with drainage area than do the basins reported here.

Time-area histograms may also be of value for comparisons across basins. For example, the peak of the time-area histogram may be useful as an independent variable in regional regression equations. Our preliminary results, however, have been discouraging. The principal problem is that the value of the coefficient K in the travel-time equation which we used appears to vary substantially from basin to basin. In fact, we believe that a better understanding of the physical factors affecting this variation will lead to improvements in the regional regression equations.

Should time-area histograms prove useful, there will be a need for more efficient methods for their construction, possibly exploiting digitized topography. Standard methods are extremely tedious and time-consuming.

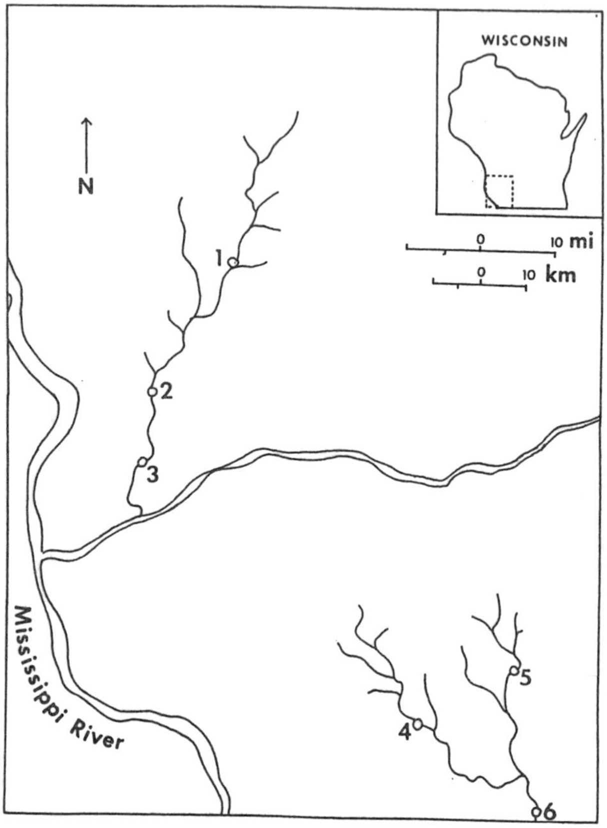

Figure 1 Location of six gages on Kickapoo River (1,2,3) and Pecatonica River (4,5,6)

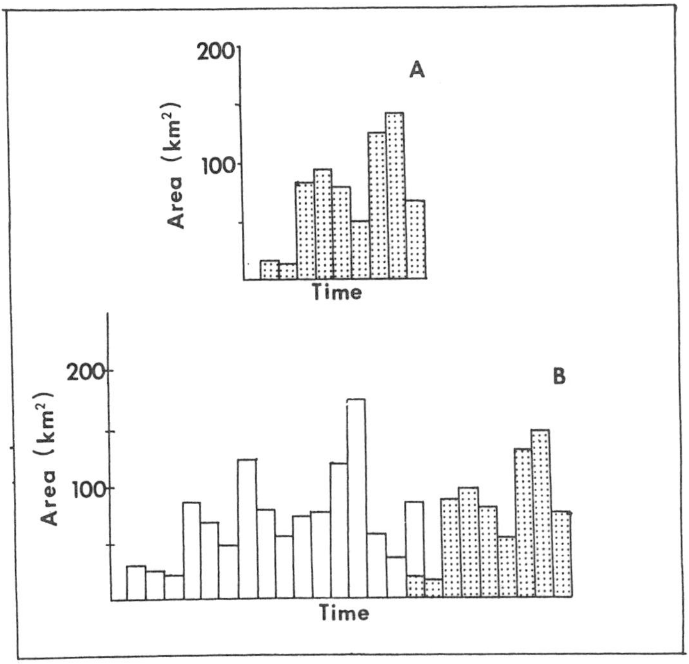

Figure 2 Time-area histograms for Kickapoo River
A. Upper gage (5408000)
B. Lower gage (5410500)

Note: The size of the time increments is unknown, but probably is on the order of two to four hours. Shading shows contribution of upper basin.

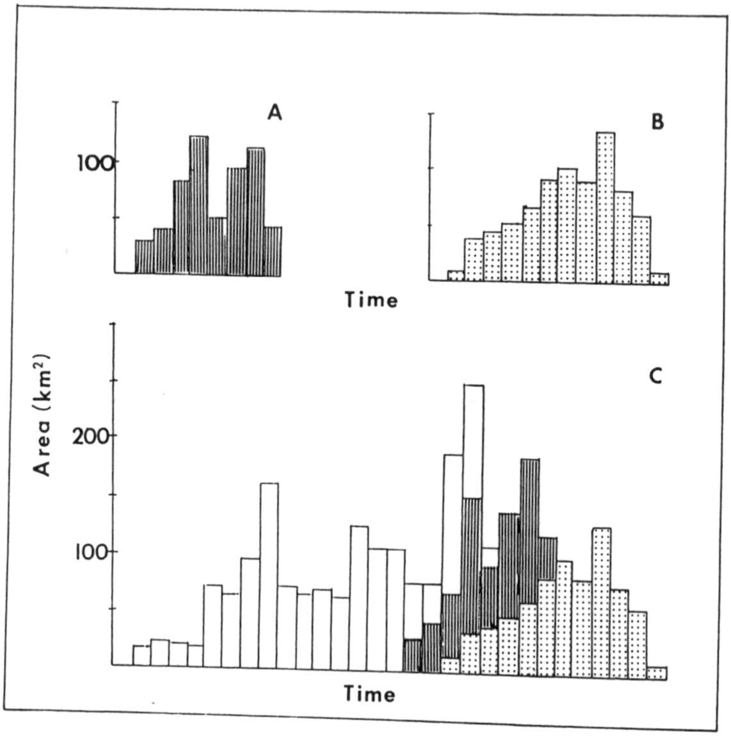

Figure 3 Time-area histograms for Pecatonica River
A. Upper gage, east branch (5433000)
B. Upper gage, main branch (5432500)
C. Lower gage (5434500)

Note: The size of the time increments is unknown but is probably on the order of two to four hours. Shading shows contributions of upper basins.

5. ACKNOWLEDGEMENTS

This work was funded by a grant from the Wisconsin Department of Transportation.

6. REFERENCES

Conger, D.H, 'Techniques for estimating magnitudes and frequency of floods for Wisconsin streams,' United States Geological Survey Open-File Report No. 80-1214, 1981.

Newton, D.W. and Herrin, J.C, 'Assessment of commonly used flood frequency methods', Transportation Research Record Series TRR89, 10-30, 1983.

Pilgrim, P.H., 'Isochrones of travel time and distribution of flood storage from a tracer study on a small watershed, Water Resources Research, **13**(3), 587-595, 1977.

A REGIONAL PEAK FLOW STUDY IN SOUTHERN BRITISH COLUMBIA, CANADA

J.D. Cheng[1] and D.E. Reksten[2]
[1]Research Section, Ministry of Forests, Kamloops, B.C. and [2]Hydrology Section, Ministry of Environment, Victoria, B.C.

ABSTRACT. To provide assistance in estimating peak flows for hydraulic structure design associated with road development, drainage and flood control works, this study was undertaken in two areas in southern British Columbia (B.C.) with different peak flow generation mechanisms. The two study areas are: (A) the Lower Fraser Valley in southwestern B.C. near Vancouver where prolonged storms with high rainfall amounts are the major factor in causing annual peak flows; and (B) the South Thompson Basin in south central B.C. near Kamloops where snowmelt is the dominant cause to annual peak flows. A mean annual daily peak flow map was first developed for both areas using peak flow data from gauged watersheds, topographic characteristics plus regionalized mean annual maximum 24-hr. precipitation data for area A and regionalized mean annual precipitation data for area B. The regional frequency curves relating return period peak flows to mean annual daily peak flows were determined for identified subregions within each study area. The regionalized precipitation frequency data were derived from results of earlier studies using available climatic station records and a precipitation model based on orographic and frictionally induced vertical moisture flux parameters.

INTRODUCTION

To design hydraulic structures associated with road drainage and flood control works, estimates of selected return period peak flows are often required at crossings of ungauged streams. Regional analysis techniques are often used to correlate peak flows with one or more watershed characteristics (Leith, 1976), most frequently drainage area only (Benson, 1962). In many parts of British Columbia (B.C.) peak flow cannot be adequately estimated using only drainage area because the magnitudes of precipitation and evapotranspiration vary substantially with factors such as elevation and location relative to the ocean and topographic barriers (Waylen and Woo, 1981). For small watersheds (< 100 km^2), the application of many empirical or semi-empirical formulas, originally developed elsewhere with data from specific areas and conditions, has often failed to produce satisfactory results. For example, formulas such as Talbot's formula and the Rational Formula contain a coefficient to adjust the basic equation for local conditions. However, the selection of a proper coefficient often proves difficult due to the lack of observed data.

V. P. Singh (ed.), Regional Flood Frequency Analysis, 77–90.

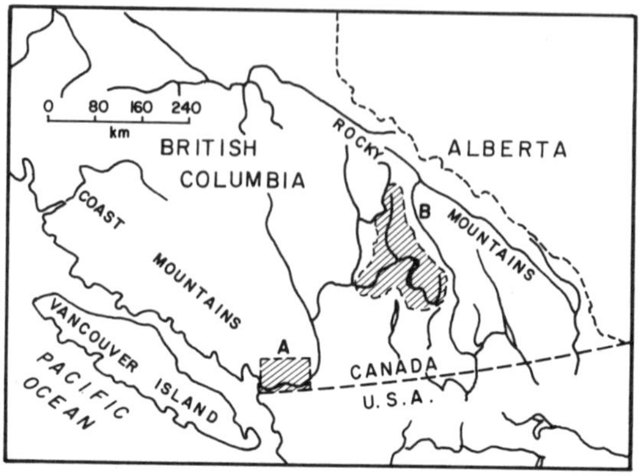

Figure 1. Location map of the Lower Fraser Valley (A) and South Thompson Basin (B).

The objective of this study was to regionalize peak flow characteristics for small to medium sized watersheds in two areas of B.C. with different peak flow generation mechanisms. The two study areas are A) the Lower Fraser Valley in southwestern B.C. near Vancouver, where prolonged storms with high rainfall amounts are the major factor causing annual peak flows, and B) the South Thompson River Basin in southern B.C. near Kamloops, where snowmelt is the dominant contributor to annual peak flows (Figure 1). These two areas were selected because adequate streamflow and regionalized precipitation information are available. This study makes use of these available streamflow and precipitation data as well as watershed topographic characteristics to develop regional mean annual daily peak flow maps and frequency curves for use in estimating peak flows of various return periods on ungauged streams.

STUDY AREA

Lower Fraser Valley

The study area is comprised of two main physiographic zones: the relatively flat, low elevation valley bottom south of the Fraser River and the steep valley side in the northern portion of the study area formed by the southern end of the Coast Range. The Fraser River floodplain, composed of fine sands and silts, has superimposed on it low rolling hills (100 - 200m) formed by morainal deposits and the alluvial fans of outwash gravels deposited by the steep mountain streams entering the flat valley floodplain. The flat portion of the valley bottom is predominantly farmland with some residential and industrial areas. The rolling hills are mainly forested with some areas cleared for pasture. The steep, mountainous zone in the northern portion of the study area is characterized by frequent bedrock exposures and thin morainal and colluvial deposits.

The climate of the Lower Fraser Valley is typically maritime with October to December being the wettest period and June to August being the driest. Annual precipitation, ranging from 750 mm to over 3500 mm, increases northward from and eastward along the valley, as well as with elevation. Snowfall usually occurs from late November to February with its relative contribution to total annual precipitation increasing significantly with increasing elevation. Rainfall intensities are usually moderate but total ranfall amounts over durations of one day or more can be well over 100 mm (Figure 2A). The annual streamflow distribution pattern follows closely the seasonal precipitation with a winter maximum and a summer minimum. Peak flows generally occur as a result of long duration, moderately intense winter storms with high total rainfall amounts, sometimes enhanced by melting snow.

South Thompson Basin

The South Thompson (River) Basin above Kamloops is approximately 16,400 km^2 in area with mountain slopes ranging from moderate to steep. Variations in annual precipitation amounts in the basin result mainly from large elevation differences (750 to 2400 m) and the orientation of its mountain ranges perpendicular to the prevailing Pacific air masses. Mean annual precipitation, mainly in the form of snow, varies greatly, ranging from 200 mm near Kamloops to over 2 000 mm along the ridge of the Monashee Mountains.

The streamflow characteristics in this region are mainly governed by snowmelt and, to a lesser degree, by rainstorms. Streams start to rise in late March or April and reach their peak in May, June, or early July, depending on watershed elevation, aspect and weather pattern and then recede to low flows in the late summer, fall and the winter periods.

ANALYSIS AND RESULTS

The analysis mainly consists of a) developing a map of mean annual daily peak flow, q_{pd}, for each region, b) determining the annual daily and instantaneous peak flows for various return periods, $(q_{pd})_r$ and $(q_{pi})_r$, for gauged streams with adequate data, and c) constructing regional peak flow frequency curves for each identified peak flow sub-region by grouping the ratios of peak flows of various return periods $(q_{pd})_r$ and $(q_{pi})_r$, to mean annual daily peak flows, q_{pdm}.

Development of Mean Annual Daily Peak Flow Map

Lower Fraser Valley

The mean annual maximum 24-h precipitation map (Figure 2A) was developed using regionalized data for 18.5 km x 12.1 km grid squares estimated by Danard (1975). The regionalized mean annual maximum 24-h precipitation data from Danard (1975) were derived from available climatic stations and a model based on orographic and frictionally induced moisture flux parameters (Danard, 1976). A preliminary

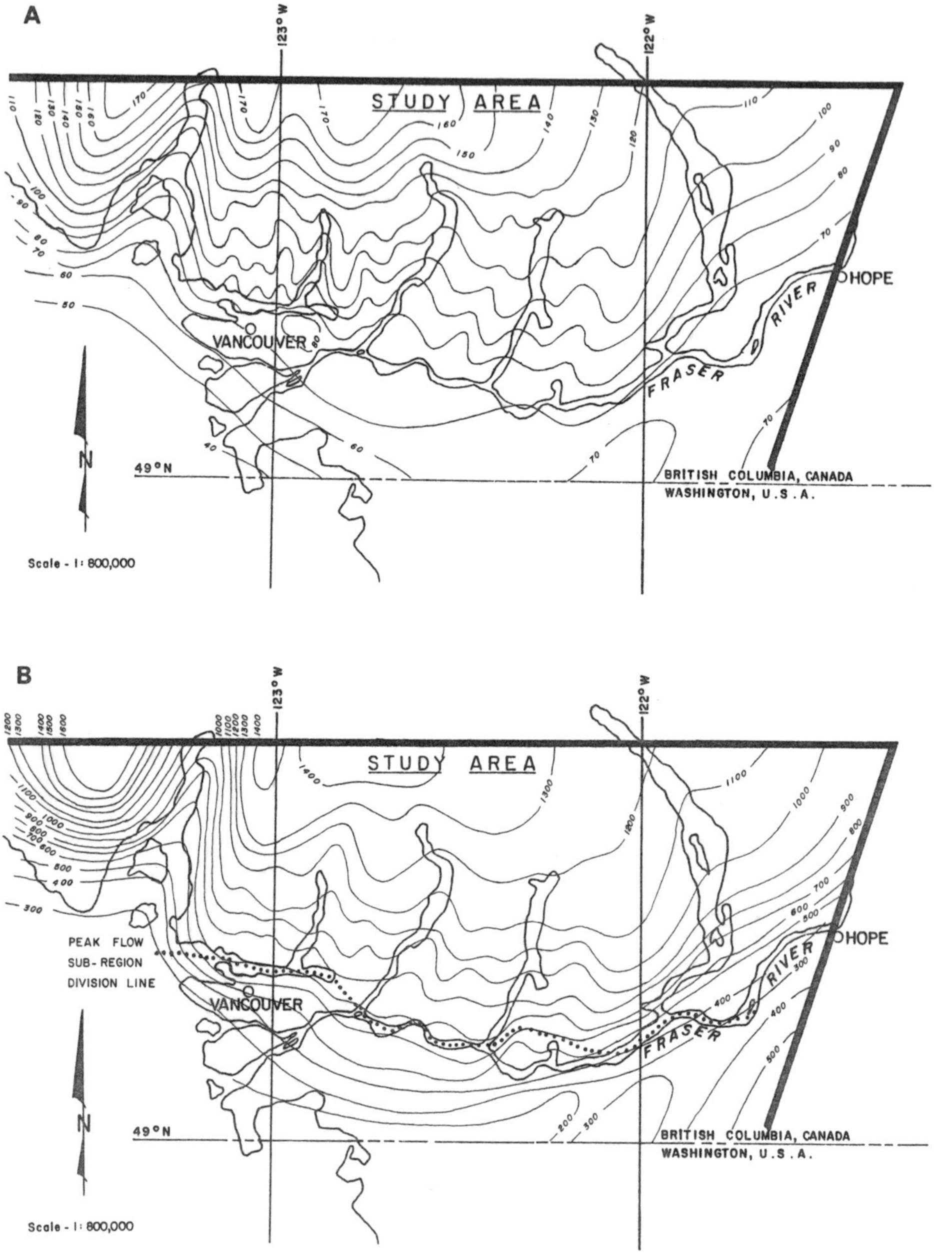

Figure 2. A map of (A) mean annual maximum 24-h precipitation (mm) and (B) mean annual daily peak flow ($1/s/km^2$) for the Lower Fraser Valley.

analysis based on data from 18 gauged watersheds with at least six years of data in this study area indicates a good correlation between the mean annual daily peak flow, l/s/km^2, and mean annual maximum 24-h precipitation (Figure 3A). An earlier peak flow study in this region (Cheng, 1977) also obtained a good correlation between annual maximum daily peak flow and annual maximum 24-h precipitation of 25-year return period. Therefore, a map of mean annual daily peak flow, q_{pdm}, was developed for this study region (Figure 2B) using 1) the mean annual maximum 24-h precipitation map (Figure 2A) derived from Danard (1975), and 2) peak flow data from 40 gauged streams having at least four years of record. Mean annual precipitation information from Wright and Trenholm (1969), Schaefer and Niklevia (1973) and elevation distribution of the area are also used as supplementary information. Isopleths (lines of equal mean annual peak flow) were drawn and adjusted to give reasonably correct averages for the drainage areas above each gauging station.

South Thompson Basin

The isopleth map of mean annual precipitation (Figure 4A) was developed based on data presented in Danard (1971). Danard (1971) used an objective technique to compute orographic effects on precipitation and generated regionalized annual precipitation amounts for the South Thompson River Basin. The mean annual precipitation map was then used, together with mean annual peak flow data from 55 stations with at least four years of data and a topographic map, to derive a mean annual daily peak flow map for this region (Figure 4B).

The decision to use the mean annual precipitation map as the major guide for developing a mean annual peak flow map was mainly due to good correlation between mean annual precipitation and mean annual daily peak flow for 22 gauged watersheds with at least six years of data in the South Thompson Basin (Figure 3B). This good correlation is probably due to the fact that for snowmelt streams, mean annual precipitation is usually related very closely to mean annual water yield. The mean annual water yield of a watershed is, in turn, a good index to the magnitude of its peak flow (Bethlamy, 1971).

Peak Flow Frequency Analysis

Published annual daily and instantaneous peak flow data of gauged streams in the two study areas, with at least 8 years of record, were used to derive values of the mean, 10, 25, 50 and 100-year return period daily and instantaneous peak flows, $(q_{pd})_r$ and $(q_{pi})_r$, through frequency analysis. An upper drainage area limit of 125 km^2 (50 mi^2) and 1250 km^2 (500 mi^2) was selected for gauged streams in the Lower Fraser Valley and the South Thompson Basin respectively . These upper limits were selected in order to have a sufficient number of gauged watersheds for analysis. Figure 5 shows the location of study watersheds whose data are used in frequency analyses and are referred to as index watersheds. Four frequency distributions, Log-Normal, Gumbel, Pearson Type III and

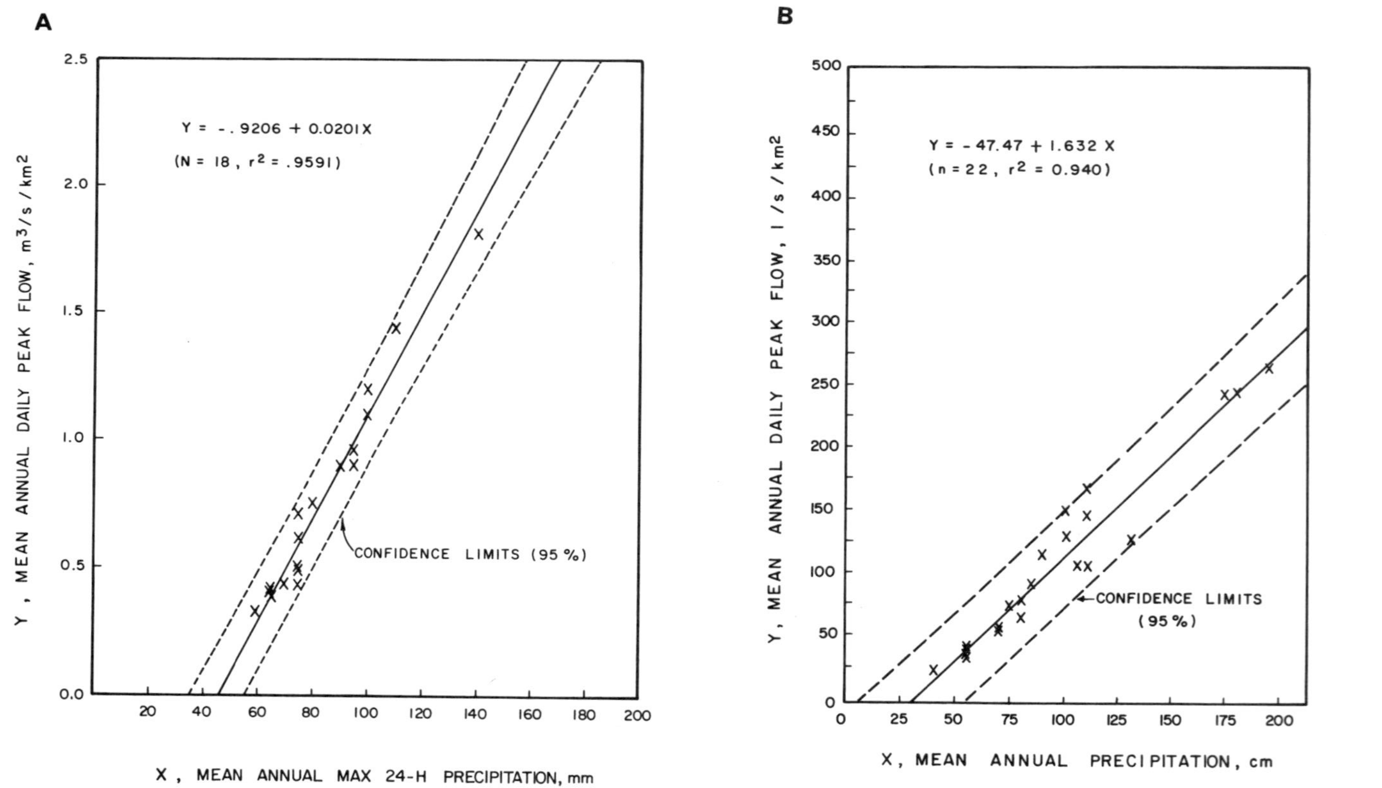

Figure 3. (A) Relationship between mean annual maximum 24-h precipitation and mean annual daily peak flows for 18 gauged watersheds in the Lower Fraser Valley, and (B) relationship between mean annual precipitation and mean annual daily peak flow for 22 gauged watersheds in the South Thompson Basin.

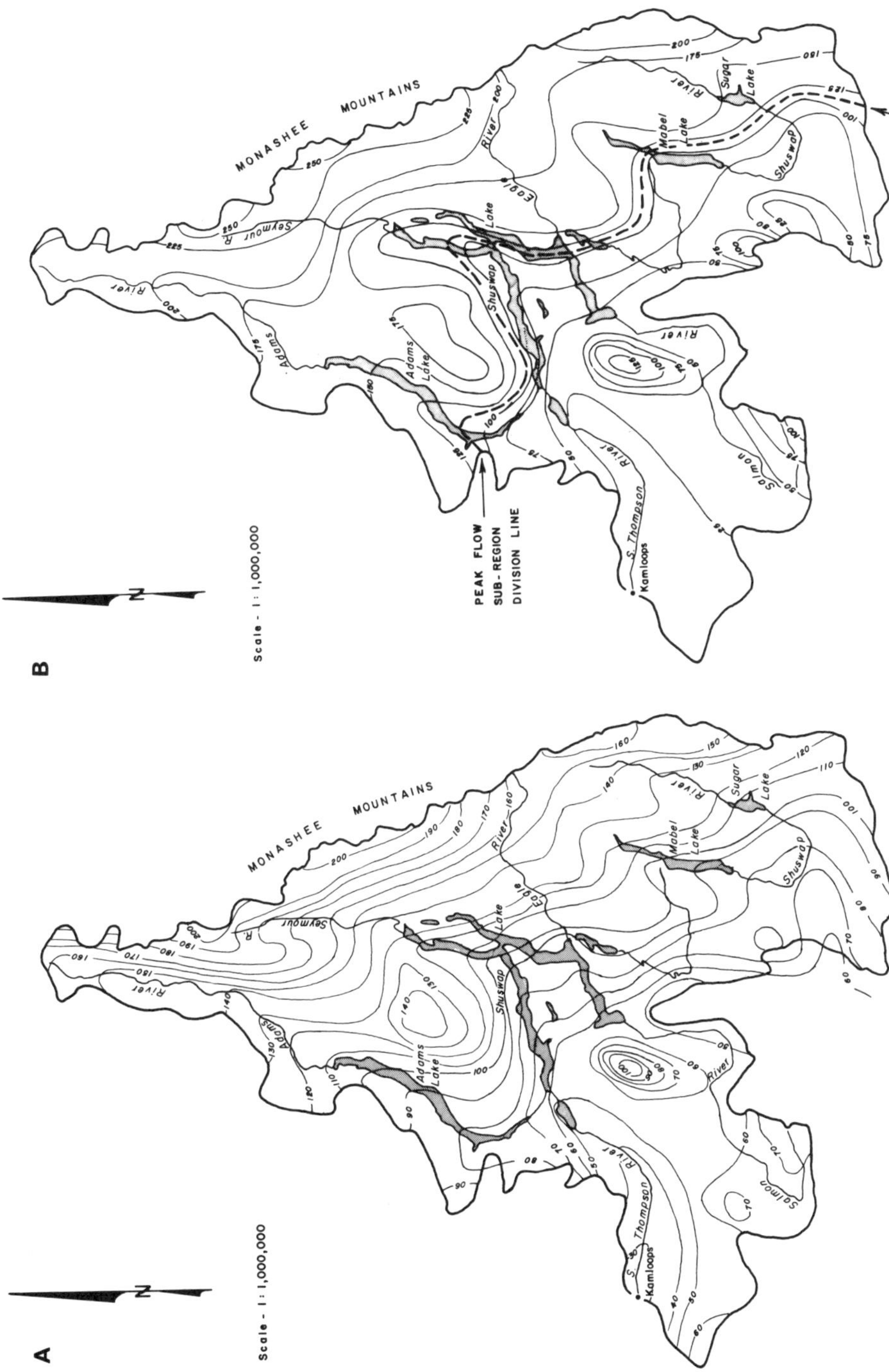

Figure 4. A map of (A) mean annual precipitation (cm) and (B) mean annual daily peak flow ($1/s/km^2$) for the South Thompson Basin.

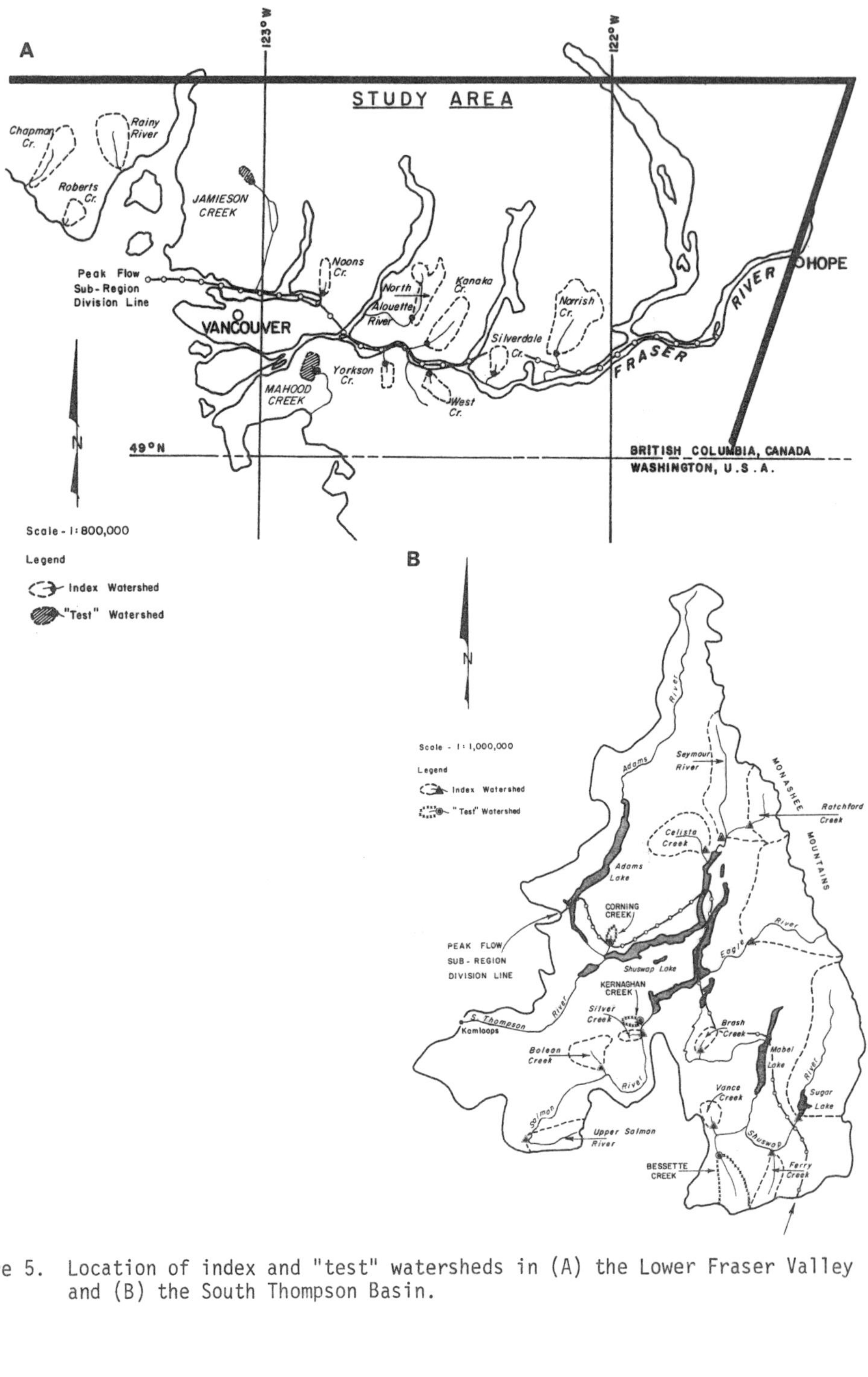

Figure 5. Location of index and "test" watersheds in (A) the Lower Fraser Valley and (B) the South Thompson Basin.

Log-Pearson Type III, were used to obtain return period peak flow estimates in the frequency analysis (Benson, 1968). Since differences among the four results for most watersheds were relatively small and the criteria for selecting the best estimate are not well-defined, the average of the four estimates was used in the analysis. The mean, 10, 25, 50 and 100-year recurrence period annual peak flows for the index watersheds are given in Table 1.

Regional Frequency Curves

The values of Rd and Ri, i.e. ratios of peak flows of selected return periods, $(q_{pd})_r$ and $(q_{pi})_r$, to mean annual daily peak flows, q_{pdm}, respectively, were computed for each of the index watersheds in both the Lower Fraser Valley and in the South Thompson Basin. A homogeneity test (Dalrymple, 1960) applied to ratios of 10-year return period daily peak flows over the mean annual daily peak flows indicated a need to further sub-divide each of the two study regions into two peak flow sub-regions. The Lower Fraser Valley was subdivided into two peak flow sub-regions, separated approximately by the Fraser River, due to the differences in topographic and precipitation characteristics (Figure 2B). The South Thompson Basin was sub-divided into two peak flow sub-regions separated approximately by the 100 $l/s/km^2$ mean annual peak flow isopleth (Figure 4B). These peak flow sub-region division lines follow approximately the limits of the coastal and interior Douglas fir biogeoclimatic zones respectively (Farley, 1979).

The development of regional frequency curves representing the ratios of 10, 25, 50 and 100-year return period peak flows to mean annual daily peak flows for each of the four peak flow sub-regions involves the following steps: (1) the peak flows of selected return period at each index stream are first divided by the mean annual peak daily flow at the station, to reduce them to dimensionless ratios; and (2) individual curves, plotted by use of the dimensionless peak flow ratios for each station are combined to obtain the regional frequency curve for each sub-region. For each of selected return periods, the median of the individual ratios to the mean annual daily peak flow, q_{pdm}, is determined. These median values define the regional composite frequency curve. Figure 6 shows the composite peak flow frequency curves for each of the peak flow sub-regions. The steeper slope of a composite frequency curve for a peak flow sub-region usually reflects a higher degree of variation in peak flow causative factors, i.e., maximum snowmelts or rainstorms, from year to year.

Estimation of Instantaneous Peak Flows

Daily peak flow data are usually available at many more stations and for many more years than instantaneous peak flow data. However, for hydraulic structures, instantaneous peak flows rather than daily peak flows will be the critical flows for design. There are several ways to estimate instantaneous peak flows of selected return

TABLE 1. PEAK FLOW FREQUENCY CHARACTERISTICS FOR INDEX WATERSHEDS IN THE LOWER FRASER VALLEY AND SOUTH THOMPSON BASIN

Watershed	Area (km^2)	q_{pdm} ($l/s/km^2$)	Mean	Return Periods (Yrs.) 10	25	50	100
LOWER FRASER VALLEY							
Northern Sub-Region							
Chapman Cr.	64.5	757	a. 1	1.53	1.83	2.05	2.27
			b. 1.70	2.53	2.96	3.25	3.55
Kanaka Cr.	47.7	899	a. 1	1.59	1.91	2.16	2.41
			b. 1.84	2.77	3.16	3.44	3.69
Noons Cr.	2.59	770	a. 1	1.54	1.96	2.31	2.70
			b. 1.84	2.63	3.13	3.51	3.90
Norrish Cr.	117	957	a. 1	1.54	1.86	2.09	2.33
			b. 1.86	2.95	3.56	4.01	4.48
N. Alouett	37.3	1,200	a. 1	1.50	1.76	1.94	2.13
R.			b. 1.87	2.73	3.12	3.40	3.64
Roberts Cr.	32.6	429	a. 1	1.62	2.08	2.40	2.78
Rainy Cr.	69.4	1,815	a. 1	1.59	2.00	2.72	3.30
Southern Sub-Region							
Silverdale	17.1	488	a. 1	1.59	1.91	2.15	2.38
Cr.			b. 1.48	2.29	2.71	3.02	3.33
West Cr.	11.4	715	a. 1	1.63	1.95	2.18	2.40
			b. 1.48	2.39	2.84	3.16	3.46
Yorkson Cr.	6.0	510	a. 1	1.52	1.85	2.11	2.38
SOUTH THOMPSON BASIN							
Northern Sub-Region							
Seymour R.	805	239	a. 1	1.31	1.48	1.59	1.71
			b. 1.20	1.57	1.84	2.05	2.28
Ratchford Cr.	253	240	a. 1	1.36	1.67	1.94	2.24
Eagle R.	904	242	a. 1	1.31	1.46	1.56	1.66
			b. 1.12	1.44	1.66	1.85	2.04
Shuswap R.	1130	174	a. 1	1.44	1.71	1.93	2.16
Celista Cr.	282	126	a. 1	1.50	1.65	1.75	1.84
Southwestern Sub-Region							
Silver Cr.	25.9	79	a. 1	1.65	2.04	2.34	2.65
Vance Cr.	73.3	37	a. 1	1.58	1.92	2.17	2.43
Ferry Cr.	145	65.5	a. 1	1.56	2.04	2.44	2.89
Brash Cr.	32.6	93.3	a. 1	1.57	1.97	2.30	2.64
Salmon R.	143	53.6	a. 1	1.53	1.78	1.95	2.11
			b. 1.21	1.89	2.27	2.46	2.64

a. $(q_{pd})_r/q_{pdm}$
b. $(q_{pi})_r/q_{pdm}$

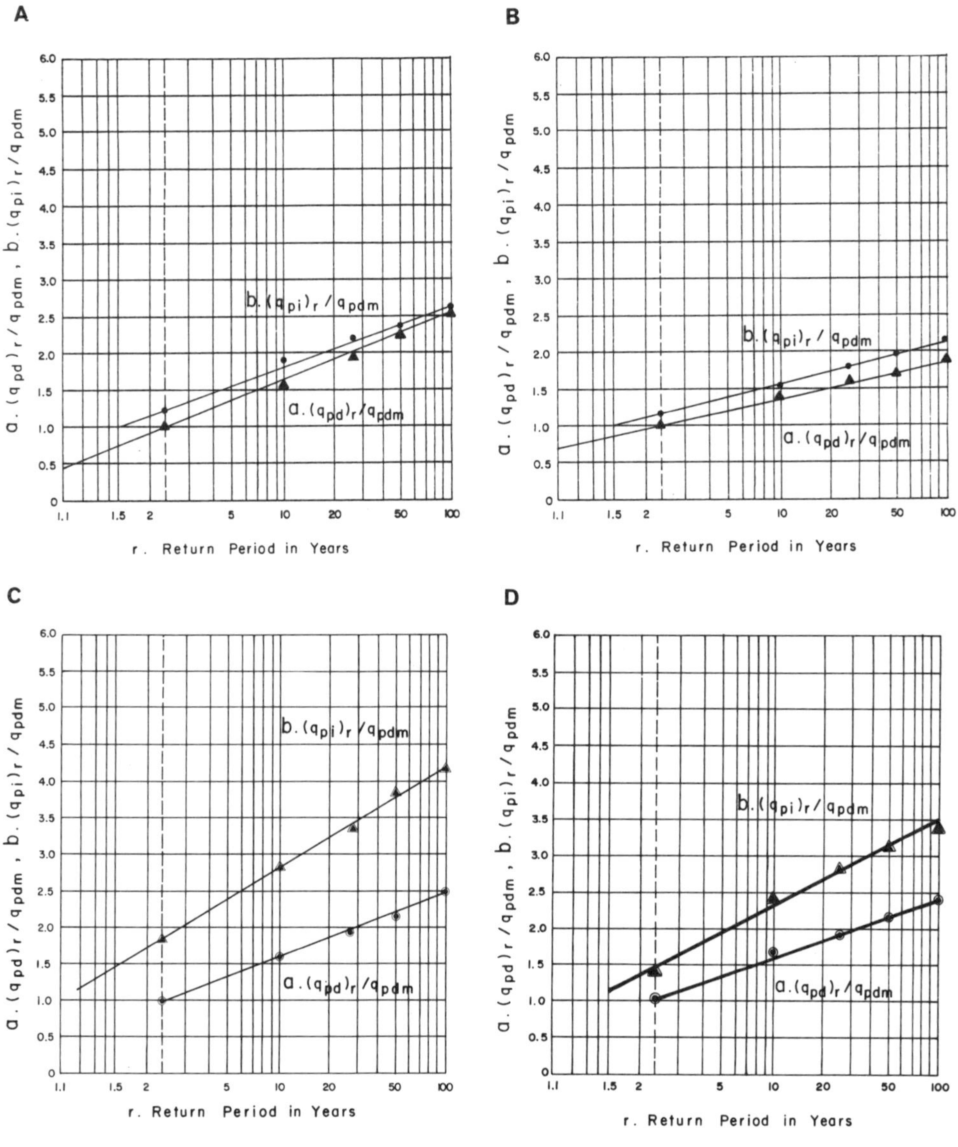

Figure 6. Regional frequency curves for four peak flow sub-regions. (A) South-western South Thompson Basin, (B) Northeastern South Thompson Basin, (C) Northern Lower Fraser Valley and (D) Southern Lower Fraser Valley.

periods, $(q_{pi})r$, for ungauged streams or gauged streams without instantaneous peak flow data. For example, streams with only daily peak flow values in both study areas a practical method recently proposed by Sangal (1981) may be used to estimate instantaneous peak flows from recorded daily flows, of three consecutive days with the annual maximum daily peak flow occupying the middle position. This method can be used to reconstruct annual q_{pi} values from annual q_{pd} values. The $(q_{pi})_r$ can then be determined by frequency analysis of their reconstructed annual q_{pi} values. For ungauged streams, $(q_{pi})r$ is often estimated by applying a ratio of instantaneous to daily peak flow derived from data obtained at gauged locations which are equipped with automatic water-level recorder. If C_r represents the ratio of $(q_{pi})_r$ over $(q_{pd})_r$, the instantaneous peak flow, $(q_{pi})_r$ can be calculated by multiplying the daily peak flow, $(q_{pd})_r$ with the C_r value determined from these gauged streams with q_{pi} data.

For this study, the estimate of $(q_{pi})_r$ for ungauged watersheds and gauged watersheds without q_{pi} data can be determined from the ratio R_i, of instantaneous peak flows of a desired return period $(q_{pi})_r$ over mean daily peak flow q_{pdm} from the regional instantaneous frequency curve for a peak flow sub-region (Figure 6). This method is used because $(q_{pi})_r$ can be easily calculated once q_{pdm} is determined. It is interesting to note that the magnitudes and range for ratios of $(q_{pi})_r$ over q_{pdm} are smaller for watersheds in the South Thompson Basin than for streams in the Lower Fraser Valley. This is mainly due to their differences in peak flow causative factors, i.e. rainstorms vs snowmelts.

DISCUSSION AND CONCLUSION

There are no formulas or methods that can be universally applied to accurately estimate design peak flows for ungauged streams. Therefore, many regional peak flow studies have been carried out to define regional peak flow characteristics and develop maps and equations according to local conditions. The development of a regional peak flow frequency curve is based on the empiricism that for a large region of homogeneous meteorologic and physiographic conditions, individual watersheds within it should have peak flow frequency curves of similar slopes (Chow, 1964).

The present study was conducted to provide regional peak flow maps and equations as well as regional peak flow frequency curves by using factors and relationships according to the peak flow generation mechanisms at each of the two study areas. The data used include regionalized precipitation and peak flow records. The study produces maps and regression equations for determining mean annual daily peak flow for any watershed within respective study areas. It also generalizes peak flow frequency relations over a region by combining the results of several gauged watersheds to produce a regional frequency curve.

The peak flow of a desired return period for a watershed can be estimated from the appropriate regional composite frequency curve after the q_{pdm} value has been determined from the mean annual peak flow maps (Figures 2B and 4B) or the regression equations relating q_{pdm} to mean annual maximum 24-h precipitation or the mean annual precipitation (Figures 3A and 3B). The estimates of 25-year return period peak flows using these procedures for 5 test watersheds in the study areas (Figure 5) are compared in Table 2 with those determined from frequency analysis of observed peak flow data. The peak flow data of these five watersheds had not been used in the previous analysis. There are no significant differences between the two results. This indicates that dependable peak flow estimates can be made for watersheds in the study areas using maps and equations developed in this study. It should be noted that the applicability of the results presented in this paper is limited by the range of data used to develop them. As well, because the results reflect only the regional hydrologic conditions, they should not be used to calculate peak flows for streams outside the study regions.

Improvements will be made to the maps and equations developed in this study as more data become available in the future. The results from this study can be used to estimate with a reasonable degree of accuracy and a minimum amount of subjective judgments design peak flows for watersheds in the two study areas. The procedures to use the study results for peak flow estimates are simple and straightforward, features that are important for direct field application. The peak flow estimates using this practical method can be used as a check for estimates made by other methods and to serve as a guide for more detailed evaluation.

TABLE 2. VALUES OF 25-YEAR RETURN PERIOD PEAK FLOW, $(q_{pi})25$, ESTIMATED FROM THE PRESENT REGIONAL ANALYSIS FOR FIVE "TEST" WATERSHEDS AS COMPARED TO THOSE CALCULATED FROM FREQUENCY ANALYSIS OF OBSERVED PEAK FLOW DATA

Watershed	Drainage Area km^2	$(q_{pi})25$, $l/s/km^2$ Regional Analysis a	Regional Analysis b	Frequency Analysis
Lower Fraser Valley				
Jamieson Cr.	3	4,080	4,920	3,910
Mahood Cr.	18	1,540	1,640	1,720
South Thompson Basin				
Kernaghan Cr.	9	270	250	380
Corning Cr.	26	330	360	320
Bessett Cr.	600	110	130	70

a) q_{pdm} determined from the mean annual daily peak flow maps.
b) q_{pdm} determined from regression equations.

ACKNOWLEDGEMENTS

The review of this paper by Paul Doyle and Rita Eremko is gratefully appreciated. Special thanks are due Doug Golding, Kuo-chi Rae, Larry Barr, Patricia Frehlick, Heather Gell, Ruth Morley and Sharon Laitinen for their contribution toward the completion of this paper.

REFERENCES

Benson, M.A. (1962). Evolution of methods for evaluating the occurrence of floods. U.S. Geol. Water Supply Paper 1580-A.

Benson, M. (1968). Uniform flood frequency estimating methods for federal agencies. Water Resources Research 4:891-908.

Bethlamy, N. (1971). Maximum peak flows for selected return periods for watersheds west of the continental divide in Idaho and Montana. USDA For. Serv. Res. paper INT-113. 12 p.

Cheng, J.D. (1977). Peak flow estimates from regionalized 24-hour rainfall frequency data. Pro. Second Conference on Hydrometeorology, Oct. 24-26, 1977, Toronto, Ontario. pp 119-123.

Chow, V.T. (ed.-in-chief), (1964). Handbook of Applied Hydrology. McGraw-Hill Book Co., New York.

Coulson, C.H. (1975). Regional peak flow analysis - southwest mainland. Hydrology Division Report. B.C. Water Resources Service, Victoria, B.C.

Dalrymple, T. (1960). Flood frequency analysis. U.S. Geological Survey. Water Supply Paper, 1543-A, 77 pp.

Danard, M.B. (1971). A simple method of computing the variation of annual precipitation over mountainous terrain. Boundary-Layer Meteorology. 2, 188-206.

Danard, M. (1975). Twenty-four hour rainfall amounts for various return periods over southwestern British Columbia. Water Investigations Branch, B.C. Water Resources Service. Victoria, B.C.

Danard, M. (1976). On frictional and orographic effects on precipitation in coastal areas. Boundary-Layer Meteorology 10, 409-422.

Farley, A.L. (1979). Atlas of British Columbia. University of British Columbia Press. Vancouver, British Columbia, 136 pp.

Leith, R.M. (1976). Streamflow regionalization in British Columbia. No. 1: Regression of mean annual floods to physiographic parameters. Environment Canada. Inland Waters Directorate. Report Series No. 40. p. 21.

Sangal, B.P. (1983). Practical method of estimating peak flow. Journal of Hydraulic Engineering 109(4):549-563.

Wright, J.B. & Trenholm, C.H. (1969). Greater Vancouver precipitation. Tech. Memo. TEC 722, Canada Dept. of Transport, Meteorological Branch, Toronto, Ont., 36 pp.

Waylen, P. & Woo, M.K. (1981). Regionalization and prediction of annual floods in the Fraser River catchment, British Columbia. Water Resource Bulletin. 17(4):655-661.

REGIONAL TREATMENT OF FLOOD DATA

Maria Mimikou
Division of Water Resources, Hydraulic and Maritime Engineering, Civil Engineering Department, Technical University of Athens
5, Iroon Polytechniou, 157 72 Athens, Greece

ABSTRACT. The maximum observed peak discharges in stations of the northwestern and western regions of Greece are plotted against drainage areas and are found to be power functions of them. The curves enveloping the plotted points are regionally varying according to the climatic regime. When the peak discharges are plotted against a morphoclimatic index, which is the product of the expected storm duration by the maximum observed average storm intensity for this duration by the area of the drainage basin, a single linear envelope curve is drawn for both regions that explains 95% of the regional variation in the dependent variable. The envelope curves may give evidence as to the flood magnitude to be expected from ungauged basins of the area, with a return period at least equal to the return period of the analysed flood data. The use of a "cookbook" method can give information for even less frequent floodflows.

INTRODUCTION

Floodflows estimation is very important for hydrologic design. Floodflows are determined by the interactions of basin and precipitation characteristics. Much research has been done in relating flood-flows of specified probability to the causative factors such as precipitation, duration and intensity, basin sizes and various hydraulic characteristics. Because of the complexity of the physical interactions that produce floodflows hydrologic knowledge is not sufficiently advanced as yet to justify a precise mathematical expression for the definition of the maximum flow to be expected from a given basin. Thus, only empirical equations have been developed relating extreme floods of observed occurrences mainly to basin sizes (Creager et al., 1945; Matthai, 1969; Crippen and Bue, 1977; Grippen, 1982).

In this paper maximum flood experience in two hydrologically characteristic regions of Greece is depicted in graphs on which the maximum observed peak discharges are plotted against drainage areas. Regional curves enveloping the plotted points are developed and studied. The Greek envelope curves are compared to envelope curves developed in the

V. P. Singh (ed.), Regional Flood Frequency Analysis, 91–101.

United States (Crippen, 1982) and conclusions are drawn. A successful attempt is made to develop a single envelope curve for the whole area under study. This is accomplished by using as independent variable, not the drainage area alone but a morphoclimatic index containing not only morphological but also hydrometeorological information from the drainage basin. This morphoclimatic index is the product of the expected storm duration by the maximum observed average storm intensity for this duration by the area of the drainage basin and explains a significant amount of the regional variation in the dependent variable. The single envelope curve is free from several drawbacks of floodflow-drainage area curves, i.e. it does not need the determination of region boundaries, etc. The envelope curves may give evidence as to the flood magnitude to be expected from ungauged basins of the area at certain probability levels. Thus, the envelope curves can be used in specific engineering applications.

ENVELOPE CURVES DEVELOPED IN THE U.S.A.

Crippen (1982) presents envelope curves for extreme flood events recorded through the water-year 1975, which have been developed for 17 regions in the conterminous U.S. and are described by equations of the following form:

$$Q_p = K_1 A^{K_2} (A^{C_1} + C_2)^{K_3} \tag{1}$$

The equations have been developed by using the U.S. Customary units, namely with the peak discharge Q_p is cubit feet per second (0.0283 m^3/s) and the drainage area A in square miles (2.5931 sq.kilom.). The form of the equations and the constant parameters C_1 and C_2 were derived by a trial-and-error procedure; the constant K_1 and the exponents K_2 and K_3 of the curves were computed from sets of points selected from postulated envelope curves. Another curve was developed by Matthai (1969) to define an upper limit of peak flood discharges observed through 1965 in the U.S.A. This curve was based on nationwide data from basins less than 200 sq. miles (520 km^2) in area and is described by equation of the following form:

$$Q_p = aA^b \tag{2}$$

The forms of Eqs (1) and (2) have been used as well for describing envelope curves for floodflows observed in regions of Greece.

DATA USED

The floodflows Q_p which have been used are the maximum observed peak discharges (instantaneous) measured at eleven flow measuring stations on six rivers of the area. The flow data collection program has started in 1964, but investigations have shown that these peak discharges are actually the largest events to have occured during the past N=29 years. No reliable earlier historical flood information could be found. Thus, using the plotting position: T=N+1 for the return period T of the largest even (Gumbel, 1958), the return period which could be now assigned to the floodflows Q_p of the analysis is: T=30 years. Apparently, this return period increases with increasing N. Besides, the maximum observed peak discharges need to be modified and thus, their envelope curves presented in the following to be appropriately adapted, whenever new floodflow observations exceed the analyzed flood data. The names and the general location of the rivers of the measuring stations are given in Fig. 1. The peak discharges Q_p and the drainage areas A for all stations, covering a wide range between 200 to 5000 sq. kilom. are given in Table I. The majority of the peak discharges have not been measured directly due to equipment inefficiency in very high discharges, but they have been estimated from well-established extensions of stage-discharge rating curves at the measuring sites by using the maximum observed water-level readings from stage recorders

Fig. 1 General location of the measuring stations
Scale: 1:2,000,000

TABLE I. Morphological and hydrometeorological characteristics of the drainage basins

Region	River	Measuring station	Floodflows Q_p (m^3/s)	Drainage area A (sq.km)	Expected storm duration $\tilde{d}$ (days)	Max.observed average storm intensity $\tilde{i}_m$ (mm/day)	Characteristic extreme precip. volume V_c ($m^3 \times 10^6$)
Western	Acheloos	Avlaki	2148	1349	2.70	115.2	419.59
	Acheloos	Kremasta	4400	3570	2.80	108.0	1079.57
	Arachthos	Tsimovo	896	640	2.71	93.3	161.82
	Arachthos	Plaka	1360	970	2.82	90.6	247.83
	Arachthos	Arta	2100	1855	2.83	83.1	436.25
	Kalamas	Vrossina	859	1031	2.60	74.4	199.44
	Aoos	Vovoussa	380	202	2.50	105.6	53.33
North-western	Aoos	Konitsa	700	665	2.56	100.8	171.60
	Aliakmon	Siatista	1358	2725	2.30	57.6	361.01
	Aliakmon	Ilarion	2100	5005	2.40	52.9	635.43
	Venetikos	Venetikos	950	818	2.10	64.8	111.31

operating at the sites. The data were provided by the Public Power Corp. (P.P.C.).

REGIONAL ENVELOPE CURVES FOR FLOODFLOWS AGAINST DRAINAGE AREAS

The maximum observed peak discharges in cubic meters per second plotted against drainage areas in square kilometers on a log-log scale are shown in Fig. 2. The plotted points are clearly classified into two groups.

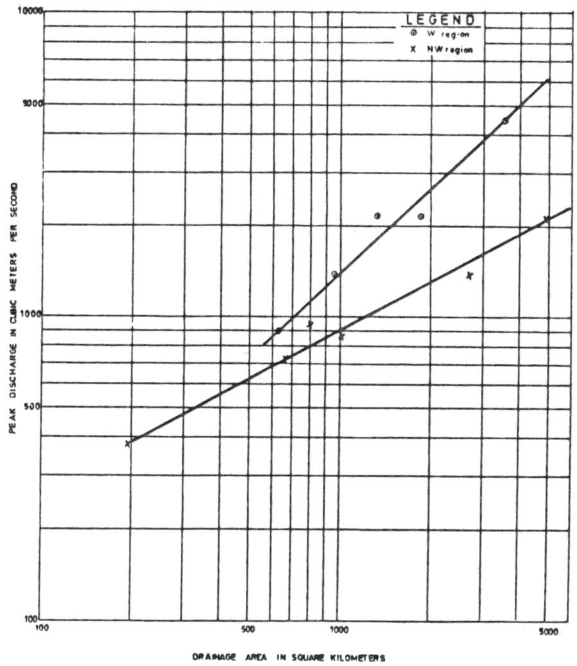

Fig. 2 Envelope curves for maximum peak discharges against drainage areas

The upper group includes the points corresponding to the Arachthos and the Acheloos river basins of the western region and the lower group the points corresponding to the Kalamas, the Aoos, the Aliakmon and the Venetikos river basins of the northwestern region. This separation is in accordance with the climatic conditions and the hydrological experience from the area, which expects higher floodflows for the same drainage area in the western than in the northwestern Greece. This is because the western region of Greece accepts the highest precipitation depths. It appears from Fig. 2 that for decreasing basin sizes the two groups of points approach each other. Thus, one can reasonably assume that for relatively small basin sizes, i.e. for $A < 300$ sq. kilom. the flood runoff rates per unit area of the two regions are equalized. It was found that Equ. (2) calibrated by the least squares method gives better results than Equ. (1) for both groups of points. The parameters of the latter equation have been estimated by a trial and error procedure, such that error σ in the residuals of the equation is the minimum. This error σ, which is given by:

$$\sigma = \left[\sum (Q_p - Q_p)^2 / v\right]^{\frac{1}{2}}$$

with Q_p the estimated by the empirical equation peak discharge and v the number of associated points, is also the criterion for selecting one of the two empirical Equs. (1) and (2) for representing the floodflow-drainage area relationship. The equation with the minimum σ is the best. The values of σ for Equ. (2) are: $\sigma = 87.4$ m^3/s and $\sigma = 99.6$ m^3/s for the western and northwestern group respectively, whereas the corresponding σ-values for Equ. (1) are $\sigma = 207.3$ m^3/s and $\sigma = 346.5$ m^3/s. Therefore, Equ. (2) fits better on both groups of points. The equations describing the envelope curves which are given as straight lines in Fig. 2 and their coefficients of determination r^2 are the following:

$$\text{W region:}\quad Q_p = 2.948\ A^{0.982},\ r^2 = 0.96 \qquad (4)$$

$$\text{NW region:}\quad Q_p = 26.140\ A^{0.511},\ r^2 = 0.98 \qquad (5)$$

Since additional flood data are not available for basin sizes less than 300 sq. kilom., where the flood runoff rates per unit area of both regions are assumed to become equal, Equ. (5) can reasonably be taken to express the floodflow-drainage area relationship for these basin sizes and for both regions.

Even though the envelope curves developed in the U.S. by Matthai (1969) and Crippen (1982) refer to flood data recorded through only one year, a comparison is made between these curves and the Greek envelope curves in order to draw conclusions on the floodflow-drainage area covariation for different climatic regimes. The empirical equations of the form of Eq. (1) calibrated with the Greek flood data and estimated in U.S. Customary units are the following:

$$\text{W region:}\quad Q_p = 2.0 \times 10^3 A^{1.1} (A^{0.5} + 5)^{-1.0} \qquad (6)$$

$$\text{NW region:}\quad Q_p = 3.8 \times 10^3 A^{1.0} (A^{0.5} + 5)^{-1.2} \qquad (7)$$

One the other hand, the range of values of the constant K_1 and of the exponents K_2 and K_3 and the parameters C_1 and C_2 of the curves described by Equ. (1) for all 17 regions analyzed in the U.S. (Crippen 1982, p.1210) are given along with the parameters of Eqs. (6) and (7) in Table II. It appears that the parameters C_1 and C_2 of the curves estimated in the U.S.A. are appropriate for the Greek data as well, the K_2 and K_3 values for the Greek data are in the range of K_2 and K_3 values.

TABLE II. Envelope curves' parameters estimated in Greece and in the U.S.A.

Developed in	Range of values				
	$K_1(\times 10^3)$	K	K	C	C
U.S.A.	10∿121	0.710∿1.116	-1.572∿-0.844	0.5	5.0
Greece	2.0∿3.8	1.0∿1.1	-1.2 ∿ -1.0	0.5	5.0

for the U.S. data, but the K_1 values are much lower for the former than for the latter data. Consequently, the exponents of the maximum observed floodflow-drainage area envelope curves developed in the Greek Mediterranean basins are in the range of values of the exponents of yearly floodflow-drainage area envelope curves developed in conterminous U.S. basins, but the former curves' constants are much lower than the latter. This conclusion seems reasonable since the rate at which a change in the drainage basin size is converted to a floodflow change is not expected to be affected by the differences between Mediterranean and continental climate, which for a given basin size certainly affect the floodflow magnitude. The same conclusion is drawn when comparison is made with the envelope curves described by Eq. (2). Because of the drainage area upper limit in Matthai's (1969) analysis comparison is made with Eq. (5) only, whose expression in U.S. Customary units is the following:

$$Q_p = 1{,}503\ A^{0.511} \tag{8}$$

The constant a and exponent b estimated in the U.S. are: a = 11,000 and b = 0.61 respectively. Thus, the difference in the exponent is not significant but the constant a is much lower for the Greek than for the U.S. flood data.

SINGLE ENVELOPE CURVE

In order to establish a single envelope curve for the maximum observed peak discharges of the total area under study, a combination of several factors in addition to basin area has been tried. The additional factors used describe hydrometeorological characteristics of the drainage basin like precipitation duration and intensity, which are directly related to maximum floodflows. More specifically, it was found that a single envelope curve for all analyzed extreme flood events was obtained when the drainage area variable of the developed envelope curves was replaced by another independent variable, which is the product of the expected storm duration by the maximum observed average storm intensity for this duration by the area of the drainage basin. This product constitutes a drainage basin morphoclimatic index and can be estimated in units of volume.

The expected storm duration $\tilde{d}$ is the average of all storm events in the mean precipitation record of the drainage basin. A storm event is an uninterrupted sequence of wet (rainy) periods bounded by dry ones. Since there are very few continuously recording rain gages in the area under study and the majority are gages with daily readings, it was found more convenient and reliable to estimate duration d in days. The mean daily precipitation records (containing rainfall and snow water-equivalent values) of all analyzed basins have been estimated from point precipitation data measured by P.P.C. by using the Thiessen polygon method and cover a 29-year period. A day was defined as wet if the water equivalent of the precipitation estimated on that day was at least 0.25 mm otherwise it was defined as a dry day (Mimikou 1983). Since one day is a sufficient time to guarantee independency between successive storm events, all uninterrupted sequences of wet days bounded on both sides by one or more dry days were defined as storm events. For each mean daily precipitation record the storms have been numbered consecutively: $i=1,2,...,n$ and a duration in days d_i has been assigned to each of them. The expected value of variable d_i: $E[d_i]$, is defined as the expected storm duration $\tilde{d}$ of the drainage basin as follows:

$$d = E[d_i] = \sum_{i=1}^{n} d_i/n \tag{9}$$

Then, the maximum observed average storm intensity i_m for an event of duration d within the precipitation record time length of each basin is estimated. Since the precipitation data used in this study are daily and duration d is in general a decimal number, intensities i_m cannot be estimated directly from the records. For this purpose the maximum observed average intensities i_m (mm/day) for precipitation durations: $d = 1$ day, $d = 2$ days, $d = 3$ days and $d = 4$ days have been estimated for each basin precipitation record. By using a duration correction factor δ for reducing the scattering of points, the $[i_m, (d+\delta)]$ pairs have been plotted for each basin on a log-log scale and a least square linear curve has been drawn through them (Nemec 1972, p. 193). The maximum observed average storm intensity $\tilde{i}_m$ (mm/day) for the expected storm duration of each basin is estimated by entering the linear curve with a duration $(\tilde{d}+\delta)$. The whole procedure for the estimation of the $\tilde{d}$- and $\tilde{i}_m$-values has been implemented in computer code, whereas these values are given for all basins in Table I. Besides the previously mentioned eventual modification of the maximum observed peak discharges, the $\tilde{d}$ and $\tilde{i}_m$ values are also subject to eventual modifications as new precipitation data become available for each basin.

The product:

$$V_c(m^3) = 10^3 \times \tilde{i}_m \text{ (mm/day)} \times \tilde{d}\text{(days)} \times A\text{(sq. kilom.)} \qquad (10)$$

can be defined as a characteristic extreme precipitation volume of the drainage basin and is given for all basins in Table I. Using the volumes $V_c(m^3)$ as independent and the maximum observed peak discharges $Q_p(m^3/s)$ as dependent variables, a new analysis has been undertaken relating the variables. The plotting of the associated variables in Fig. 3, clearly shows that a single curve is now sufficient to envelope all points.

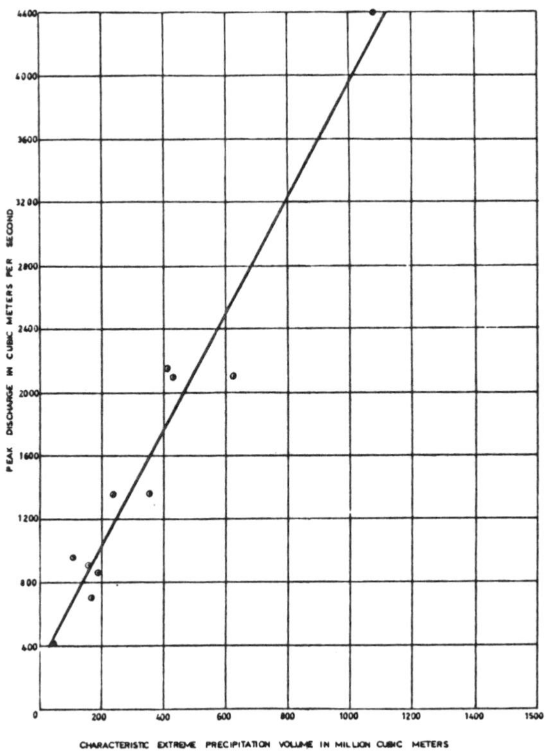

Fig. 3 Single envelope curve for maximum peak discharges against characteristic extreme precipitation volumes

Two linear curves have been calibrated by the least squares method: one with the physical variables and one with their logarithms. The former is found to be more appropriate according to the minimum residual error criterion as given by Equ. (3). The equation describing the single linear envelope curve in Fig. 3 and its coefficient of determination r^2 are the following:

$$Q_p = 269.5 + 3.685 \quad 10^{-6} V_c, \; r^2 = 0.95 \qquad (11)$$

In other words, 95% of the regional variation in the maximum observed peak discharges Q_p is explained by Equ. (11)

APPLICATION

The developed envelope curves for extreme flood events may give evidence as to the flood magnitude that has been occurred or might occur at other ungauged basins of the area. They may give regional information for the flood magnitude to be expected from ungauged basins of the area with a return period at least equal to the return period of the analyzed flood data: T = N+1. Therefore, they can be used in specific engineering applications, like the hydrologic design of safety structures built for a relatively short-term protection, e.g. in the design of diversion works at ungauged river sites of an area where a dam is to be built and the prediction of a diversion design floodflow is needed. The single envelope curve in Fig. 3 is preferable for predicting peak discharges compared to the regionally varying curves in Fig. 2, since it does not need the determination of region boundaries and is free from any limits on drainage basin sizes and is more complete in expressing the physical interdependence between floodflows and drainage basin characteristics. In such a case, one has to estimate in addition to the drainage area A the expected storm duration $\tilde{d}$ and the maximum observed average storm intensity $\tilde{i}_m$, for this duration of the drainage basin, as explained previously. Then, the characteristic extreme precipitation volume V_c is estimated from Eq. (10) and the design peak discharge is predicted from the single envelope curve of the area. This discharge is not expected to be exceeded for at least (N+1) years. In case where the needed precipitation data are not available, peak discharges can be predicted from the regionally appropriate floodflow drainage area envelope curve by using the drainage area of the ungauged basin only.

The use of a "cookbook" method like the one proposed by Gumbel (1958) can give information for even less frequent floodflows. Gumbel (1958, p. 239) proves, with some simplifying assumptions that the return period corresponding to double the largest flood observed within the last N years is the square of the return period of the largest flood multiplied by the factor e^{au}, which is always larger than unity, with a and u parameters of the Gumbel's first asymptote for extremes. Consequently, by doubling the peak discharge estimated from the envelope curves for extreme flood events with a return period: T=N+1, one can predict peak discharges with a return period at least equal to $(N+1)^2$ years, which can be also used in specific engineering applications, like the hydrologic design of spillways, etc.

CONCLUSIONS

The conclusions drawn from this research are the following:

(1) The maximum observed peak discharges in the northwestern and western regions of Greece are found to be power functions of the drainage

areas; these functions differ from region to region in accordance with the climatic regime. For relatively small basin sizes the flood runoff rates per unit area of the two regions can be reasonably assumed to become equal.

(2) The exponents of the maximum observed floodflow-drainage area envelope curves developed in the Greek Mediterranean basins are into the range of values of the exponents of yearly floodflow-drainage area envelope curves developed in conterminous U.S. basins, but the former curves' constants are much lower than the latter.

(3) When the peak discharges are plotted against a morphoclimatic index, which is the product of the expected storm duration by the maximum observed average storm intensity for this duration by the area of the drainage basin, a single linear envelope curve is drawn for both regions, explaining 95% of the regional variation in the dependent variable. This curve is advantageous compared to the regionally varying floodflow-drainage area envelope curves since it does not need the determination of region boundaries, is free from any limits on drainage basin sizes and is more complete in expressing the physical interdependence between floodflows and drainage basin characteristics.

(4) The developed envelope curves can predict peak discharges at certain probability levels for ungauged basins of the area and can be used in specific engineering applications like the hydrologic design of diversion works, spillways, etc.

ACKNOWLEDGEMENTS

The author wishes to thank the Hydrology Section of the Greek Public Power Corporation for providing the flood and point precipitation data used in this paper.

REFERENCES

Creager, W.P.; Justin, J.D. and Hinds, J., 1945. Engineering for Dams, General Design, Vol. I, John Wiley and Sons, Inc., New York, N.Y.

Crippen, J.R.; 1982. Envelope Curves for Extreme Flood Events. Journal of Hydraulic Division, Am. Soc. Civ. Eng., Proc. Rap. 17391, 108 (HY10), pp. 1208-1212.

Crippen, J.R. and Bue, C.D., 1977. 'Maximum Floodflows in the Conterminous United States'. Water Supply Paper 1887, U.S. Geological Survey, Washington, D.C.

Gumbel, E.J., 1958. Statistics of Extremes. Columbia University Press, New York, N.Y.

Matthai, H.F., 1969. 'Floods of June 1965 in South Platte River Basin, Colorado'. Water Supply Paper 1850-B, U.S. Geological Survey, Washington, D.C.

Mimikou, M., 1983. 'Daily Precipitation Occurrences Modelling with Chain of Seasonal Order'. Hydrological Sciences Journal, **28**, 2, 6, pp. 221-232.

Nemec, J., 1972. Engineering Hydrology. McGraw-Hill, London.

STATISTICAL ANALYSIS OF FLOODS AND REGIONAL RELATIONS

Maria Ozga-Zielinska
Institute of Environmental Engineering
Warsaw Technical University
ul. Nowowiejska 20
00-653 Warsaw
Poland

ABSTRACT. The paper is concerned with several basic issues and problems that pertain to the use of mathematical statistics for flood analysis, including development of regional relations used in many countries for estimation of specific flood events, e.g. for the design purposes. The following issues and problems are discussed in the paper: maximum discharges and floods, nonhomogeneity of the observation series, flood as a multivariate phenomenon, probability in a set vs. annual probability, and regional statistical relations.

1. MAXIMUM DISCHARGES VS. FLOODS

The terms "maximum discharge" and "flood" are often used as synonyms. In fact they refer to different phenomena and have different definitions.

The maximum discharge is the largest discharge in a given time interval. Most often discharges are analysed in monthly, half-a-year or annual intervals and maximum discharges refer to these intervals.

Concerning flood, its definition is not that clear and it can be defined from different viewpoints as seen in the hydrological literature where one may encounter several different definitions. The definition proposed in this paper is sufficiently general to embrace all definitions formulated so far, and at the same time it allows for an easy and univocal separation of flood from the discharge hydrograph. Thus, flood is a period of time when due to increase of supply, river discharges increase above certain discharge Q_o to such extent, that the peak discharge reaches or exceed certain threshold value Q_f.

The value of this threshold discharge is of arbitrary character and it can be defined on the basis of different criteria. The criteria may be of economic or hydrological character. The most common economic criterion is flood control and in such case $Q_f = Q_{acc}$, where Q_{acc} is the largest river flow which still does not cause flood losses. Other factors may also be taken into account, for example: water supply, when water excess periods are taken as a flood; adequate (safe) operation of various hydraulic equipment, with discharge above which operational

V. P. Singh (ed.), Regional Flood Frequency Analysis, 103–118.

difficulties being taken as Q_f.

The hydrological criteria are most often determined on the basis of statistical analysis. The statistically and hydrologically most justified value of Q_f is the smallest of the annual maximum discharges. Such a discharge has occured every year in the period of record, therefore, from the hydrological point of view, it can be taken as a measure of a minimum "flood potential" of a given river. Taking into account, however, that the maximum discharge in exceptionally dry year may be relatively small, the Q_f discharge may be adopted higher than the above mentioned value, for example, the mean or the median of annual maximum discharges may be taken as Q_f.

2. NONHOMOGENEITY OF OBSERVATION SERIES

The discharge observation series (hydrographs) must be homogeneous if they are to be used for statistical analysis.

In a general sense, given phenomenon is considered homogeneous if a set of factors that cause and decide on the course of this phenomenon does not change (is constant). But homogeneity of the measurement series depends also on the conditions of an experiment (measurement). Providing we can identify and define, even qualitatively, factors which are conditioning homogeneity and factors that cause nonhomogeneity of the measurement series, we can say that we know the genesis (origin) of the phenomenon being analysed and origin of its nonhomogeneity. Such nonhomogeneity that can be identified is called in this paper the genetic nonhomogeneity. But the phenomena under analysis and the conditions of measurement are also under influence of many factors that cannot be defined and separated -- their impact can only be seen through some changes in the statistical properties of the measurement series. This type nonhomogeneity, contrary to the genetic one, is called the statistical nonhomogeneity, i.e. the one that can be identified by the statistical methods only (Fig.1).

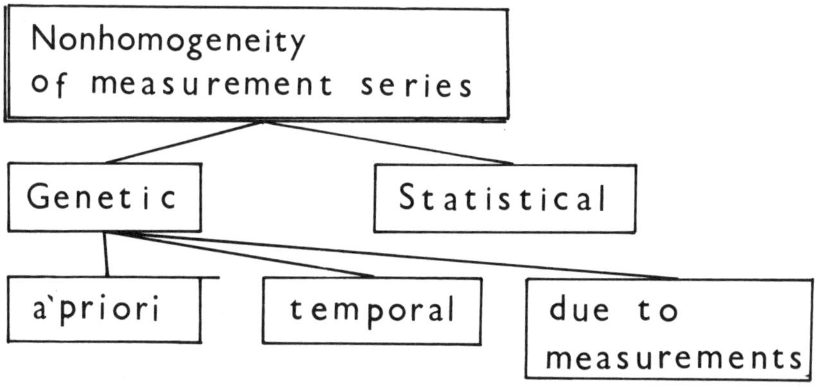

Figure 1. Classification of nonhomogeneity of measurement series.

Taking into account the above, one can distinguish three classes of genetic nonhomogeneity: (i) a priori, (ii) temporal, and (iii) due to measurements (Fig.2). The a priori nonhomogeneity results from the fact, that the same phenomena, or those that exhibit very similar properties, are caused by completely different factors. In hydrology, most typical examples of such phenomena provide floods and droughts. Floods as well as droughts may occur in different seasons of a year, but their origin is not always the same. For example, different factors are responsible for initiation and the course of a snow-melt flood than those responsible for the rainfall flood. This type of nonhomogeneity is called a priori, since it may be identified a priori, based on the analysis of the origin of the phenomena under investigation. Consequently, snow-melt floods and rainfall floods must be treated separately in statistical analysis, forecasting, etc.

Temporal homogeneity requires that factors causing the phenomenon under analysis do not change in time. But more and more intensive human activities in river catchments and channels continuously change runoff conditions as well as conditions of other hydrological phenomena -- this consequently causes a loss of homogeneity in the measurement series.

The third type of genetic nonhomogeneity is the one due to measurements -- i.e. due to the nonhomogeneous conditions in which the experiment was carried out. This type of nonhomogeneity is not related to the phenomenon itself; the phenomenon may be homogeneous both a priori and temporally, but measurement series may still be nonhomogeneous. It may be caused by a change in the measurement method, change of measurement instrument, and/or change in the time of measurement and sampling frequency. For the spatially varying phenomena it may also be due to a change of sampling density in the geographic space. It may be caused as well by the methods applied for averaging and interpolation, both in time and in space. The nonhomogeneity due to measurements is generally small in comparison to the temporal one, especially in the records from the last few decades. In some cases, however, it may be significant and has to be taken into account.

To summarize, the hydrological measurement series may exhibit all three types of genetic nonhomogeneity, but most often encountered is the temporal one. Thus, the term "nonhomogeneity" is used quite often with a specific reference to the temporal nonhomogeneity.

It may happen, however, that analysis of a given phenomenon and examination of the measurement conditions do not allow for identification of any concrete cause of nonhomogeneity. In such case, only the methods of statistical analysis can be used for examination of the series nonhomogeneity.

Some methods which may be used for testing the homogeneity of measurement series are discussed in a farther section of this paper.

3. FLOOD AS A MULTIVARIATE PHENOMENON

The floods, as phenomena which occur in certain time intervals, may have different courses which can be characterized in different ways. One possibility is to provide an equation of flood hydrograph $Q = f(t)$. The other one is to give certain number of characteristics (parameters) which describe the magnitude and the shape of the hydrograph. In this section we are concerned with statistical analysis of floods, thus we shall characterize them in the other way as the multivariate phenomena.

The characteristics used most often are: flood duration - t_f, peak discharge - $Q_{max,f}$, flood volume - V_f, mean flood discharge - $Q_{av,f}$, duration of flood rise (time from the beginning of flood till the peak discharge) - t_r, duration of flood fall (time from the peak till the end of flood) - t_a (Fig.2).

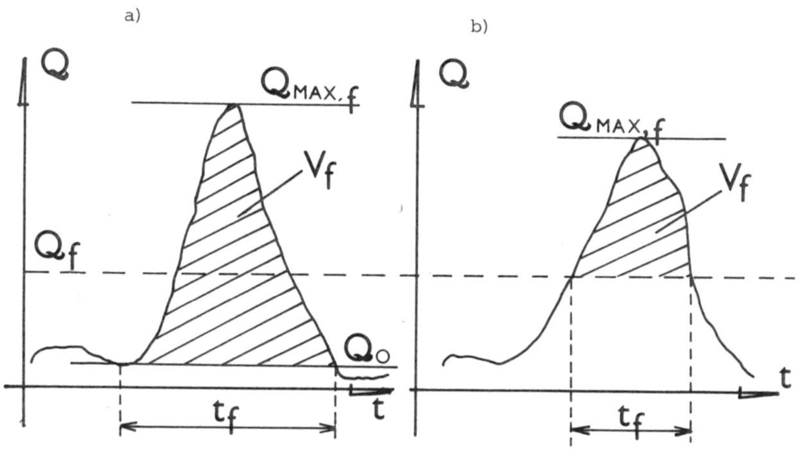

Figure 2. Flood characteristics

Depending on the needs of the analysis, the above characteristics can be estimated for the entire flood above Q_o (see Fig.2a), or for its upper part above the threshold discharge Q_f (see Fig.2b).

Assume, that during N years we have recorded n homogeneous floods, characterized by k characteristics each. Thus, we have a k-dimensional random sample of n elements, which can be described by a multivariate normal distribution which requires that distributions of all individual variables are also normal. Since generally distributions of flood characteristics do not satisfy this requirement, they must be transformed into the normally distributed variables.

The probability determined on the basis of n-element random sample defines probability in a set which must be next converted into the annual probability (see equation (4) in the next section).

The above presented way of multivariate characterization of hydrological phenomena was proposed in Poland in 1963 for low flow analysis (Zielinska, 1963 , 1964), and in 1967 for flood analysis (Strupczewski, 1967).

4. PROBABILITY IN A SET VS. ANNUAL PROBABILITY

The way of defining floods and extreme discharges on the basis of some threshold values leads to n-element random samples of these variables, where n does not necessarily coincides with the number of years in the period of record N. Hence, it is necessary to introduce the notions of a probability in a set and the annual probability.

Providing only one event is chosen per year and the sample size n is equal to the number of years N in the period of record, the probability in a set:

$$p_s = \frac{m}{n + 1} \quad (1)$$

is equal to the probability of occurence of the m-th value in a year:

$$p_y = \frac{m}{N + 1} \quad (2)$$

In case of $n \neq N$ these probabilities are not equal, equation (2) loses its sense, and annual probability must be determined in a different way.

When the phenomenon under analysis occurs at least once every year (e.g. for small Q_f), the probability of m-th event being the highest in a year can be determined on the basis of Bernoulli scheme (Langbein, 1949; Zielinska, 1965):

$$p_y = 1 - (1 - p_s)^{\lambda} \quad (3)$$

where: p_s - probability of the m-th event in the n-element sample, computed acc. to (1),
λ - mean number of occurencies of a given phenomenon in a year, $\lambda = n/N \geqslant 1$.

When the phenomenon under analysis does not occur every year, the probability of m-th event being the highest in a year is:

$$p_y = \frac{r}{N}\left[1 - (1 - p_s)^{\lambda}\right] \quad (4)$$

where: r - number of years in which the phenomenon has occured at least once,
λ - mean number of occurencies of the phenomenon in a year, based on the r years of record, $\lambda = n/r \geqslant 1$,
- other symbols are the same as in (3).

It may be seen that equation (4) is a product (conjunction)of the probability defined by (3) and the probability (r/N) of occurence of a year in which the phenomenon under analysis has occured at least once.

Equation (4) is of a general character and it comprises all possible situations. In other words, it fits all patterns of annual occurence of the phenomenon under analysis. If r = N and $\lambda = 1$, than n = N, i.e. given phenomenon occurs once every year, and equation (4) is reduced to (1) or (2) which in this case are the same.

5. REGIONAL STATISTICAL RELATIONS

The statistical methods of flood analysis presented in this paper are applied for analysis of observations recorded at a gauging station. However, hydrological information may be needed as well for a specific profile of the uncontrolled river, i.e. the river which has no gauging stations at all. In such cases most often various types of regional relations or formulas are used - their coefficients are adopted on the basis of physiographic characteristics of river catchment, characteristics of precipitation, or other meteorological characteristics of various type. There are many such formulas and regional relations abound in the hydrological literature of every country. In recent years several new regional relations have been developed, taking full advantage of the powerful contemporary statistical methods and based on large amount of measurement data. Unfortunately development of these relations is not always preceded by detailed analysis of the measurement data, especially from the point of view of their representativeness, areal independence and synchronism in time. Thus, some relations (developed in the past as well as some of the more recent) raise doubts concerning their formal and substantive correctness.

Determination of a correct regional statistical relation requires:
(a) adequate definition of the explained variable,
(b) homogeneity analysis of all measurement series used in the computations,
(c) analysis of areal (spatial) independence of the explained variable and of the synchronism of measurement series,
(d) selection of a specific form of regional relation,

All these requirements are discussed one bv one in the remaining part of the paper.

5.1. Definition of explained variable

The characteristic for which the statistical relation is being thought most often is called explained variable - the one being explained by the relation (model). The variables on the values of which depends the value of the characteristic in question are called explanatory variables.

According to the actual needs, any characteristic of maximum discharges may be adopted as the explained variable. This may be the mean maximum discharge, the largest of maximum discharges or a maximum discharge corresponding to the specific frequency of recurrence $Q_{max,p}$. In the last case it is necessary to determine the probability distribution function (pdf) of maximum discharges at each gauging station and take as the explained variable either:
- maximum discharges of specific probabilities p (separate regional relation must be developed for each of them) or,
- parameters of the pdf, i.e. $\bar{Q}_{max}$, c_v and c_s, with regional relation developed separately for each parameter. Parameter values based on these relations are used for determination of pdf for a given river profile in the region.

The characteristics relative to catchment area, e.g. Q_{max}/A or $Q_{max}/\bar{Q}_{max}$, where $\bar{Q}_{max}$ is the mean maximum discharge, may also be used as explained variables.

Other flood characteristics, such as V_f, t_f, $Q_{max,f}$ may also be used as explained variables and in such cases one deals with the multivariate explained variable.

Each of the above described ways of defining the explained variable leads to a regional relation of a specific class (Table I).

5.2. Analysis of the Homogeneity of Measurement Series

The analysis of the homogeneity of measurement series is carried out by sequential elimination of individual nonhomogeneity types. First phase is devoted to separation of the series which are a priori homogeneous, for example in case of floods or maximum discharges, preparation of separate series taking into account whether each specific event is of a snow-melt or rainfall origin. In case of one of those series being insignificant, for example snow-melt floods may be small and rare, all further analysis may be limited to the rainfall floods only. When floods of both origin are more or less equally significant, all further analysis and computations must be carried out separately for each series. If one has to determine what is the probability of a highest flood in a year irrespective of the year's season, i.e. irrespective whether it is caused by snow-melt and/or rainfall, it may be computed as the probability of an alternative:

$$p(Q_{max,y}) = p(Q_{max,s}) + p(Q_{max,r}) - p(Q_{max,s})\cdot p(Q_{max,r}) \qquad (5)$$

where: $p(Q_{max,y})$ - probability of a flood with the peak equal or higher than Q_{max}, occuring as a highest one in a year, irrespective whether it is caused by snow-melt and/or rainfall,

$p(Q_{max,s})$ - probability of a snow-melt flood with the peak equal or higher than Q_{max}, occuring as a highest snow-melt flood in a year,

$p(Q_{max,r})$ - probability of a rainfall flood with the peak equal or higher than Q_{max}, occuring as a highest rainfall flood in a year.

The second phase of the analysis is devoted to testing the homogeneous series for nonhomogeneity due to measurements. It is necessary to check whether the type of measurement equipment, measurement methods, etc. have not been subject to some changes in the period of record. If there was a change, it is necessary to determine a correction factor (in percentage or absolute values), which applied to the appropriate elements of the series will render it homogeneous. In case we know that the series is nonhomogeneous due to measurements but the correction factor cannot be determined, one has to conclude that this nonhomogeneity is not identifiable by the methods of genetaic analysis and must be treated as the statistical one.

The factors which influenced in the period of record the phenomenon under investigation must be carefully analysed to check temporal homogeneity of the series. If some significant changes in these factors are identified, they must be taken into account by correction of specific

Table 1 Regional statistical relations

explained variable u: dimensionality	explained variable u: unit	class	sub-class	explanatory variables x: choice method	explanatory variables x: dimensionality	explanatory variables x: type	regional relation: form	regional relation: method of development	the surface area and character of the region
one-dimensional	non-dimensional or related to the unit area	I	a	is not done	0		$\bar{u}$ = const. in the region	averaged in the region	small regions homogeneous in P and γ_i
			b	-"-	2	φ, λ	graphical	charting-isolines	unlimited
			c	-"-	2	φ, λ	$u = f(\varphi, \lambda)$	surface trend	-"-
	any	II	a	arbitrary	1	A	$u = f(A)$	II-type regression	small regions homogeneous in P and γ_i
			b	-"-	2	A, P	$u = f(A, P)$	-"-	small regions homogeneous in γ_i
			c	-"-	k	A, P, α_i	$u = f(\alpha_1 .. \alpha_i, A, P)$	-"-	unlimited
	any	III		statistical choice of the optimal set of explanatory variables	k	$x_i = A, P, \underline{\gamma}^T, \underline{z}^T$	$u = f(\underline{x})$	I-type regression	-"-
multivariate	any	IV	a	-"-	k	$x_i = A, P, \underline{\gamma}^T, \underline{z}^T$	$\underline{u} = F(\underline{x})$	-"- multiple regression	-"-
			b	-"-	k	$x_i = A, P, \underline{\gamma}^T, \underline{z}^T$	$u_j = f_j(\underline{x})$, $j = 1, 2, .., m$	-"- multi-equation regression	-"-

Legend:

φ, λ - geographical coordinates
A - catchment area
P - rainfall
α_i - coefficients depending on physiographical properties
γ - vector of physiographical properties
$\underline{z}$ - vector of additional explanatory variables
$\bar{k}$ - the number of explanatory variables
m - the number of explained variables
$\underline{u}$ - the m-dimensional vector of explained variables

elements of the series, what usually is not an easy task. The alternative procedure is subdivision of the series into homogeneous sub-series, but in this case all further computations must be carried out separately for each sub-series. Undefined temporal nonhomogoneity is treated as the statistical one.

The last phase of the analysis is identification of statistical nonhomogeneity of the series. As mentioned before, various statistical tests are used for verification of hypotheses concerning specific characteristics of the measurement series. Among the characteristics which most often are subject to such tests are mean values, standard deviations and serial correlation coefficients; analysis of multivariate variables requires also tests of cross-correlation coefficients.

Testing the means is most often accomplished by the t-Student test, variances - by the F-Snedecor test, and correlation coefficients - by the z-Fisher test. These parametric tests require samples to be distributed normally, while nonparametric tests may be used for verification of the homogeneity hypotheses concerning random variables of any distribution. Since nonparametric tests use less statistical information about the sample, they are less powerful than the parametric tests and should be used only when the latter cannot be employed (Gren, 1975). The nonparametric tests which can be used for analysis of nonhomogeneity of measurement series include such as Kruskal-Wallis test (series test), Spearman rank order correlation coefficient test for trend and others.

5.3. Analysis of Areal Independence of the Explained Variable and of the Synchronism of Measurement Series

The areal independence of a regional explained variable is understood here to be such a situation, that the areas assigned to specific hydrometric stations in the region do not overlap. In such case, individual stations represent the mutually exclusive areas. By areal dependence we understand partial overlapping of areas assigned to specific stations To satisfy the requirement of areal independence of flow characteristics assumed as explained variables, the areas assigned to individual stations must be the source or differential catchments. The flow characteristics considered for the partially overlapping catchments are areally dependent variables (Fig.3).

For some types of regional statistical relations explained variable must be satisfying the requirement of independence (Table I - Class I), while this requirement is not binding for other types (Class II, III III and IV).

The other significant factor which must be taken into account in development of a regional statistical relation is the synchronism of measurement series, i.e. series overlapping in time. The series which do not overlap in time and originate from different time periods are asynchronic. In development of regional relation, the requirement of all series being synchronic should always be satisfied. The only exception is such a situation, that the series available for all stations in the region are not completely synchronic but they are very long and homogeneous. In such case relatively small asynchronism does not influence the values of explained variable estimated at each particular station and these values are comparable among themselves.

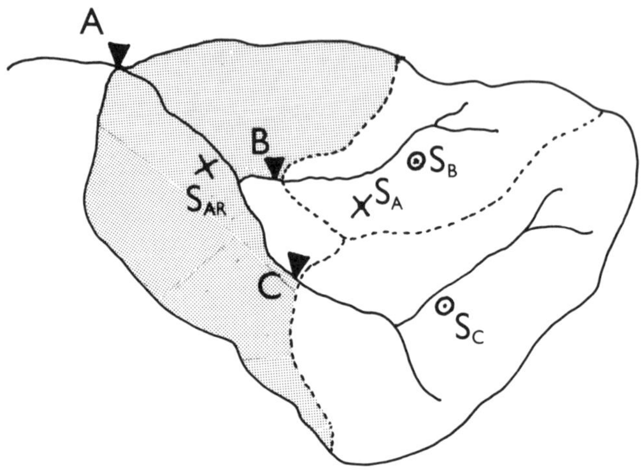

Figure 3. River catchment scheme
Note: catchments of stations B and C and differential catchment of station A (darkened area) are areally independent. Catchments of station A and B or A and C are areally dependent. S_A, S_B, S_C, S_{AR} - location of gravity centres of stations A, B, C and differential catchment AR respectively.

5.4. Selection of a Specific Form of Regional Relation

Regional relations may be of different form and different computational methods may be used for their development. Thus, depending on the adopted classification criteria, several classes (types) of these relations may be distinguished.

The most fundamental classification criterion is the time factor. It means that the explained variable, and explanatory variables as well, are either functions of time or are defined as some static values representing certain time interval (e.g. month, year, several years long period). In the first one of these cases we deal with the variables and relations of a dynamic nature; in the second one, they are static. Dynamic regional relations are based on time series analysis and analysis of dynamic random fields. They will not be considered in this paper, although it should be recognized that in the recent years dynamic models and characteristics are used quite often, especially in water management.

To introduce some order, it is proposed to classify static regional relations according to such criteria as a number and character of explained variables as well as a number, type and selection method of explanatory variables. Based on these criteria, classification of regional relations to be used for determination of maximum discharges and floods is presented in Table I. Four classes are distinguished. Three for a single-dimensional explained variable that differ in the selection method and the number of explanatory variables. The fourth class is for the multivariate explained variable.

First class includes relations in which the explained variable u is not explained by any other variable than the coordinates of geographical location (φ, λ). Determination of such regional relation is accomplished by calculation of u values for gauging stations located in this region, based on measurement series recorded at each of these stations. Next, representative values of u are established for that region. Representative values of u may be defined in different ways, which serve as a basis for separation of sub-classes.

In sub-class Ia, the region is represented by a constant value $\bar{u}$ which is considered to be average for the region, but this average may be determined in different ways, not necessarily as an arithmetic mean (e.g. it may be a weighted mean). It should be underlined that explained variable may correspond not necessarily to a single concrete value of this variable but it may represent certain "mean" relation characteristic for a given region. For example, it may be the dimensionless probability curve of maximum discharges (this case will be discussed in more detail in the last part of the paper), or dimensionless flow duration curve. These relations are determined for individual gauging stations in the region under analysis, and next the relation (curve) "average" for the region is computed.

Class Ib and Ic relations, contrary to class Ia, take into account spatial variability of explained variable. Class Ib includes relations in which this variability is represented in the graphical form of isolines established by interpolation. Subject to interpolation are the values of given variable determined for certain number of points for which their geographical coordinates (φ_i, λ_i) are known. Thus, geographical coordinates play the role of explanatory variables.

In the last decades, relations describing spatial variability of the explained characteristics in the analytical form (contrary to graphical isolines) are used quite often in several countries, including Poland. These relations are called surface regressions or trends. The explanatory variables, similar to Class Ib relations, are geographical coordinates (φ, λ). Relations of this type are distinguished in Table I as Class Ic.

In Class II regional relations, explanatory variables are chosen arbitrarily by the analyst, therefore, he applies his own hierarchy of variables resulting from his analysis of the origin of a given phenomenon. If he finds that there is a single dominating variable, it may be sufficient to develop the simplest relation with one explanatory variable only (sub-class IIa). To improve or generalize the already defined relation (this may be needed for broadening of its applicability), additional explanatory variables are sequentially introduced (sub-classes IIb and IIc). It is known that factors which decide on the river flow are sequentially: catchment area A, specific precipitation P which decides on the magnitude of a given characteristic flow, physiographical characteristics of a catchment γ_i, various less significant climatic factors, and others. According to this sequence, the above mentioned variables are introduced to the individual sub-classes of Class II, although instead of physiographical characteristics γ_i they are coefficients α_i which depend on these characteristics. This procedure has been widely used in the past for selection of explanatory variables, but now the procedure is usually replaced by more objective methods. These methods are based on correlation analysis and they lead to the selection of the optimal (in the statistical sense) set of explanatory variables (Class III and IV relations). Detailed discussion of different methods used for selection of the optimal variables may be found in Kaczmarek (1969) and Nowak (1984). Moreover, it should be underlined, that Class II relations have been defined on the basis of short measurement series; they also employ α_i coefficients instead of physiographic characteristics γ_i. These coefficients are related to the physiographical characteristics in the descriptive and not quite univocal way (usually in form of various tables). Under such circumstances, objective computations and attainment of fairly representative (real) results is rather impossible. Class III and IV relations are free from these faults. They are determined as type I multiple regressions (multivariate conditional distributions) for explanatory variable selected from the potential set by the well developed and scientifically sound procedure. Moreover, relations being developed now are based on much longer measurement series than those of the past.

Neverthless it must be remembered, that the set of potential variables selected by the analyst is to some extent of a subjective nature - selection is made on the basis of analyst's experience and knowledge of the phenomenon being analysed. Thus, selection of the optimal set depends not only on the statistical method applied but, what is even more important, also on the choice of a potential set. The optimal set includes only variables chosen by the analyst for the potential set, therefore, although optimal set is statistically the best, it always contains variables a priori postulated by the analyst. To summarize, correctness of the regional relation being thought depends above all on the appropriate set of potential explanatory variables as well as on the measurement series available for these variables which must be satisfying the requirements discussed above.

Class III relations are now widely used in hydrology (e.g. in Poland Ozga-Zielinska, 1975; Fal, 1979), while Class IV relations are applied very rarely so far.

5.5. Comments Concerning Some Regional Relations

The comments presented herewith are concerned with the Class I relations only, since their development sometimes raises certain doubts.

Developing any type relation of this class, the measurement series of explained variable must be spatially independent and synchronic. The second one of these requirements is usually satisfied, although there are some cases that it is given insufficient attention. The spatial (areal) independence requirement usually is not observed.

Class Ia relations are generally quite rare, since hydrological phenomena exhibit considerable spatial variability and there are not too many regions where the assumption of a regionally constant value of explained variable could be satisfied. Neverthless, this concept has been used for determination of regional curves describing the probabilities of maximum discharges in Great Britain (Flood..., 1975). The procedure and assumptions adopted in this study raise some doubts concerning:
- synchronism of the series of explained variable (dimensionless chara-

cteristic $Q_{max}/\overline{Q}_{max}$),
- computation of probabilities on the basis of observations from less than 10 years,
- assumption concerning statistical independence of events (floods) recorded at the stations which are not directly neighbouring each other in the region,
- the way of extrapolating regional probability curve for high recurrence intervals.

The synchronism of Q_{max} series at individual stations in the region is not analysed in the study. Region 5 used as a numerical example (Flood..., 1975, pages 173-176) includes 47 stations with observations available from 4 to 30 years (Table 2.40), but it is not indicated which years. It is unknown, therefore, whether these observations coincide in time and consequently whether they can be averaged in specific y intervals, i.e. intervals of recurrence period T (y is a function of T).

The second reservation is due to the fact that computations employ observations from stations with the period of record less than 10 years, and there are 29 such stations in the region (more than 60% of all stations). The use of the 4-element series - even if they are used for averaging $Q_{max}/\overline{Q}_{max}$ values in specific y intervals only - raises doubts from the statistical viewpoint.

Next doubt is the assumption concerning the statistical independence of floods recorded at the stations located in the region but not close one to another. If regional relation is being thought, moreover average relation for the region, it means that the region is characterized by some specific common properties that distinguish this region from the neighbouring ones. It means that from the point of view of the explained variable, the region is to some extent treated as a homogeneous one. In light of the above, assumption of independence of floods recorded at different stations in the region - even stations which are not directly neighbouring each other - also raises some doubts. Moreover, independence of events does not depend on the distances among the stations but first of all on their relative location in the catchment. It all depends whether the stations are representative for the non-overlapping catchments - in other words, whether stations are areally independent.

Special doubts are caused by the procedure used for extrapolation of the regional probability curve. This procedure is based on the assumption of independence of observations recorded at these stations which are neighbouring each other. Accordingly, stations located in the region are divided into few groups (4 to 5), and each group includes stations located far away one from another. Next, four highest $Q_{max}/\overline{Q}_{max}$ values are selected from each group, irrespective of the station at which these discharges have been recorded. Thus, in the extreme situation, these may be four values from a single station or each value from a different station. It is assumed next, that these four values are the highest values in the M-element sample, where M is the total number of station-years in the group, i.e.

$$M_i = \sum_{j=1}^{k_i} N_j \qquad (6)$$

where N_j is the number of years of record at the j-th station, and k is the number of stations in the i-th group.

This way, four highest values of the explained variable in each i-th group are related to the M_i- element random samples. In the example given above, these are 100-element samples in spite of the fact that they origin from the stations with the period of record less than 30 years. It may happen, moreover, that all four highest values have been recorded in the same year but at different stations. In such case, the condition of independence of events is not satisfied, and if the stations are located along the same river, areal independence is not satisfied as well. Such a way of extrapolating the regional probability curve seems to be unjustified. It must be said that the notion of "station-years" does not improve anything, although it makes an impression that large amount of statistical information is available. However, in a time series sense, this is kind of a duplicated information. The fact of having observations available from a single year from 50 stations does not mean that 50-element observation series is available for estimation of the probability of occurence of a given phenomena in time, since this series describe distribution in space but not in time. Such a procedure would be justified only if observations at each station would be recorded in different years, and the region in question would behydrologically fully homogeneous. One may only presume, that the notion of "station-years" was introduced under assumption that spatial and temporal distributions are the same, and consequently time series and cross-sectional series constitute together one homogeneous random sample. Unfortunately, there is no such assumption mentioned in the Flood Studies Report (1975). In reality, these distributions are underlied by entirely different random mechanisms and they are generated by different factors. Thus, they cannot be exchanged one by another; their pooling is methodologically unjustified. Temporal and spatial density functions of maximum annual flow, their parameters and specially their variability, are quite different.

The report (Flood..., 1975, page 173) presents probability curves developed this way for each of 10 regions covering together the whole territory of Great Britain. The parameters of these regional curves are next averaged to produce a single curve for the entire Great Britain treated as a single region with 420 stations and 7760 station-years. This allowed determination of the curve points for the values of random sample $Q_{max}/\bar{Q}_{max}$ corresponding to recurrence intervals as high as $T > 1000$ years, in spite of the fact that observation series at individual stations have been 18 years long on the average (Fig. 2.16 in Flood..., 1975). Thus, several questions seems to be in order:

- Is it possible to consider all Great Britain to be a single homogeneous region?
- Is relation determined in such a way of any practical significance? It means, are there any possibilities of its use in e.g. design of hydraulic structures?
- Do the values of $Q_{max}/\bar{Q}_{max}$ given in Fig.2.16 (Flood..., 1975) really correspond to such a long time intervals exceeding 1000 years?
- The notion of "station-years", which is not justified neither in substance nor statistically, is practically understood as a number of years of record (length of the series). Is there no danger that this

artificially increases the recurrence intervals for the flows which in fact occur much more often? The consequences may be very serious when it turns out that the structure has been designed in fact not for the flow occuring e.g. once in 1000years, but once in 50 years or less!

Developing Class Ib relations for hydrological characteristics, i.e. the isolines, it is very often that the requirement of areal independence of explained variable is not satisfied. The values of explained variable are defined for the partially overlapping catchments and next they are referred to by the geometrical centres of these catchments (Dynowska, 1973; Gutry-Korycka, 1984). Geometrical centres of the partially overlapping catchments may be located very close one to another, but the corresponding values of explained variables may differ quite substantially (Fig.3). It is clear, therefore, that construction of isolines for the partially overlapping catchments is inappropriate, and this is true not only because of the proximity of the geometrical centres of these catchments. The value referring to the geometrical centre of a larger catchment always includes information about each catchment located within it, therefore, these data repeat the same information to some extent - information is partially duplicated and consequently is nonhomogeneous and noncomparable. Unfortunately such a procedure is almost a rule in drafting hydrological isolines in Poland.

It should be noticed in general, that drafting hydrological characteristics in form of isolines raises several doubts. First, hydrological phenomena are not continuous spatially. Second, there is a problem of considering catchment boundaries in the process of drafting isolines (in practice these boundaries are disregarded - isolines ignore catchment boundaries). Third, hydrological characteristics refer to the areal units such as catchments - thus, how to assign the catchment to a particular point in the geographic space? Fourth, is it justified that the values of some areal characteristics (catchment characteristics) are referred to some specific points and next interpolated between these points? Is such a procedure formally correct? If isolines are to serve engineering computations or design needs as a certain form of spatial relation, and their role is to be more than just that of a general illustration, such a methodology should be worked out for drafting isolines as to avoid all of the doubts expressed above.

In the process of drafting the isolines, the requirement of synchronism of the explained variable series is generally always satisfied - observations from the entire region come from the same time period.

All comments presented above for the Class Ib relations are equally valid for the relations determined in form of a surface trend (Class Ic). For Class Ic, the values of explained variable are also referred to the gravity centres of catchments partially overlapping, hence areally dependent. In such situations, the choice of a relation and of numerical values of its parameters is a difficult task. This is mostly due to small distances between geometrical centres of partially overlapping catchments, which cause large differences in the respective values of explained variable. Regression (trend) surface has many "folds" in this case, and their approximation, even with the use of high degree polynomials, is very difficult. But the problem extends beyond the difficulties associated with the choice of a function - the essence of the problem is the nonhomogeneity of data and the fact that they are not comparable among themselves. The situation is similar to that of relations developed in form of isolines.

REFERENCES

Dynowska, I. (1973) 'Problems of work on runoff maps' (in Polish), Folia Geographica, Series Geographica-Physica, Vol.VII

Fal., B. (1979) 'Spatial variability of maximum discharges in the lowland part of Poland' (in Polish), Research Papers of the Institute of Meteorology and Water Management, Hydrology and Oceanology Series, Warsaw.

'Flood Studies Report', (1975), Hydrological Studies, Vol.I, Natural Environment Research Council.

Gren, J. (1975) 'Mathematical Statistics' (in Polish), PWN, Warsaw

Gutry-Korycka, M. (1984) 'Ana analysis and models of Poland's hydrological structure', Dissertationes Universitatis Varsoviensis, No.246

Kaczmarek, Z. (1969) 'On methods of selection of predictors set', Bulletin of Hydrological and Meteorological Service, No.3, Warsaw.

Langbein, W. (1949) 'Annual floods and the partial duration flood series', Transactions of the American Geophysical Union, 30,6.

Nowak, E. (1984) ' Problems in selecting variables for an econometric model' (in Polish), PWN, Warsaw

Ozga-Zielinska, M. (1975) ' Statistical analysis of relationships among characteristics of streamflow and physiographic characteristics of river basin', Symposium of Tokyo, IAHS Publication No. 117

Ozga-Zielinska, M. (1984) 'Hydrology as a source of information for the the national economy' (in Polish), Review of Geophysics, No.4

'Rukovodstvo po opredeleniju rascetnych gidrologiczeskich characteristic', (1973), Leningrad (in Russian).

Strupczewski, W. (1966) 'Statistical analysis of flood shapes' (in Polish), Ph.D. Dissertation, Warsaw Technical University.

Zielinska, M. (1963,1964) 'Statistical methods of defining low-water stages' (in Polish) , Review of Geophysics, Part I-1963, Part II-1964. Republished in English by the NSF, Washington, D.C. in 1967.

Zielinska, M. (1965) 'Methods of the arrangement of hydro-meteorological data for statistical purposes (in Polish), Water Management, No.7

COMBINED REGIONAL FLOOD FREQUENCY ANALYSIS AND REGRESSION ON CATCHMENT CHARACTERISTICS BY MAXIMUM LIKELIHOOD ESTIMATION

R.J. Moore
Institute of Hydrology
Wallingford, Oxon,
OX10 8BB, United Kingdom.

ABSTRACT. Regional flood frequency procedures are commonly based on the assumption that a single distribution obtains over a region after standardisation by some statistic (for example, division by the mean annual flood at each site). The standardising statistic is subsequently related to catchment characteristics to permit estimation of flood quantiles at ungauged sites within the region. Here, a maximum likelihood approach is developed to jointly estimate the parameters of the flood frequency distribution and the parameters relating them to catchment characteristics. Explicit standardisation is thus avoided in this unified approach. Application to U.K. data illustrates the utility of the unified approach as a means of estimating flood quantiles, and their standard errors, at ungauged sites.

1. INTRODUCTION

The frequency distribution of annual flood maxima varies spatially over a region because of changing vegetation, soil, physiography and climate. However it is desirable to seek levels of homogeneity in a region in order that estimates of flood quantiles may be made at ungauged sites using information at gauged sites in the region, or to improve the precision of estimates at gauged sites where only short records exist. Homogeneity may be sought by characterising drainage basins within the region by indices which attempt to quantify those vegetation, soil, physiographic and climatic characteristics which most affect flood frequency. The most commonly sought form of homogeneity is homogeneity in the frequency distribution after standardising by an index flood, which is commonly chosen to be the mean annual flood $\bar{Y}$. Thus the n year annual maximum flood sequence, $Y_{i1}, Y_{i2}, \ldots Y_{in}$, at site i, has mean $\bar{Y}_i$, and $Y^*_{ij} = Y_{ij}/\bar{Y}_i$ is assumed to have a common distribution across all m sites, $f(Y^*_{ij};\underline{\theta})$, where $\underline{\theta}$ is a parameter vector defining the shape of the distribution f(.).

V. P. Singh (ed.), Regional Flood Frequency Analysis, 119–131.

2. REGIONAL HOMOGENEITY

Homogeneity of the flood frequency distribution for a region may be specified at different levels. For example, we may consider that a common parametric distribution applies over a region and that it is of the generalised extreme value (GEV) form

$$F(x) = \exp[-\{1-k(x-u)/\alpha\}^{1/k}] , \quad (1)$$

with location parameter, u, scale parameter, α, and shape parameter, k (Jenkinson, 1955; Natural Environment Research Council, 1975). Within this level of homogeneity we may consider that the most homogeneous is one where all points in the region share the same u, α and k values. The most heterogeneous would be where u, α, and k differ at all points within the region. For the 3 parameter GEV distribution the following 4 levels of homogeneity may be defined:

(a) Perfect homogeneity

$$Y_{ij} \sim GEV(u, \alpha, k) \quad (2.a)$$

(b) Homogeneity in scale and shape

$$Y_{ij} \sim GEV\ (u_i, \alpha, k) \quad (2.b)$$

(c) Homogeneity in shape

$$Y_{ij} \sim GEV\ (u_i, \alpha_i, k) \quad (2.c)$$

(d) Total heterogeneity

$$Y_{ij} \sim GEV\ (u_i, \alpha_i, k_i)\ . \quad (2.d)$$

The location, scale or shape parameters will vary spatially over the region in response to the spatial variation in variables which shape the flood hydrograph, such as physiographic, climatic, soil and vegetation characteristics. By indexing such characteristics in a quantitative way and treating them as covariate information, the variation in u, α and k may be accounted for. This requires the establishment of a relation between the chosen GEV parameter(s), or some function of them, and the covariate information. Often dependence of the mean annual flood on catchment characteristics is assumed, where the mean annual flood corresponding to the GEV (u,α,k) distribution is a function of all 3 parameters ie

$$\bar{Y} = u + \frac{\alpha}{k}(1 - \Gamma(1+k)) \quad k \neq 0, \quad (3.a)$$

$$\bar{Y} = u + .5772\,\alpha \quad\quad k = 0\ . \quad (3.b)$$

The mean annual flood is used as a standardising variable such that if $\bar{Y}_i$ denotes the mean annual flood for site i, then $Y^*_{ij} = Y_{ij}/\bar{Y}_i$ is often assumed to be distributed GEV (u,α,k). This is a different level of homogeneity to those discussed previously although it is similar to the homogeneity in scale and shape level.

Here we will consider dependence relations between u, α, or k, and the covariate information, C_{1i}, C_{2i}, ... C_{pi}, i=1, 2,...,m, a set of p catchment characteristics for each of the m sites. A combined maximum likelihood estimation procedure for the parameters defining the dependence relations and the regional flood frequency distribution will be developed, using as an example the case where homogeneity in shape is assumed appropriate for a region.

3. AN EXAMPLE: HOMOGENEITY IN SHAPE

Consider a sequence of nm annual maxima Y_{ij}, i=1,2,...,m, j=1, 2 ... n, from m sites in a region which are distributed

$$Y_{ij} \sim \text{GEV}\ (u_i,\ \alpha_i,\ k) \quad i=1,2,...,m; \quad j=1,2,...,n \tag{4}$$

so that they share a common shape parameter, k. The location and scale parameters, u_i and α_i, vary from site to site as a log linear function of covariate information, in the form of catchment characteristics, C_{1i}, C_{2i}, ... C_{pi}, so that

$$u_i = C_0^{\beta_0}\ C_{1i}^{\beta_1}\ C_{2i}^{\beta_2} \ldots C_{pi}^{\beta_p} \tag{5.a}$$

$$\alpha_i = C_0^{\gamma_0}\ C_{1i}^{\gamma_1}\ C_{2i}^{\gamma_2} \ldots C_{qi}^{\gamma_q} \tag{5.b}$$

where $C_0 \equiv e$, the Napierian constant.

Denoting the GEV density by $f(Y_{ij}; \underline{\theta})$ where $\underline{\theta}^T = [k,\beta_o,\beta_1,...,\beta_p,\gamma_o,\gamma_1,...,\gamma_q]$ then the likelihood function is

$$L(\underline{\theta}) = \prod_{i=1}^{m} \prod_{j=1}^{n} f(Y_{ij} : \underline{\theta}). \tag{6}$$

Maximum likelihood estimates of $\underline{\theta}$ may be obtained by maximising the log-likelihood

$$\ell(\underline{\theta}) = \log L(\underline{\theta})$$

$$= -n \sum_{i=1}^{m} \log \alpha_i - (1-k) \sum_{i=1}^{m} \sum_{j=1}^{n} y_{ij} - \sum_{i=1}^{m} \sum_{j=1}^{n} e^{-y_{ij}} \tag{7}$$

where

$$y_{ij} = - k^{-1} \log \{ 1-k(Y_{ij}- u_i)/\alpha_i\} ; \tag{8}$$

this may be carried out by Newton or quasi-Newton function minimisation methods. To simplify notation, only the log-likelihood at any one of the m sites, $\ell_i(\underline{\theta})$ will be considered; also the site suffix i will be dropped so $\ell \equiv \ell(\underline{\theta}) \equiv \ell_i(\underline{\theta})$ below. Defining the scores evaluated at $\hat{\underline{\theta}}$ as

$$S_r(\hat{\underline{\theta}}) = \frac{\partial \ell(\hat{\underline{\theta}})}{\partial \theta_r} \qquad r = 1,2,\ldots,p+q+3 \tag{9}$$

then expanding this as a Taylor series about a preliminary estimate $\underline{\theta}_0 = (\theta_{10}\ \theta_{20}\ldots\theta_{p+q+3,0})^T$, up to the first power of $\hat{\theta}_r - \theta_{r0}$, and setting the result equal to zero gives

$$S_r(\hat{\underline{\theta}}) \simeq S_r(\underline{\theta}_0) + \sum_s (\hat{\theta}_s - \theta_{s0}) \frac{\partial S_r(\underline{\theta}_0)}{\partial \theta_s} = 0 \tag{10}$$

which may be solved for $\hat{\underline{\theta}}$. This is then used as the new preliminary estimate and the solution procedure repeated, repeating these steps until a satisfactory convergence towards $\hat{\underline{\theta}}$ is attained. In matrix notation

$$\underline{\theta}_{t+1} = \underline{\theta}_t - \left[B(\underline{\theta}_t)\right]^{-1} S(\underline{\theta}_t) \qquad t=0,1,2\ \ldots \tag{11}$$

where

$$S(\underline{\theta}_t) = \left[S_1(\underline{\theta}_t)\ S_2(\underline{\theta}_t)\ldots S_{p+q+3}(\underline{\theta}_t)\right]^T$$

$$\left[B(\underline{\theta}_t)\right]_{rs} = \partial S_r(\underline{\theta}_t)/\partial\theta_s \qquad r,s = 1,2,\ldots,p+q+3$$

and $\underline{\theta}_t$ denotes the estimate $\hat{\underline{\theta}}$ at the t'th iteration.

Since the observed information matrix, $B(\underline{\theta}_t)$, depends on the likelihood $L(\underline{\theta}_t)$ and therefore on the sample observations, it is sometimes preferable to replace $B(\underline{\theta}_t)$ by its expectation, the negative of the Fisher information matrix, $A(\underline{\theta}_t)$, so

$$E(B(\underline{\theta}_t)) = -A(\underline{\theta}_t) = E\left[\frac{\partial^2 \log L(\underline{\theta}_t)}{\partial\theta_r\,\partial\theta_s}\right]$$

$$= n\,E\left[\frac{\partial^2 \log f(Y;\,\underline{\theta}_t)}{\partial\theta_r\,\partial\theta_s}\right]$$

$$= -n\,E\left[\frac{\partial \log f(Y;\,\underline{\theta}_t)}{\partial\theta_r}\;\frac{\partial \log f(Y;\,\underline{\theta}_t)}{\partial\theta_s}\right]. \qquad (12)$$

Consequently

$$\underline{\theta}_{t+1} = \underline{\theta}_t + [A(\underline{\theta}_t)]^{-1}\; S(\theta_t), \quad t = 0,1,2, \ldots . \qquad (13)$$

However, Prescott and Walden (1983) have performed simulations which suggest that the observed information matrix provides more rapid convergence to $\hat{\underline{\theta}}$. Analytical expressions for the first and second derivatives of the log likelihood function with respect to u, α and k have been obtained by Prescott and Walden (1983) and are presented here for completeness

$$\frac{\partial \ell}{\partial \alpha} = -\frac{P+Q}{\alpha k}$$

$$\frac{\partial \ell}{\partial u} = -\frac{Q}{\alpha}$$

$$\frac{\partial \ell}{\partial k} = -\frac{1}{k}\left(R - \frac{P+Q}{k}\right)$$

with
$$P = n - \Sigma\, e^{-y_i}$$
$$Q = \Sigma\, e^{-y_i+ky_i} - (1-k)\Sigma\, e^{ky_i}$$
$$R = n - \Sigma y_i + \Sigma y_i\, e^{-y_i}$$
$$y_i = -k^{-1} \log\{1 - k(Y_i-u)\}$$

$$\frac{\partial^2 \ell}{\partial \alpha^2} = -\frac{1}{\alpha}\frac{\partial \ell}{\partial \alpha} + \frac{1}{\alpha k}\left(\frac{\partial P}{\partial \alpha} + \frac{\partial Q}{\partial \alpha}\right); \quad \frac{\partial^2 \ell}{\partial u^2} = -\frac{1}{\alpha}\frac{\partial Q}{\partial u}$$

$$\frac{\partial^2 \ell}{\partial \alpha \partial k} = -\frac{1}{k}\left(\frac{\partial R}{\partial \alpha} + \frac{\partial \ell}{\partial \alpha} + \alpha\frac{\partial^2 \ell}{\partial \alpha^2}\right); \quad \frac{\partial^2 \ell}{\partial u\,\partial \alpha} = -\frac{1}{\alpha k}\left(\frac{\partial P}{\partial u} + \frac{\partial Q}{\partial u}\right)$$

$$\frac{\partial^2 \ell}{\partial k^2} = -\frac{1}{k}\left[\frac{\partial R}{\partial k} + \frac{\partial \ell}{\partial k} + \alpha\frac{\partial^2 \ell}{\partial \alpha\,\partial k}\right]; \quad \frac{\partial^2 \ell}{\partial u\,\partial k} = -\frac{1}{k}\left(\frac{\partial R}{\partial u} + \alpha\frac{\partial^2 \ell}{\partial u\,\partial \alpha}\right).$$

$$\frac{\partial P}{\partial u} = -\frac{1}{\alpha} \Sigma e^{-y_i(1-k)}, \quad \frac{\partial P}{\partial \alpha} = \frac{1}{\alpha k} \Sigma e^{-y_i} + \frac{1}{k}\frac{\partial P}{\partial u}$$

$$\frac{\partial Q}{\partial u} = \frac{(1-k)}{\alpha} [\Sigma e^{-y_i(1-2k)} + k \Sigma e^{2ky_i}]$$

$$\frac{\partial Q}{\partial \alpha} = \frac{1}{k}\frac{\partial Q}{\partial u} - \frac{(1-k)}{\alpha k} [\Sigma e^{-y_i(1-k)} + k \Sigma e^{ky_i}]$$

$$\frac{\partial R}{\partial u} = \frac{1}{\alpha} [\Sigma e^{ky_i} (1-e^{-y_i} + y_i e^{-y_i})]$$

$$\frac{\partial R}{\partial \alpha} = -\frac{1}{\alpha k} [(m-r) - \Sigma e^{ky_i} + \Sigma y_i e^{-y_i} - \Sigma e^{-y_i} + \Sigma e^{-y_i(1-k)} - \Sigma y_i e^{-y_i(1-k)}]$$

$$\frac{\partial R}{\partial k} = \frac{1}{k} [\Sigma y_i - \Sigma y_i e^{-y_i} + \Sigma y_i^2 e^{-y_i} - \alpha \frac{\partial R}{\partial \alpha}] .$$

In the present case we require derivatives with respect to the regression parameters, $\{\beta_r\}$ and $\{\gamma_r\}$, and not with respect to u and α since these vary from site to site. The derivatives needed may be expressed in terms of those for u and α, or more properly u_i and a_i for site i, using the usual chain rule as follows:

$$\frac{\partial \ell}{\partial \beta_r} = \frac{\partial \ell}{\partial u_i}\frac{\partial u_i}{\partial \beta_r} = u_i \log C_{ri} \frac{\partial \ell}{\partial u_i} \qquad r = 0,1,2,\ldots,p$$

$$\frac{\partial^2 \ell}{\partial \beta_r \partial \beta_s} = u_i \log C_{ri} \log C_{si} \left(2 \frac{\partial \ell}{\partial u_i} + u_i \frac{\partial^2 \ell}{\partial u_i^2}\right) \qquad r,s=0,1,2,\ldots,p$$

$$\frac{\partial^2 \ell}{\partial k \partial \beta_r} = u_i \log C_{ri} \frac{\partial^2 \ell}{\partial u_i \partial k} \qquad r = 0,1,2,\ldots,p.$$

The derivatives with respect to γ_r (in place of β_r) are essentially the same with u_i replaced by α_i. Also

$$\frac{\partial^2 \ell}{\partial \beta_r \partial \gamma_s} = u_i \log C_{ri}\, \alpha_i \log C_{si} \frac{\partial^2 \ell}{\partial u_i \partial \alpha_i}$$

for $r = 0,1,2,\ldots,p$ and $s = 0,1,2,\ldots,q$.

The observed information matrix (symmetric), $B(\underline{\theta})$, is:

$$\left[\begin{array}{c|cccc|cccc}
-\frac{\partial^2 \ell}{\partial k^2} & & & & & & & & \\ \hline
-\frac{\partial^2 \ell}{\partial k\, \partial \beta_0} & -\frac{\partial^2 \ell}{\partial \beta_0^2} & & & & & & & \\
-\frac{\partial^2 \ell}{\partial k\, \partial \beta_1} & -\frac{\partial^2 \ell}{\partial \beta_0 \partial \beta_1} & -\frac{\partial^2 \ell}{\partial \beta_1^2} & & & & & & \\
\vdots & \vdots & \vdots & & & & & & \\
-\frac{\partial^2 \ell}{\partial k\, \partial \beta_p} & -\frac{\partial^2 \ell}{\partial \beta_0 \partial \beta_p} & -\frac{\partial^2 \ell}{\partial \beta_1 \partial \beta_p} & \cdots & -\frac{\partial^2 \ell}{\partial \beta_p^2} & & & & \\ \hline
-\frac{\partial^2 \ell}{\partial k\, \partial \gamma_0} & -\frac{\partial^2 \ell}{\partial \beta_0 \partial \gamma_0} & -\frac{\partial^2 \ell}{\partial \beta_1 \partial \gamma_0} & \cdots & -\frac{\partial^2 \ell}{\partial \beta_p \partial \gamma_0} & -\frac{\partial^2 \ell}{\partial \gamma_0^2} & & & \\
-\frac{\partial^2 \ell}{\partial k\, \partial \gamma_1} & -\frac{\partial^2 \ell}{\partial \beta_0 \partial \gamma_1} & -\frac{\partial^2 \ell}{\partial \beta_1 \partial \gamma_1} & \cdots & -\frac{\partial^2 \ell}{\partial \beta_p \partial \gamma_1} & -\frac{\partial^2 \ell}{\partial \gamma_0 \partial \gamma_1} & -\frac{\partial^2 \ell}{\partial \gamma_1^2} & & \\
\vdots & \vdots & \vdots & & \vdots & \vdots & & & \\
-\frac{\partial^2 \ell}{\partial k\, \partial \gamma_q} & -\frac{\partial^2 \ell}{\partial \beta_0 \partial \gamma_q} & -\frac{\partial^2 \ell}{\partial \beta_1 \partial \gamma_q} & \cdots & -\frac{\partial^2 \ell}{\partial \beta_p \partial \gamma_q} & -\frac{\partial^2 \ell}{\partial \gamma_0 \partial \gamma_q} & & \cdots & -\frac{\partial^2 \ell}{\partial \gamma_q^2}
\end{array}\right]$$

4. APPLICATION

The regional flood frequency estimation scheme has been applied to the Severn and Trent river basins in the United Kingdom. The data set of annual maxima comprised 833 values from 39 stations with a minimum permissible record length of 5 years. Initial estimates of the parameter vector $\underline{\theta} = (k, \beta_o, \beta_1, \ldots, \beta_p, \gamma_o, \gamma_1, \ldots, \gamma_q)$, required as part of the Newton iterative estimation scheme, were obtained as follows. At site estimates of u_i and α_i with k=0 were obtained for each station by maximum likelihood estimation of the Gumbel distribution (or by the method of moments if this failed). Regressions were then performed of u_i and α_i on catchment characteristics to establish estimates for $\{\beta_j\}$ and $\{\gamma_j\}$. This scheme was adopted because it provided a simple and fast procedure for estimating at site values of u_i and α_i whilst constraining k to be constant for the region. For numerical reasons the catchment characteristics, $\{C_{ji}\}$, used in the at-site regression analysis and in the subsequent regional analysis, are the measured values divided by their geometric mean; note that this is equivalent to subtracting the mean of the logarithm of the catchment characteristic from each logged catchment characteristic when performing the log-linear regression. Three catchment characteristics were considered in the analysis (i) AREA, denoting basin area in km^2, (ii) SAAR, the annual average rainfall for the basin in mm for the standard period 1916-1950, and (iii) SL1085, the slope of the main stream in m/km between the 10 and 85 percentiles of mainstream length. Geometric means of AREA, SAAR and SL1085 for the 39 stations are 430, 857 and 2.60 respectively. The results of these regression analyses are summarised in Table I.

TABLE I Parameter estimates and their standard errors (in brackets) obtained from regression of at-site estimates of u_i and α_i (for k=0) on catchment characteristics.

(a) Location parameter, u_i

β_0 (Intercept)	β_1 (AREA)	β_2 (SAAR)	β_3 (SL1085)	R^2
3.77	.65	-	-	.94
(.12)	(.076)			
3.77	.80	2.2	-	.98
(.06)	(.043)	(.22)		
3.77	.89	1.9	.18	.99
(.06)	(.099)	(.34)	(.18)	

(b) Scale parameter, α_i

γ_0 (Intercept)	γ_1 (AREA)	γ_2 (SAAR)	γ_3 (SL1085)	R^2
2.68	.61	-	-	.90
(.13)	(.09)			
2.68	.76	2.2	-	.96
(.09)	(.06)	(.3)		
2.68	1.03	1.4	.5	.96
(.09)	(.13)	(.5)	(.24)	

The parameter estimates of Table I were used as initial estimates in the Newton iterative scheme to obtain the results presented in Table II. Particularly interesting is the tendency for the regional

TABLE II Parameter estimates and their standard errors (in brackets) and log likelihood and log likelihood ratios obtained for dependence of the location and scale parameters, u_i and α_i, on catchment characteristics.

Catchment characteristics	k	β_0	γ_0	β_j	γ_j	- log-likelihood	log-likelihood ratio
AREA	-.31 (.03)	3.77 (.03)	3.43 (.04)	.75 (.01)	.58 (.02)	4429	-
AREA	-.12 (.03)	3.76 (.02)	3.07 (.03)	.82 (.01)	.71 (.02)	4115	314
SAAR				2.09 (.07)	1.63 (.11)		
AREA	-.12 (.03)	3.75 (.02)	3.04 (.03)	.94 (.03)	.98 (.06)	4099	16
SAAR				1.80 (.10)	.83 (.26)		
SL1085				.21 (.05)	.50 (.12)		

value of k to tend towards zero, and hence a Gumbel distribution, as more within region variation of u_i and α_i is accounted for through incorporation of additional catchment characteristics. The Newton scheme employed includes modifications to choose a large step in the steepest descent direction if the Hessian, $B(\underline{\theta})$, is not positive definite, and to reduce the step length by a factor if the adjustment fails to reduce the function value. A large function value is returned if the parameters move outside their permissible bounds, specified as $u_i + \alpha_i/k > y_{max,i}$ for $k > 0$ and $u_i + \alpha_i/k < y_{min,i}$ for $k < 0$, where $y_{min,i}$ and $y_{max,i}$ are the minimum and maximum of the annual maxima values for site i.

The observed information matrix evaluated at the final set of parameters may be used to obtain approximate standard errors of the parameter estimates, since its inverse approximates the covariance matrix of parameter estimates:

$$\text{cov}(\hat{\theta}) \simeq B(\theta)^{-1}. \tag{14}$$

Consequently the required standard errors are

$$\text{s.e.}\ (\hat{\theta}_r) \simeq ([B(\theta)^{-1}]_{rr})^{1/2}, \tag{15}$$

where the $[\cdot]_{jj}$ denotes the jj th element: the values obtained are presented in Table II in brackets. The covariance matrix of parameter estimates is presented below as a correlation matrix for the case when two catchment characteristics, AREA and SAAR, are included:-

k	1.0						
β_0	.28	1.0					
β_1	-.08	-.47	1.0				
β_2	.00	-.38	.42	1.0			
γ_0	.13	.48	-.15	-.13	1.0		
γ_1	-.15	-.10	.30	.10	-.25	1.0	
γ_2	.05	-.11	.12	.57	-.17	.44	1.0
	k	β_0	β_1	β_2	γ_0	γ_1	γ_2

The highest correlations are seen to occur between parameters defining the two relations with catchment characteristics, particularly the exponents on SAAR, β_2 and γ_2.

An estimate of the T-year flood at site i , $\hat{Q}_i(T)$, is obtained from the regional estimation procedure using the parameter estimates u_i, $\hat{\alpha}_i$ and k in

$$Q_i(T) = \begin{cases} u_i + \alpha_i[1 - \{-\log(1-T^{-1})\}^k]/k & k \neq 0 \\ u_i - \alpha_i \log\{-\log(1-T^{-1})\} & k = 0 \end{cases} \tag{16}$$

This derives from the relation $T^{-1} = 1 - F(Q_i(T);u_i,\alpha_i,k)$. In addition an approximate standard error for $\hat{Q}_i(T)$ may be calculated using

$$\text{var}(\hat{Q}_i(T)) \simeq \sum_{r=1}^{N} \sum_{s=1}^{N} \frac{\partial Q_i(T)}{\partial \theta_r} \frac{\partial Q_i(T)}{\partial \theta_s} \text{cov}(\hat{\theta}_r, \hat{\theta}_s) \tag{17}$$

where $N = p + q + 3$, and $\text{cov}(\hat{\theta}_r, \hat{\theta}_s) \simeq [B(\theta)^{-1}]_{rs}$. Here, the derivatives are evaluated using the final parameter estimates in the following expressions:

$$\frac{\partial (Q_i(T))}{\partial k} = \frac{-\alpha_i}{k^2} [1 - \{-\log(1-T^{-1})\}^k (1 - \log\{-\log(1-T^{-1})\}^k)]$$

$$\frac{\partial (Q_i(T))}{\partial \beta_r} = u_i \log C_{ri} , \qquad r = 0,1,2,...,p. \tag{18}$$

$$\frac{\partial (Q_i(T))}{\partial \gamma_r} = (Q_i(T) - u_i) \log C_{ri} \qquad r = 0,1,2,...,q.$$

The approximate variance formulae for parameter and T-year flood estimates presented above are based on asymptotic theory. These may be poor for small samples, in circumstances where the likelihood function may be highly non-quadratic, and when the Q-T relationship is highly nonlinear, e.g. $k < 0$, T large.

The regional flood frequency scheme is based on the assumption that annual flood maxima for a region share a GEV(u_i, α_i, k) distribution having a common shape parameter k for the region but with location and scale parameters, u_i and α_i, varying as a log-linear function of catchment characteristics. The validity of this assumption may be investigated through the use of probability plots of residuals. Here we will define a residual, r_{ij}, for the j th year at site i as $r_{ij} = (Y_{ij} - u_i)/\alpha_i$. Ranking these residuals for each site and plotting them against $(1-(-\log p_r)^k)/k$, $r=1,2,...,n$, where p_r denotes the plotting position of the r th ranked value, would for each site yield a straight line plot through the origin with a slope of unity if the assumption was correct. Figure 1 presents probability plots of residuals obtained for the Vyrnwy annual maxima data set when dependence on AREA, SAAR, and SL1085 are progressively incorporated. Note the closer correspondence to the theoretical line as more heterogeneity is accounted for through incorporation of further catchment characteristics.

Particularly important is the assumption of independence of annual maxima between sites made in forming the likelihood function. The scheme may be shown to provide consistent estimates of both parameters and floods of given return periods, in the presence of inter-site dependence. However, inter-site dependence will tend to mean that the standard errors of parameter and flood estimates obtained using the variance formulae presented above understate their true value.

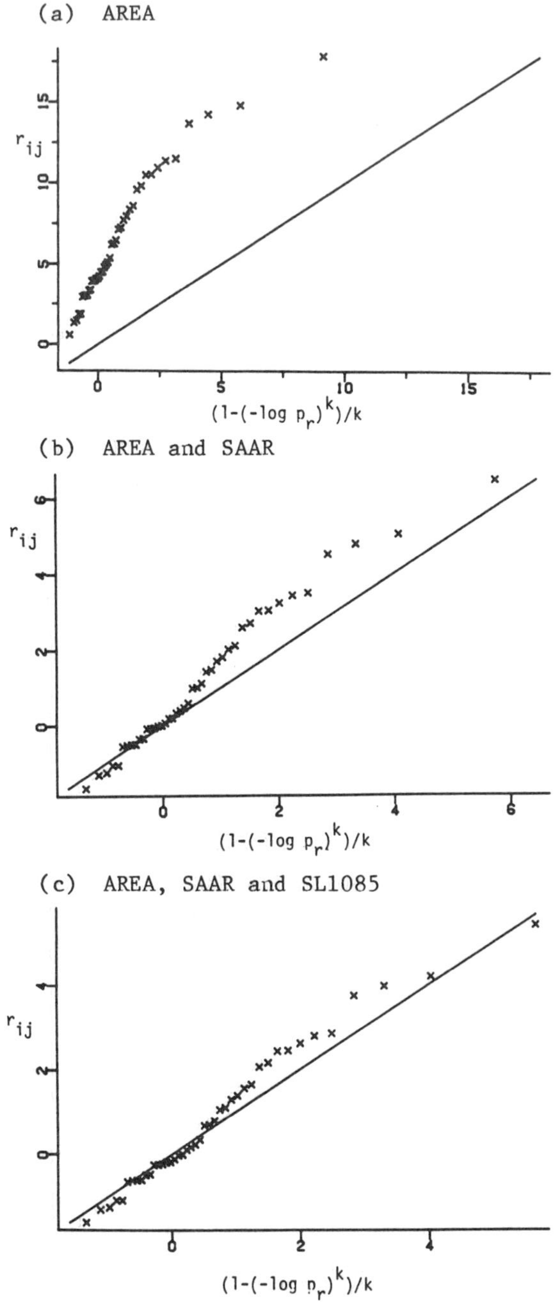

Figure 1. Probability plots of the residuals for different dependence relations on catchment characteristics: Vyrnwy at Vyrnwy Reservoir.

5. CONCLUSION

The method presented provides a means whereby regional flood frequency analysis, including the estimation of floods of a given return period at ungauged sites, may be accomplished as a single procedure. Explicit standardisation is avoided. In the example provided, where the shape parameter k is constant for the region, it also provides a novel method of obtaining a regional estimate of the parameter k in the GEV distribution. Approximate expressions for the standard errors of parameter and T-year flood estimates have also been presented. The general technique in which the procedure is embedded is applicable to distributions other than GEV, and to other forms of dependence between the parameters, or functions of them, and catchment characteristics. The present paper has demonstrated that the procedure is capable of practical implementation, but further work is required in order to assess its merits relative to existing regional flood frequency estimation procedures. In particular, the effect of inter-site dependence requires further examination.

6. ACKNOWLEDGEMENT

This contribution was undertaken as part of a research project commissioned by the Ministry of Agriculture, Fisheries and Food.

7. REFERENCES

Natural Environment Research Council, Flood Studies Report, Volume I, Hydrological Studies, London, 1975.

Jenkinson, A.F. 'The frequency distribution of the annual maximum (or minimum) values of meteorological elements', Quart. J. Roy. Met. Soc., **81**, 158-171, 1955.

Prescott, P., & Walden, A.T. 'Maximum likelihood estimation of the parameters of the three-parameter generalised extreme-value distribution from censored samples', J. Statist. Comput. Simul., **16**, 241-250, 1983.

MULTIVARIATE TECHNIQUES FOR THE IDENTIFICATION OF HOMOGENEOUS FLOOD FREQUENCY REGIONS

Stephen Wiltshire
Geomorphological Services Ltd
16 Bassett Court, The Green
Newport Pagnell
Buckinghamshire, England

and

Max Beran
Institute of Hydrology
Wallingford
Oxfordshire, England

ABSTRACT. Homogeneous collections of basins can be formed by clustering in a bivariate dataspace of coefficient of variation and specific mean annual flood. The allocation of an ungauged basin to a cluster can be achieved through the application of multivariate discriminant analysis to the basin characteristic data within the clusters. A variation of the Bayes decision rule of allocation preserves the notion of fractional cluster membership of an ungauged basin. The posterior probabilities of cluster membership are used as weighting factors in estimating a flood quantile at an ungauged site. This ensures a smooth variation of quantile values across cluster boundaries and minimises problems arising from misallocation of basins to clusters.

1. INTRODUCTION

The index flood method of regional flood frequency analysis has proved to be a popular way of providing flood information at ungauged sites since it was first described by Dalrymple (1960). The derivation of regional distributions consists of three stages:

(1) Identification of one or more homogeneous regions;
(2) Selection of an index flood and derivation of regressions relating the index flood to basin characteristics;
(3) Selection and fitting of a regional flood frequency distribution to the standardised flood data of each region.

The literature contains plentiful examples of index flood regressions, and is also replete with discussions on the choice and method of fitting of a regional distribution. However the first stage of the index flood method, the identification of homogeneous regions, is generally given

V. P. Singh (ed.), Regional Flood Frequency Analysis, 133–145.

scant attention by practitioners of the method, and yet it is no less important in terms of its effect on the reliability of quantile estimates. It has been demonstrated (Wallis and Wood, 1985) that regions should consist of basins having similar flood frequency characteristics if the variance of quantile estimates is not to be excessive.

There are two requirements for the successful identification of homogeneous regions:

(1) a method of quantifying regional homogeneity;
(2) a method of grouping hydrologically similar basins into homogeneous 'regions'.

This paper addresses the second of these two issues and then describes a robust procedure for flood quantile estimation at ungauged sites. A preceeding paper (Wiltshire and Beran, 1986) discusses the problem of testing for homogeneity.

2. REGION IDENTIFICATION

2.1 Previous studies

Heterogeneity of regions is undesirable as information is lost and the error of quantile estimation will be inflated. There have been some attempts to identify flood regions not in geographical space but in a dataspace defined by either flood statistics or basin characteristics. In New Zealand Mosley (1981) used cluster analysis to form groups of basins characterised by specific mean annual flood and coefficient of variation, while Waylen and Woo (1984) formed groups of stations in British Columbia on the basis of five different parameters of the Peaks-over-Threshold (POT) flood series and differentiated between them using discriminant analysis. White (1975) used a factor analysis of a set of basin characteristic data to identify collections of physically similar basins in Pennsylvania and Acreman and Sinclair (1986) have used cluster analysis on a matrix of basin characteristics to identify similar basins in Scotland. The present writers have employed an iterative search procedure to locate optimum divisions of the basin characteristic dataspace for flood regionalisation (Wiltshire, 1985, 1986).

In Britain there is a great deal of available flood and basin characteristic data and best use of this can be made with a multivariate analysis of the relationship between the flood frequency properties of homogeneous collections of drainage basins and the basin characteristics themselves. The following section describes such an analysis.

2.2 Derivation of Basin Clusters

It is very reasonable to follow the initiative of Mosley and to describe the location and scale of individual basin flood series by two standardised statistics, the specific mean annual flood, QBARsp, and the coefficient of variation, CV. Between them these statistics describe much that can be said of a basin frequency curve; QBARsp describes the spatial intensity of the average maximum flood while CV characterises the year-to-year variability of the annual maxima.

Figure 1 shows the distribution of the 376 British basins within this bivariate dataspace. Basins which plot close to each other in this figure can be combined together to form groups for the purposes of flood frequency analysis. The process of grouping is readily performed by a clustering algorithm which automatically allocates basins to groups according to a specified nearness criterion. The method of clustering employed here is an iterative relocation algorithm, as described by Gordon (1981). In this, the required number of groups M is arbitrarily chosen and the collection of basins is randomly divided into M similar-sized groups. Basins are successively exchanged from group to group until the total sum of squares of distances from the group centroid within the M groups is minimised. The values of QBARsp and CV for each basin are normalised prior to clustering so that neither variable has a disproportionate influence on the 'distance' between basins. This algorithm is preferred to those of hierarchical clustering schemes which tend to produce one large dominant group or cluster with small 'satellite' clusters located towards the margins of the dataspace.

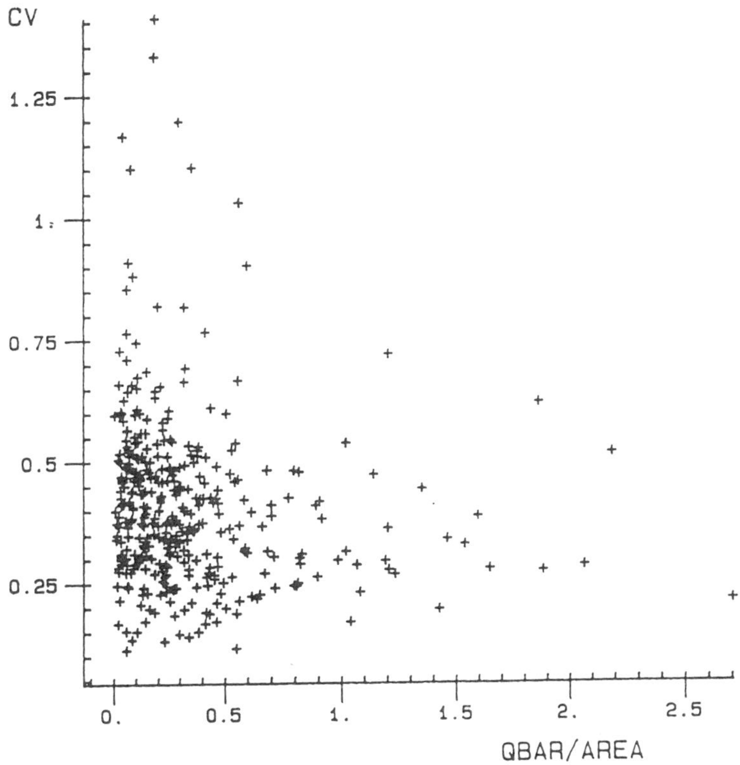

Figure 1. Plot of 376 basins in a dataspace of coefficient of variation and specific mean annual flood.

Figure 2 summarises the result of applying this clustering process to the data shown in Figure 1, with the prerequisite number of clusters specified as ten. A GEV-PWM growth curve is fitted to each cluster and these are also shown in this figure. Table I lists group sizes, mean CV's and the observed and critical values of the homogeneity statistic.

TABLE I. Homogeneity statistics of 10 clusters of basins formed in CV-QSP space

Cluster Number	Number of Sites	Number of Site-Yrs	Mean CV	Calculated R	Critica R at 1%
1	38	761	0.24	80.5	59.9
2	19	369	0.28	24.5	34.8
3	8	160	0.38	7.3	18.5
4	67	1499	0.26	135.7	95.6
5	30	603	0.65	47.4	49.6
6	13	294	0.99	24.8	26.2
7	84	2001	0.46	100.1	115.9
8	49	1156	0.48	53.6	73.7
9	14	324	0.46	23.9	27.7
10	54	1239	0.35	51.6	79.9

The majority of clusters are seen from Table I to be homogeneous at the 1% level of significance with only clusters 1 and 4 displaying heterogeneity. The set of frequency curves in Figure 2 is very diverse compared with those of the geographic regions of Figure 6 in Wiltshire and Beran (1986), with extremely steep (cluster 6) and very flat (cluster 1) curves being exhibited. Cluster 6 has a GEV shape parameter of -0.46 indicating an undefined skewness and this reflects the difficulty of fitting a regional distribution to a small number of sites whose data include some very large floods. Clusters 1 and 4 have positive shape parameters (0.1 and 0.13) indicating an upper bound to flood magnitude to which the frequency curves are asymptotic. It should be possible to interpret this effect in terms of basin characteristic and this is considered in a following section. The diversity of these cluster frequency curves is an important result as it shows the range of curve types which is inherent in the flood data but which is concealed by the heterogeneous nature of geographic regions.

The position of some of the cluster boundaries in Figure 2 is rather arbitrary and this is so for two reasons: firstly the decision to form ten clusters was an arbitrary one, and secondly there are few well-defined gaps in the data of Figure 1 which the clustering algorithm would be expected to find. Cluster boundaries could be subjectively adjusted to enlarge a particularly small group or to subdivide a heterogeneous group, but such adjustments have not been applied here.

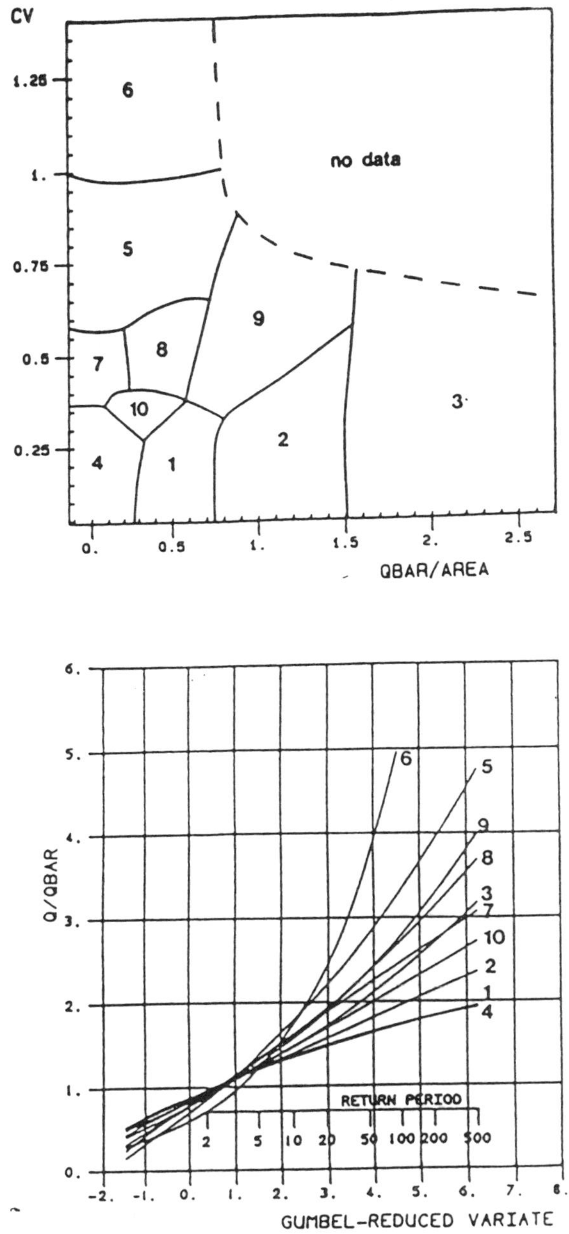

Figure 2. Cluster flood frequency curves.

2.3 Discriminant Analysis

The clustering analysis of the previous section has produced a set of ten groups, most of which are homogeneous, and each of which is defined in terms of QBARsp and CV. A practical flood frequency estimation technique must of course be geared to the ungauged case, and an ungauged basin will not have QBARsp and CV statistics available to indicate in which group or cluster it naturally belongs. However basin characteristic data can be derived for any basin within the overall study area and these data can be used to infer cluster membership, given that each cluster can itself be described by a set of average basin characteristic values.

Suppose that data for K basin characteristics are available for each drainage basin. In this K-dimensional dataspace an ungauged basin is represented by a vector X having K elements. To make an allocation of X to one of M groups or clusters of basins it is necessary to know the posterior probabilities of the basin belonging to each group, given that the basin has known characteristic values. This probability, $p(i|X)$, is given by Bayes rule

$$p(i|X) = p(X|i)p(i)/p(X) \tag{1}$$

where $$p(X) = \sum_{i=1}^{M} p(X|i)p(i)$$

The posterior probabilities allow Bayes Decision Rules to be formed. For example, if there are two groups (M=2) then the decision rules are as follows:

the basin belongs to group 1 if $p(X|1)p(1) > p(X|2)p(2)$
the basin belongs to group 2 if $p(X|2)p(2) > p(X|1)p(1)$

The decision rules can be restated as

the basin belongs to group 1 if $D_1(X) > D_2(X)$
the basin belongs to group 2 if $D_2(X) > D_1(X)$

where $D_i(X)$ is a Discriminant Function, and is given by

$$D_i(X) = p(X|i)p(i) \tag{2}$$

As its name implies, and as the decision rules indicate, the discriminant function discriminates between the groups. In a univariate problem with two groups the discriminant function would be represented as the point on the variate axis which best separates the groups, while in a multivariate problem with several clusters the function represents a hypersurface in K dimensions.

The general form of a multivariate K-dimensional normal distribution is

$$p(X|i) = \frac{1}{(|C_i|2\pi^K)^{\frac{1}{2}}} \exp((X-M_i)^T C_i^{-1} (X-M_i)/2)$$

where M_i is a vector of K means for group i, one mean for each of the K basin characteristics,
C_i is the symmetric K x K covariance matrix for group i,
$||$ denotes a determinant and a T superscript denotes a transpose.

If the covariance matrix is assumed to be common to all groups and if a log transformation of the data is applied then the K dimensional Bayes discriminant function (neglecting constant terms) becomes

$$D_i(X) = - (X-M_i)^T C^{-1} (X-M_i) + \ln(p(i)) \tag{3}$$

The posterior probability p(i) of basin X belonging to group i is

$$p_i = \exp(D_i) / \sum_{i=1}^{M} \exp(D_i) \tag{4}$$

and the Bayes decision rule will allocate X to the group having the largest value of p(i).

2.4 Application of Discriminant Analysis to Basin Data

Data for nine basin characteristics are available for each of the 376 basins used in the cluster analysis. These characteristics are listed and defined in Table II. Discriminant analysis is performed using subroutines from the International Mathematical and Statistical Language, which supply as output ten discriminant scores (the D_i's of equation 10) together with a performance matrix, S. When the discriminant function for each cluster has been computed each of the 376 basins is substituted in turn for vector X of equation 10. Each basin is then notionally allocated to the cluster which yields the highest discriminant score. This notional allocation is then compared with the actual a priori allocation of basins to clusters. The value of S(i,j) is thus the fraction of basins belonging to the ith cluster that is allocated by the discriminant analysis to the jth cluster. The S(i,j)'s therefore summarise the efficiency with which the clusters in QBARsp-CV space can be mapped into basin characteristic space.

The performance matrix derived for the ten-cluster configuration of Figure 2 is given in Table III. Entries on the diagonal of this matrix give the fraction of the basins in cluster i notionally allocated to cluster i by the Bayes decision rules. Clusters 1, 2, 3, 4, 7 and 10 are all reasonably well defined in the basin characteristic space but the others are poorly defined. Cluster 5 has a zero on the diagonal but a value of 0.64 in row 7, meaning in basin characteristic terms that all the cluster 5 basins would rather be elsewhere, 64% of them having a preference for cluster 7. The poor performance of this and some other clusters reflects the perennial difficulty of relating flow statistics to basin characteristics, a problem which is particularly evident for the smaller, more scattered clusters with higher average CV's.

TABLE II. Basin characteristics measured for 376 drainage basins in Britain

Characteristics	Definition	Units
AREA	basin plan area	km^2
AAR	average annual rainfall	mm
RSMD	5 year return period 1 day effective rainfall	mm
SMD	average annual soil moisture deficit	mm
STMFRQ	stream frequency	junctions km^{-2}
S1085	stream slope	m km^{-1}
SOIL	soil type	-
LAKE	fraction of basin draining through a lake	-
URBAN	urbanised fraction of basin	-

TABLE III. Performance matrix summarising the rate of misallocations of basins to clusters. Entry (i,j) indicates the fraction of basins in cluster (i) that are notionally allocated by discriminant analysis to cluster (j).

Original Cluster Number (i)	Notional reallocation to cluster number (j)									
	1	2	3	4	5	6	7	8	9	10
1	.58	.11	.0	.0	.0	.0	.03	.08	.0	.21
2	.16	.47	.11	.0	.0	.0	.04	.0	.11	.11
3	.0	.13	.74	.0	.0	.0	.0	.0	.13	.0
4	.10	.0	.0	.48	.0	.0	.22	.06	.0	.14
5	.06	.0	.0	.10	.0	.0	.64	.17	.0	.03
6	.0	.0	.0	.14	.0	.14	.44	.14	.0	.14
7	.0	.0	.0	.20	.02	.0	.68	.06	.10	.04
8	.14	.06	.0	.20	.0	.0	.20	.14	.0	.25
9	.29	.29	.0	.0	.0	.0	.0	.07	.14	.21
10	.15	.0	.0	.09	.0	.0	.07	.24	.0	.45

3. PREDICTION OF GROWTH CURVES VIA CLUSTER AND DISCRIMINANT ANALYSIS

The preceding section has established a framework for estimating the growth curve at an ungauged site. A set of clusters of drainage basins have been formed for which average dimensionless flood frequency relationships have been derived. A multivariate discriminant analysis provides a discriminant score for each cluster and according to Bayes rule, the ungauged basin is allocated to the cluster having the largest discriminant score. However, at this last stage a departure from the conventional Bayesian wisdom is advocated, and an alternative procedure is developed which is believed to lead to more robust quantile estimates. This alternative procedure is described below.

3.1 Weighted quantile estimation

Any cluster which has a non-zero discriminant score for an ungauged basin X has a claim to the ownership of X, and the strength of its claim is in proportion to the posterior probabilities as given by equation (4). The set of non-zero posterior probabilities implies a fractional membership of the basin X, and this is an expression of the uncertainty inherent in allocation decisions. Instead of ignoring the information from clusters other than that which has the greatest discriminant score, it is considered preferable to use this information explicity in estimating a quantile or growth curve for X.

Quantile estimation incorporating fractional membership of clusters can be achieved in the following intuitively appealing way. Suppose X_T is the T year dimensionless flood to be estimated for basin X. If the ith cluster provides an estimate T_i of the T year flood then X_T can be estimated readily as a weighted average of the M values of T_i, ie

$$X_T = \sum_{i=1}^{M} p_i \, T_i \tag{5}$$

Clusters for which the posterior probability p_i of containing X is zero do not contribute to the estimate of X_T while all others contribute in proportion to their value of P_i. The calculation of a weighted quantile estimate in this way has the result that extreme values from any one of the frequency curves of Figure 3 are avoided and the consequences of allocating X to the wrong cluster are alleviated. It will also ensure that design quantile estimates will vary smoothly across both the QBARsp-CV dataspace and across the nine-dimensional basin characteristic dataspace rather than change abruptly at cluster boundaries. This system also overcomes the problem of some clusters being poorly defined in basin characteristic terms, as these clusters will have only a weak influence on X_T.

3.2 Application of weighted quantile estimation procedure

Equation 5 can be used to construct a complete dimensionless frequency curve for a new basin. The form of this predicted curve can be compared with the curve forms expected from physical reasoning, thereby providing a test of the multivariate model introduced above. In conducting such tests the use of hypothetical sets of basin characteristics is preferred for two reasons:

(1) It is much easier to assemble a diverse set of basin characteristic values from one's imagination than from a map or computer archive as the latter sources tend to contain few obvious "types" of basin to which deductive reasoning might be simply applied;

(2) The only benefit to be gained from the use of real-world basins with historical flood data would be in a comparison of the predicted growth curve with the curve fitted to the local series. Such a comparison would only be meaningful if a long (eg 50 years) flood record was available and in Britain such records are scarce (and the few that are available have been used in the clustering process).

Thus growth curves are prepared here for six hypothetical basins whose characteristics are listed in Table IV. Basin A can be described as "average" for British conditions; basin B is similarly average but is 50% urbanised; basin C is also average but with 50% lake; basin D is small, dry and permeable; basin E is small, wet, steep and impermeable; basin F is large, wet and flat. Table V lists the posterior probability of these basins being in each of the 10 clusters of Figure 2, and Figure 3 shows the frequency curves constructed by evaluating Equation 5 for several return periods.

TABLE IV. Characteristics of six hypothetical basins

Hypothetical Basin	AREA	AAR	S1085	LAKE	SOIL	STMFRQ	RSMD	SMD	URB
A	300	800	1.0	0.	.4	1.0	40	7	0.
B	300	800	1.0	0.	.4	1.0	40	7	.5
C	300	800	1.0	.5	.4	1.0	40	7	0.
D	10	600	2.0	0.	.2	.5	30	12	0.
E	10	2000	20.0	0.	.5	2.5	70	3	0.
F	3000	2000	1.0	.1	.4	2.0	40	3	0.

TABLE V. Probability that basins A-F are members of clusters 1-10

Hypothetical	Cluster Number								
Basin	1	2	3	4	5	6	7	8	9	10

Hypothetical Basin	1	2	3	4	5	6	7	8	9	10
A	.0	.0	.0	.11	.11	.02	.24	.32	.0	.20
B	.0	.0	.0	.02	.0	.0	.0	.18	.0	.80
C	.0	.0	.0	.09	.16	.32	.31	.08	.0	.04
D	.0	.0	.0	.05	.17	.16	.33	.25	.0	.04
E	.0	1.0	.0	.0	.0	.0	.0	.0	.0	.0
F	.39	.0	.0	.45	.0	.0	.01	.02	.0	.13

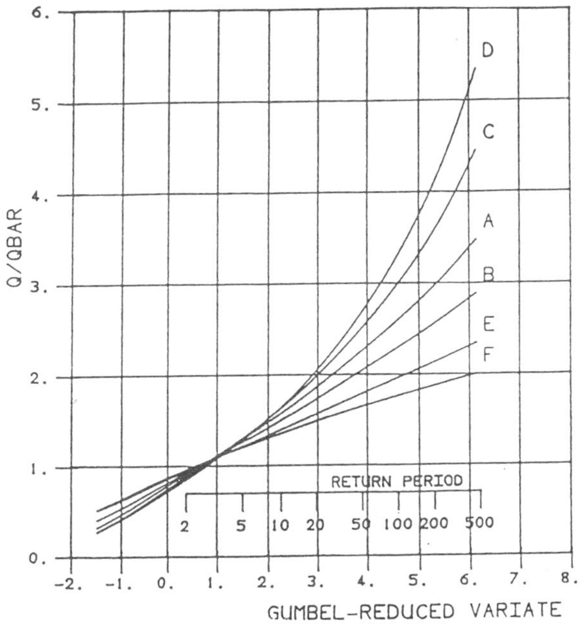

Figure 3. Frequency curves for six hypothetical drainage basins.

Basin A, true to its average basin characteristics, is not strongly drawn to any single cluster but has a fractional membership distributed among clusters lying near to the centroid of Figure 2. Basin B is identical to A except that it is 50% urbanised and Table V indicates that such a basin is very similar to the basins in cluster 10 with a membership probability of 0.8. The standardised frequency curve of B is less steep than the basin A curve as would be expected as the fixed high runoff coefficient for the urbanised part of the basin would reduce the year-to-year variability of annual maximum floods.

Basin C is the same as basin A except that half of the basin drains through a lake. This basin is seen from Table V to be quite similar to clusters 6 and 7 which has the surprising consequence of a steeper predicted basin frequency curve than that of basin A. This result is rather difficult to explain as laked basins are generally assumed to have an attenuating effect on flood peaks, although it may be that the influence diminishes as the scale of the flood event increases. However, inspection of the data reveals that one station has 83% of its area draining through a lake and yet has a CV of 1.1, albeit based on a record length of 10 years. The presence of this basin in the very small cluster 6 will have an undue attraction on basin C with the consequence of a steep frequency curve.

Basin D is small, dry and permeable and shows a kinship with the basins of clusters 7, 8, 5 and 6. Its frequency curve is correspondingly steep, reflecting the potential of such basins to vary from wet to dry and hence to produce a high degree of inter-annual variability in flood peaks. Basin E is small, wet, steep and impermeable, and such a basin would be expected to produce consistently large floods with a relatively small CV. This is demonstrated by its 100% membership of cluster 2. Basin F is large, wet and flat and 10% of the catchment drains through a lake. This combination of characteristics would be expected to produce floods of small specific runoff and small CV and indeed such are the properties of clusters 1 and 4 with which it is most strongly associated. The frequency curve for basin F is flat and is of type EV3, suggesting a finite upper bound on the maximum flood. This is a plausible result, given the combined attenuating effects of both lakes and large basins.

4. DISCUSSION AND CONCLUSIONS

A rational basis for region definition is to group together basins which are hydrologically similar and then to interpret the groups in terms of basin characteristics. This has been done for British basins using cluster analysis to group together basins having similar CV's and specific mean annual floods, and discriminant analysis to differentiate between the clusters on the basis of basin characteristics. The ten groups formed by clustering show a high degree of homogeneity, which is to be expected as the constituent basins have been grouped according to their hydrological similarity.

The clusters must be interpreted in terms of basin characteristics and the evaluation of the discriminant functions from the characteristics of a new basin typically implies a fractional membership of more than one cluster. The probability of cluster membership can then be used as a weighting factor in the calculation of quantile estimates for the new basin.

This paper attempts only to illustrate the use of these multivariate techniques in flood estimation and the methods and results are not intended to be definitive. There is considerable scope for refining the clustering process, for example by weighting observations according to the inverse of the estimation errors of CV and QBARsp. Indeed other statistics can be used to define the clustering dataspace, such as probability-weighted moments, skewness or indeed the G'-points.

The efficiency of the discriminant analysis is limited by the available basin characteristic data and there may be a requirement for additional characteristics which, for example, measure flood plain storage potential, the spatial distribution of ground characteristics within a basin or the annual variability of soil moisture deficit. The availability of such data would probably improve both the discriminant analysis described above and the regression estimation of the standardising parameter.

The judicious use of multivariate techniques has been shown to provide a framework for flood estimation that is more flexible than the conventional regional analysis and is intuitively more reasonable. Its application to a set of representative hypothetical basins shows that it gives flood predictions which accord with physical reasoning and which are responsive to quite small differences between basins.

5. REFERENCES

Acreman, M.C., and Sinclair, C.D. 1986. 'Regional flood frequency analysis from basin characteristics in Scotland.' Manuscript in preparation.

Dalrymple, T. 1960. 'Flood frequency analyses.' US Geol. Surv. Water Supply Paper 1543-A.

Gordon, A.D. 1981. Classification. Monographs on applied probability and statistics. Chapman and Hall.

Mosley, M.P. 1981. 'Delimitation of New Zealand hydrological regions.' Journal of Hydrology **49**, 173-192.

Natural Environment Research Council (NERC). 1975. Flood Studies Report. Volume I. London.

Waylen, P.R. and Woo, M. 1984. 'Regionalisation and prediction of floods in the Frazer river catchment, B.C.' Water Resources Bulletin, **20**, (6) 941-949.

Wallis, J.R. and Wood, E.F. 1985. 'Relative accuracy of Log Pearson III procedures.' Journal Hydraulic Engineering (ASCE), **111**, 1043-1056.

White, E.L. 1975. 'Factor analysis of drainage basin properties: classification of flood behaviour in terms of basin geomorphology.' Water Resources Bulletin, 11, (4) 676-686.

Wiltshire, S.E. 1985. 'Grouping basins for regional flood frequency analysis.' Hydrological Science Journal, **30**, (1), 151-159.

Wiltshire, S.E. 1986. 'Identification of Homogeneous Regions for Flood Frequency Analysis.' Journal of Hydrology (in press).

Wiltshire, S.E. and Beran, M.A. 1986. 'A Significance test for homogeneity of flood frequency regions.' Symp. flood frequency and risk analysis. Baton Rouge, May 18-21 1986. Reidel.

6. ACKNOWLEDGEMENTS

This work was conducted while the first author was employed at the Institute of Hydrology, Wallingford, and forms part of a Ministry of Agriculture, Fisheries and Food project on regional flood statistics.

A SIGNIFICANCE TEST FOR HOMOGENEITY OF FLOOD FREQUENCY REGIONS

Stephen Wiltshire
Geomorphological Services Ltd
1 Bassett Court, The Green
Newport Pagnell
Buckinghamshire, England

and

Max Beran
Institute of Hydrology
Wallingford
Oxfordshire, England

ABSTRACT. Regional flood frequency relationships are necessary for flood estimation at ungauged sites and have also gained popularity for problems on gauged rivers where they can supplement the information contained in the flood observations. A prerequisite of their development is the identification of homogeneous regions. In this paper a significance test is developed for regional homogeneity based upon the non-exceedance probabilities (termed G-points) of individual flood maxima. The power of the test is investigated in relation to region size and distribution choice. As an example of the use of the test it is used to investigate the homogeneity of the geographical regions adopted in the United Kingdom Flood Studies Report.

1. INTRODUCTION

Regional flood frequency analysis has proved to be an important tool for estimating design flood magnitudes at ungauged or inadequately gauged locations. A prerequisite of the approach is the identification of regions or data subsets that are homogeneous with respect to the flood frequency behaviour of the contained basins. This generates in turn a requirement for test statistics against which the hypothesis of homogeneity may be judged.

In this paper the development is described of such a test tailored specifically to the problem of regional flood estimation. The power and robustness of the statistic are considered in relation to number of stations and length of records in the region, and to the correct choice of parent distribution. As an example of its use the test is applied to the geographic division of the United Kingdom used in the Flood Studies

V. P. Singh (ed.), Regional Flood Frequency Analysis, 147–158.

Report (NERC, 1975) and shows most of them to be heterogeneous. A second paper (Wiltshire and Beran, 1986) describes how this may be rectified and demonstrates a procedure for quantile estimation at an ungauged site.

2. MEASUREMENT OF REGIONAL HOMOGENEITY

The approach to quantifying homogeneity adopted here begins with the a priori selection of the type of distribution applicable to the region and proceeds with an examination of the scatter of site flood data about the fitted region-average distribution.

2.1 Distribution-based Test of Homogeneity

A flood data set can be characterised by the geometry of the cumulative frequency distribution (CDF) of a flood frequency distribution which the flood data are assumed to follow.

The form of a CDF depends on the relative frequency of floods of different magnitudes and also upon the parameters of the distribution from which the flood data are assumed to derive. One property of a CDF is that the marginal distribution of the non-exceedance probabilities is uniform with lower and upper limits of zero and one. Hence, a dimensionless regional frequency distribution fitted to a set of data from N sites will have a uniform distribution of non-exceedance probability. Consider a random sample (x_1, x_2 ... x_i ... x_n) drawn from such a population. A value of non exceedance probability

$$G_j = G(x_j) \qquad (1)$$

can be calculated for each sample point x_j (ie each dimensionless annual maximum flood). These values (G_1, G_2, ... G_j ... G_n) should have, apart from deviations caused by random sampling effects, a uniform distribution.

Conversely, now consider a random sample (y_1, y_2 ... y_n) which is not drawn from the same population as above. Let the same distribution function of equation (1) be used to calculate the non-exceedance probabilities

$$G_j = G(y_j) \qquad (2)$$

Since equation (1) does not represent the distribution function of the population from which the y sample is drawn, the distribution of the set of values (G_1, G_2 ... G_n) will now differ from the uniform distribution by an amount greater than could be attributed to sampling effects alone. Figure 1 shows some possible consequences of assuming the wrong parent distribution for a particular series of floods. The true PDF is (a) but, if it is assumed to be (b) or (c), then a non-uniform distribution of the G_j's results. Thus if a region is homogeneous the dimensionless annual maximum series from each of the N sites in a region would all be expected to show a uniform distribution

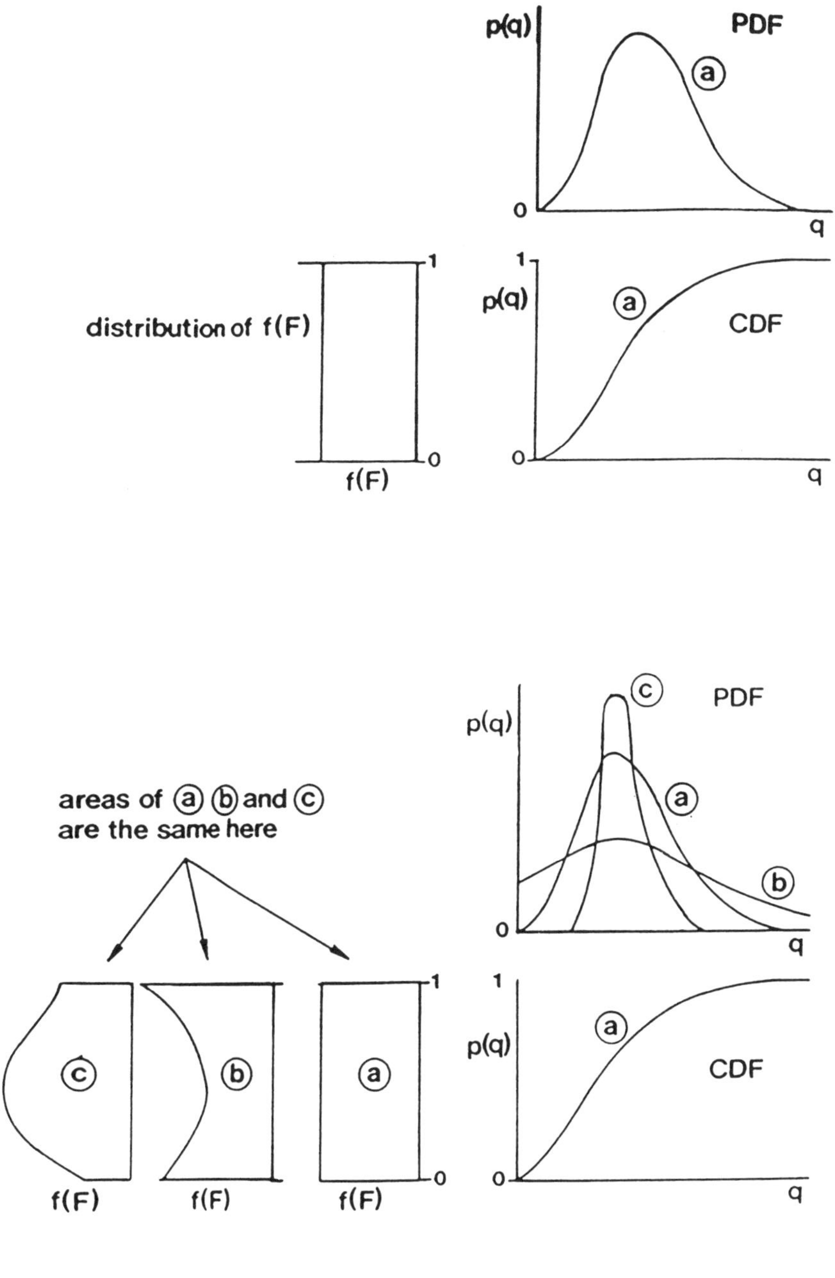

Figure 1. Distribution of f(F) of CDF when population may be either type (a), (b) or (c) but the population is assumed to be type (a).

of non-exceedance probabilities (termed here the "G-points"), with deviations from this expected geometry due solely to sampling error. Here lies the basis for a significance test of homogeneity.

The formulation of this test has a requirement for a single measure to describe the set of G-points at each site. The mean G-point is an obvious candidate, and if a flood series did follow exactly the regional (parent) distribution, the site mean should be close to 0.5. However, flood series with a high CV relative to the parent would have a greater proportion of G-points near the limits of the 0-1 range but the sample mean may still be near to 0.5. Similarly a series with a low CV relative to the parent would show a clustering of G-points around the 0.5 value and again the sample mean is likely to be at or near 0.5. In other words the mean G-point cannot distinguish between data with high CV's, low CV's and CV's comparable to that of the parent. A simple transformation can overcome this problem: The non-exceedance probability of the kth flood at the jth site, G_{jk}, can be transformed by the expression

$$G'_{jk} = 2\,|G_{jk} - 0.5| \tag{3}$$

The set of G'-points are thus the transformed G-points formed by 'folding' the distribution to the left of 0.5 point onto the right-hand part of the distribution. The expected result for sites with the same distribution as the parent is another identical uniform distribution. At sites whose data do not derive from the parent systematic departures of G-points from the expected horizontal uniform distribution will be effectively amplified by the folding process. Figure 2 shows the result of folding five different hypothetical sets of G-points. In Figure 2(a) floods follow the same distribution as the parent and so the set of G'-points is also uniform. The mean value of the G-points at site j, $G'_{j.}$ is indicated by the arrow in Figure 2(a) and would be expected to be close to 0.5. In figure 2(b) and 2(c) the effect of applying the transformation of equation (3) is respectively to increase and to decrease the value of $G'_{j.}$, indicating that the folded G-points can now distinguish between high and low CV data. Figures 2(d) and 2(e) show that the folding process will represent positively skewed and negatively skewed data in the same way, ie with a value of $G'_{j.}$ larger than 0.5.

Within a region of N sites the variability, R, of the $G'_{j.}$ values can be written as

$$R = \sum_{j=1}^{N} \frac{(G'_{j.} - G'_{..})^2}{u_j} \tag{4}$$

where $G'_{..}$ is the regional mean value of the $G'_{j.}$ s, given by

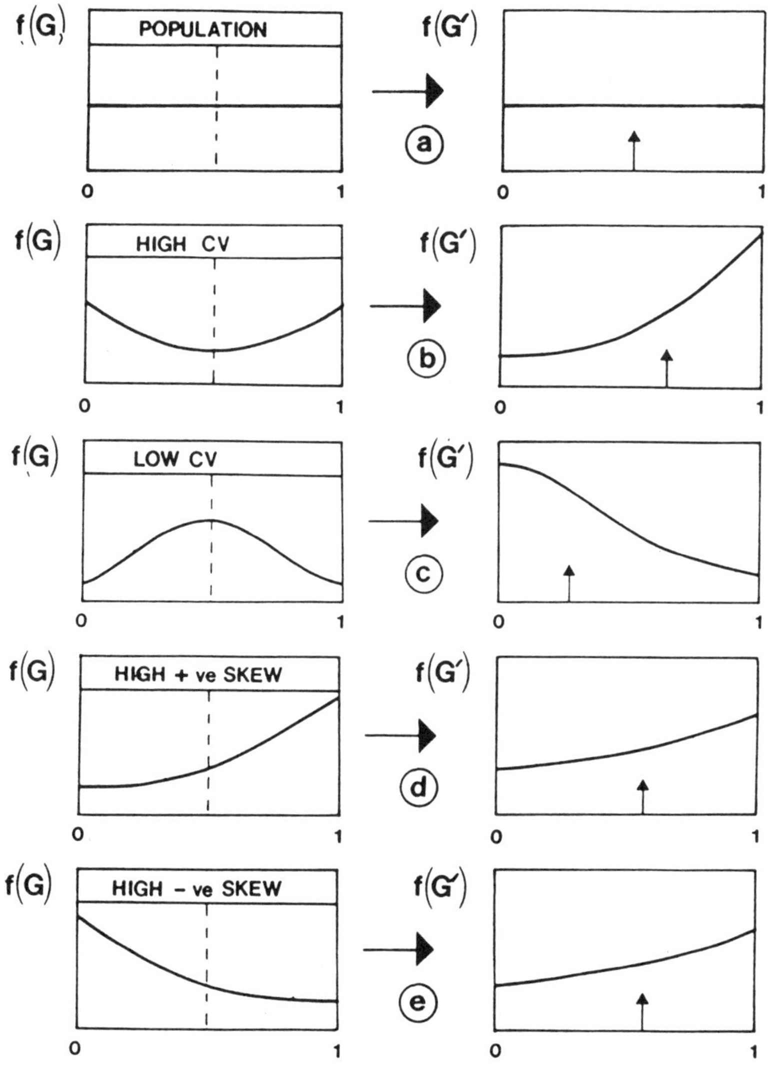

Figure 2. The effect of applying the transformation of Equation 3 on the distribution of G-points.

$$G'_{..} = (\sum_{1}^{N} n_j G'_{j.})/\sum_{1}^{N} n_j$$

and u_j is the sampling variance of $G'_{j.}$, which in turn is given by

$$u_j = V/n_j$$

V is the expected regional variance of the G'points which is a constant and equal to one-twelfth (the variance of a uniform distribution).

2.2 Distribution of the R-Statistic

The form of the R statistic of equation (4) is such that it is expected to have a χ^2 sampling distribution with N-1 degrees of freedom. The assumption of a χ^2 distribution for R can be tested using Monte Carlo simulation. Data can be randomly generated from a known flood-like distribution for an artificial region of N sites each with n years of data. For each realisation of such a region the R statistic can be calculated and, when repeated very many times, the mean and variance of R can be computed and compared with the expected χ^2 mean and variance of N-1 and 2(N-1).

Table I shows the mean and variance of R computed from 1000 realisations of each of sixteen configurations of regions. It can be seen that the calculated values of the mean and variance follow closely the expected values across the range of region configurations. The mean value of R appears to be biased slightly upward for shorter record lengths and the variance shows some downward bias for small regions, but overall this performance is quite adequate and the R statistic may be reliably assumed to have a χ^2 distribution.

TABLE I. Mean and variance of the R-statistic observed in 1000 simulations of a homogeneous region

record length (m)	Number of sites (N) 10	20	50	100
5	9.6	19.9	51.1	103.8
	17.8	34.9	96.7	195.9
10	9.5	19.3	49.5	100.4
	19.7	40.2	90.9	197.9
20	9.0	18.6	48.4	98.0
	17.2	32.1	91.0	197.7
50	8.7	18.6	48.0	97.7
	16.0	35.2	95.0	197.1

An R value can thus be evaluated for the flood data of any region and its magnitude, when compared with tabulated χ^2 values, tests the null hypothesis that the N flood series in the region do not differ significantly from the fitted regional distribution.

2.3 Power of the Homogeneity Test

2.3.1 Power curve construction. A test statistic of region homogeneity has been defined in Equation (4), but so far it is not clear how powerful this test will be when applied to real data. In other words, how heterogeneous must a collection of data be before the computed R-statistic indicates rejection of the null hypothesis of homogeneity? The power of the test may be explored through the use of simulation experiments, and in particular a power curve can be constructed which graphs the rate of H_0 rejection against a measure of region heterogeneity. Production of a power curve from simulation experiments uses the following procedure:

(1) For a region of N sites and n years of record at each site generate flood data from two parent distributions of known CV. Data for half of the sites come from the first distribution, with CV = CV_1 and the data at the other sites is generated from the second distribution with CV = CV_2.

(2) Initially $CV_1 = CV_2$ so the region is homogeneous. If a 5% level significance is adopted, then in a large number of repetitions, 5% of the R values should exceed the critical χ^2 value.

(3) Subsequently, CV_1 is held constant and CV_2 is incremented, to produce a heterogeneous region characterised by the ratio CV_1/CV_2. Data are generated from these two parents and the frequency with which R exceeds the critical value from repeated realisations is plotted against the ratio CV_1/CV_2.

(4) Stage (3) is repeated several times with successive increments in CV_2 and hence an increasing degree of regional heterogeneity.

This technique can be used to examine the effect on the R-test of the length of record, the number of sites in a region and on the choice of frequency distribution which is fitted to the regional data. In this study, data are generated from a pair of EV1 distributions (see NERC, 1975, p.I.42 for notation) with CV calculated from

$$CV = \frac{\pi\alpha/\sqrt{6}}{u + .5772\alpha}$$

The location parameter u is fixed at a value of 10.0 for both distributions; α, the scale parameter, is set to 4.0 for the first distribution but is incremented to produce increasing values of CV_2 in the second distribution. A regional EV1 distribution is fitted to the generated data using the method of probability-weighted moments (PWMs) as described by Greis and Wood (1981). A G-point is calculated from the EV1 distribution function for each generated data point as

$$G(y) = \exp(-e^{-y}) \qquad (5)$$

where y is the reduced variate given by

$$y = (x_{jk} - u)/\alpha$$

where x_{jk} is the kth flood at the jth site.

2.3.2 Sample size effect. Figure 3 shows power curves for four hypothetical regions, all with 20 sites but with respective record lengths of 5, 10, 20 and 50 years at each site. These curves are based on 1000 repetitions at each of ten CV_1/CV_2 ratios. The four curves all pass through the 5% point on the y-axis, indicating that when the two parent distributions in the region are equivalent ($CV_1/CV_2 = 1$) there is indeed a 5% chance of rejecting the null hypothesis. It can be seen that the test is more powerful when applied to a region containing sites with longer records; eg when $CV_2/CV_1 = 0.75$, p(reject H_0) is 90% for the n = 50 years region, but only 12% for the n = 5 years region. This is qualitatively the expected result as the greater sampling errors or 'noise' associated with the shorter records would be expected to obscure the genuine heterogeneity or 'signal' within the region.

In producing the power curve of Figure 4 the record lengths in each region are held constant at n = 20 years, and the four regions differ in terms of the number of their constituent sites. Again the intuitive result is seen, namely that a given degree of heterogeneity is more readily detected in a region containing a greater number of sites. A comparison of power curves in Figures 3 and 4 based on regions having the same total number of station-years (N=50, n=20 and N=20, n=50; N=20, n=10; N=10, n=20) shows that the information content of the data is greater when deployed in a small number of long records rather than in a large number of short records. Two other conclusions can be drawn from these sets of curves. Firstly, if the series available to a regional analysis are typically ten years or less in length then there will rarely be sufficient information present to detect heterogeneity even when there are gross differences among the constituent flood series. Secondly, for a typical region with N=50 and n=20, the R-test is of only moderate power, although alternative types of test are unlikely to do any better. The application of the test to real data should therefore be made with the power of such tests in mind, and the interpretation of the R-statistic as a significance test should not be too literal.

2.3.3 Distribution effect. A further set of simulation experiments can be performed to examine the effect on the power of the R-test of different distributional assumptions. In Figures 3 and 4 a two parameter EV1 distribution was fitted to the simulated regional data using the PWM fitting technique but more flexible distributions might be preferred. The three parameter General Extreme Value (GEV) distribution has been widely used in the study of flood and rainfall maxima and Hosking et al (1985) have shown how its parameters may be fitted by PWM's. The distribution function of the GEV is

$$G(y) = \exp(-(1 - ky)^{1/k}) \qquad (6)$$

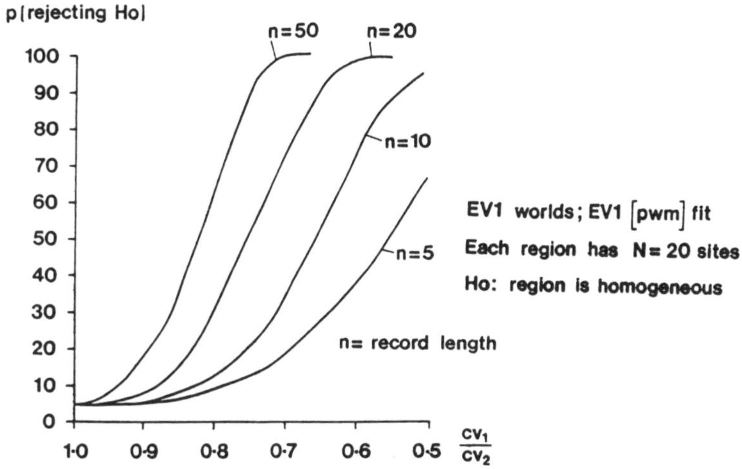

Figure 3. Power curves for the R test: variable record length.

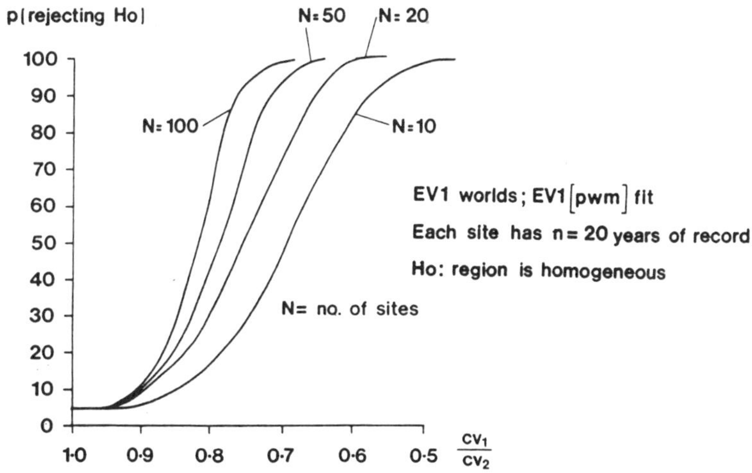

Figure 4. Power curves for the R test: variable region size.

where y is the reduced variate of Equation (5) and k is a shape parameter. The five parameter Wakeby distribution can also be fitted using PWM's (Landwehr et al 1979). The Wakeby distribution function must be expressed in inverse form

$$x = m + a(1 - (1-G)^{b}) - c(1 - (1-G)^{-d}) \quad (7)$$

so a technique such as Newton-Raphson iteration must be used to solve for G given the flood, x. Figure 5 shows power curves for the R-test, fitting regional EV1, GEV and Wakeby distributions to the two-parent EV1 regions. The EV1 curve is the same as the N=20, n=20 curves in Figures 3 and 4. The GEV and Wakeby power curves lie well to the right of the EV1 curve, signifying a lower H_0 rejection rate than the EV1 power curve for a given degree of heterogeneity. With more parameters with which to characterise the data the GEV and Wakeby distributions are more tolerant of heterogeneous effects in the data and so the region is more likely to appear homogeneous.

2.4 Application to UK geographical regions

Most applications of regional flood frequency techniques define regions to be geographic, administrative or political subdivisions of the study area. For example in the Flood Studies Report (NERC, 1975) Britain was divided into ten geographic regions as shown in Figure 6, together with their associated regional GEV-PWM dimensionless (index flood) frequency curves. These "growth" curves are based on a total of 7406 station-years of annual maxima at 376 sites, each site having a minimum of ten years of data. Table II summarises region sizes and lists the observed and critical values of the R-statistic of homogeneity described above.

TABLE II. Homogeneity statistics of ten geographic regions in Britain

Region Number	Number of Sites	Number of Site-Yrs	Mean CV	Calculated R	Critical R at 1%
1	34	960	0.39	105.5	54.8
2	53	1125	0.35	156.0	78.6
3	51	1123	0.39	150.1	76.2
4	34	801	0.43	108.9	54.8
5	45	910	0.53	75.9	68.7
6	44	961	0.46	79.4	67.5
7	18	286	0.57	27.5	33.4
8	31	653	0.48	87.5	50.9
9	35	866	0.35	94.5	56.1
10	31	721	0.34	50.8	50.9

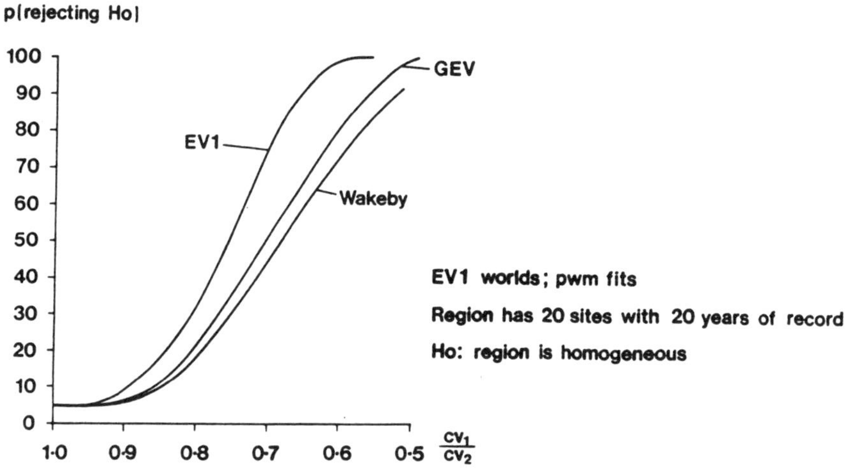

Figure 5. Power curves for the R test as a function of distribution type.

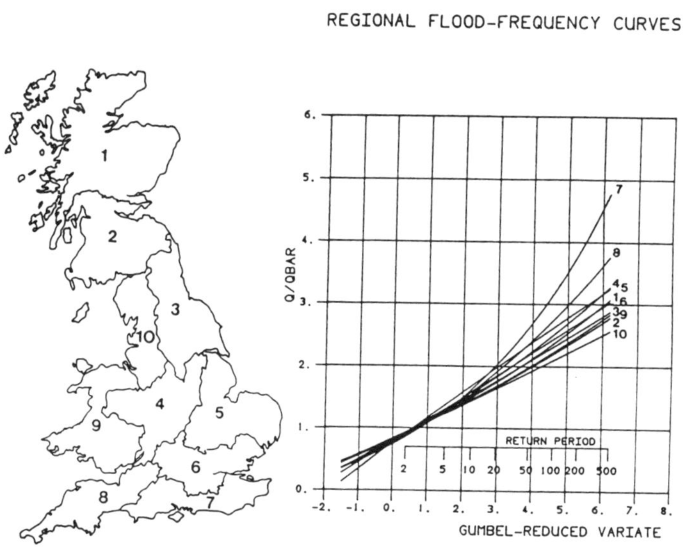

Figure 6. Regional flood frequency curves in Britain.

Only regions 1 and 10 are homogeneous at the 1% level of significance; all other regions must be interpreted as being heterogeneous in terms of their flood frequency characteristics. The steep growth curve of region 7 and the flatter curve of region 10 reflect the differences in basin geomorphology between the drier south-east and the wetter north-west of England, but overall little sense can be made of the variations as most curves are rather similar to each other. In fact this is to be expected if regions are heterogeneous as the average of diverse collections of basin growth curves will tend to be similar from region to region.

3. CONCLUSIONS

The importance of homogeneous regions has been indicated and a significance test described based upon the folded non-exceedance probabilities, termed G'-points. The test statistic is shown to be distributed as chi-square and for its operation requires the previous adoption of a flood frequency relationship for the area under test. The power and robustness properties of the test have been explored through simulation. The statistic was applied to UK geographical regions which are thereby shown, with two exceptions, to be heterogeneous.

4. ACKNOWLEDGEMENTS

This work was conducted while the first author was employed at the Institute of Hydrology, Wallingford, and forms part of a Ministry of Agriculture, Fisheries and Food project on regional flood statistics. We also acknowledge the contribution of our colleague Dr. D.A. Jones for his assistance in the development of the significance test.

5. REFERENCES

Greis, N.P., and Wood, E.F. 1981. 'Regional flood frequency estimation and network design'. Water Resources Research **17**, (4), 1167-1177.

Hosking, J.R.M., Wallis, J.R. and Wood, E.F. 1985. 'Estimation of the Generalised Extreme Value distribution by the method of probability weighted moments'. Technometrics **27**, (3), 251-261.

Landwehr, J.M. Matalas, N.C. and Wallis, J.R. 1979. 'Estimation of parameters and quantiles of Wakeby distributions'. Water Resources Research **15**, 1361-1372.

Natural Environment Research Council. 1975. Flood Studies Report, **I**. London.

Wiltshire, S.E. and Beran, M.A. 1986. 'Multivariate techniques for the identification of homogeneous flood frequency regions'. Proc. Int. Symp. Flood Frequency and Risk Analyses. Baton Rouge. May 18-21, 1986. Reidel.

TESTING THE SUITABILITY OF THE TWO-COMPONENT EXTREME VALUE DISTRIBUTION FOR REGIONAL FLOOD ESTIMATION

Nigel Arnell and Max Beran
Institute of Hydrology
Crowmarsh Gifford
Wallingford,
Oxon OX10 8BB
U.K.

ABSTRACT. The Two-Component Extreme Value distribution is characterized as the maximum of two Gumbel distributions representing different flow generating processes. This paper describes some of the tests used to evaluate the suitability of the distribution for regional flood frequency analysis.

Some geometrical properties of the distribution are outlined, and the distribution is fitted to UK flood data. Results show that the distribution is capable of reproducing the observed variability in sample skewness.

Simulation experiments show that the TCEV estimation scheme has little bias but high variability when data are drawn from TCEV and Wakeby distributions, but displays larger bias with GEV data. Variability increases significantly as record length decreases. The experiments also show that estimation procedures based on GEV and, to a lesser extent, Wakeby distributions do not perform well when applied to data drawn from a TCEV distribution.

1. INTRODUCTION

1.1 Objectives of Paper

The design of structures and strategies for flood mitigation requires extrapolation beyond the range of the short flood records normally available. The source of information for such an extrapolation will be either a physical or a statistical model. In this paper we discuss one particular statistical model, the two-component extreme value (TCEV) distribution. Earlier papers concerning the TCEV distribution (Rossi et al, 1984; Beran et al, 1986; Arnell and Gabriele, 1985) have detailed its statistical properties and parameter estimation so here we focus on its suitability for solving the practical problems of flood estimation for the UK.

Particular attention is paid to the TCEV distribution's relationship with currently accepted procedures and how its properties match flood maxima. Of particular interest is its ability to cater for

V. P. Singh (ed.), Regional Flood Frequency Analysis, 159–175.

uncertainties about the true form of the flood population and inaccuracies in the basic data. A strategy is described for drawing a final conclusion about its suitability.

1.2 UK Conditions

In the UK flood estimation tends to be pervasive in the sense that it is conducted on very many streams, large and small, and is certainly not confined to major watercourses or population centres. The majority of flood protection schemes involve structural measures and are peak rather than volume sensitive. A range of design return periods is usually tested, commonly between 10 and 200 years, before a final scheme is settled.

The pervasive nature of estimation means that the gauging station network, although dense, is seldom of direct use. The problem is compounded by the shortness of the flow records; of over 1000 gauging stations that have existed in the UK network fewer than 60 have accumulated records longer than 30 years. The response to this has been to prefer regional to site flood frequency estimates: see Figure 1 of Beran (1986) presented at this meeting for the currently recommended procedure. Many flood protection projects are quite small so cannot sustain high design costs and therefore flood estimation must not be computationally expensive.

Flood protection is publicly funded so uniformity of approach is important. This means that the estimation technique must be capable of working with the range of response types that are met in a country of varied topography, land use and geology. Uniformity is also demanded in the sense that local data and historical flood experience are used both formally and informally to improve estimation. Thus the method for an ungauged site must be continuous with the recommended procedure for a gauged site; ie the same distribution must be capable of fitting (perhaps with constraints) to local at-site data.

A final requirement concerns the need for the estimation procedures to appear reasonable to design engineers. These are the major class of user and because they carry ultimate responsibility for the integrity of their design they need to appreciate the significance of the calculation steps and how they correlate with reality even though not conversant with the underlying theory. Particularly prominent in their internal checking system is the flood experience of the locality; for example through the implied return period of overbank flow, and the regional frequency of large events.

2. DERIVATION OF THE TCEV DISTRIBUTION

In their original derivation of the TCEV distribution Rossi et al (1984) were motivated by the observation that Italian rainfall and flood series contain occasional outliers due to unusual synoptic and catchment conditions. They modelled the occurrence of maxima by two parallel Poisson processes with exceedence rates λ_1 and λ_2 for the basic and outlier events. The event magnitudes were considered to be exponentially distributed with parameters θ_1 and θ_2. The outlier exceedance rate was considerably smaller than the basic series rate, but due to its greater tail length was likely to generate larger events.

From this basis they proved that the annual maximum distribution followed the cumulative distribution function:

$$F(x) = \exp\{-\lambda_1 \exp(-x/\theta_1) - \lambda_2 \exp(-x/\theta_2)\} \qquad (1)$$

The distribution is the product, and hence the maximum, of two Gumbel or EV1 distributions. In accordance with the time series formulation the basic series is responsible for 'normal' events through its location to the right of the outlier series (Figure 1), but the latter's much heavier tail tends to throw up occasional large values.

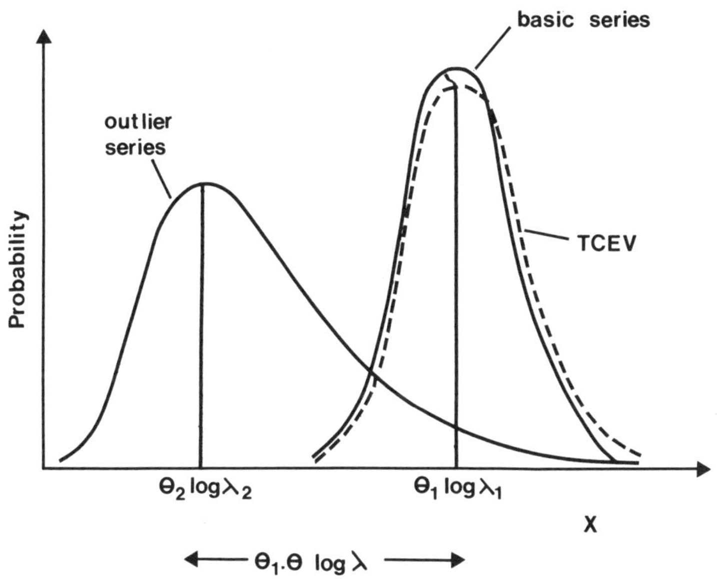

Figure 1 Gumbel and TCEV density functions

This formulation has an intuitive appeal to many hydrologists who recognise a different genesis for typical and extraordinary floods. Although such a picture is clearly justifiable in some regions, for example where spring floods are due to snowmelt and in other seasons storm rainfall is responsible, the UK seasonal pattern of flooding is not clearcut and in the trials reported here the model is treated as an empirical approximation to be fitted without regard to the specific origins of flood events.

3. PROPERTIES OF TCEV DISTRIBUTION

3.1 Relationship to Gumbel Distribution

It is useful to standardise the TCEV distribution using the Gumbel reduced variate

$$y' = (x-\theta_1 \ln\lambda_1)/\theta_1 \tag{2}$$

The parameter θ_1 is the scale and $\theta_1 \ln\lambda_1$ is the mode, ie location parameter, of the basic series Gumbel distribution. On Figure 2 the basic series appears as the line of equality and the outlier series a line of the form

$$y' = \theta y + \theta \ \ln\lambda \tag{3}$$

where: $\theta = \theta_2/\theta_1 \quad : \quad \lambda = \lambda_2/\lambda_1^{1/\theta}$

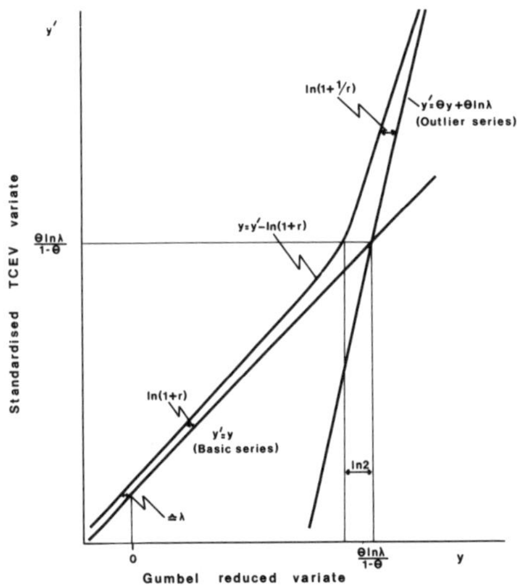

Figure 2 TCEV and Gumbel frequency curves

This further reparametrisation allows us to write equation 1 in standardised form as

$$F(y') = \exp\{-\exp(-y') - \lambda\exp(-y'/\theta)\} \tag{4}$$

Recalling that, for the Gumbel distribution,

$$y = -\log\{-\log(F(y')\}$$

we obtain:

$$y = y' - \ln\{1 + \lambda\exp(-y'/\theta)/\exp(-y)\},$$

Setting $r = \lambda\exp(-y'/\theta)/\exp(-y')$ equation (4) becomes

$$y = y' - \ln(1+r) \qquad (5)$$

The ratio, r, is capable of easy geometric interpretation on Figure 2 in terms of the abscissae of the basic and outlier lines. By consideration of log(1+r) it can readily be shown that the TCEV line is asymptotic to the y'=y line for small y (in fact at y=0 $y'\approx\lambda$). A similar process shows that the TCEV curve rapidly approaches equation (3) at the upper end. Thus Figure 2 shows the TCEV distribution bounded by the basic and outlier Gumbel lines.

The intersection of the two asymptotes has special significance in appreciating the properties of the TCEV distribution. By definition, at that point the non-exceedence probability must be equal in both limiting lines, the ratio r is unity and the relationship

$$y = y' = y_o = \theta\ln\lambda/(1-\theta)$$

holds. The corresponding return period of a TCEV variate is nearly exactly half that of the event on either parent distribution. In fact for hydrological samples this point occurs well to the right of the mode of the basic series, typically at $y\approx 3$ and flood estimation is often concerned with the region $r\approx 1.0$.

At this point $y'=y_o$ the gradient of the TCEV frequency curve is $2/(1+1/\theta)$ ie the harmonic mean of the gradients of the two asymptotes. Above this point it approaches the upper asymptote rapidly (ultimately as $1/r^2$). The relative sharpness of the change in direction gives the frequency curve the appearance of a broken line, again a feature often noticed in recorded flood data (eg Figure 3).

3.2 Relationship to Extreme Value Type 2 Distribution

The TCEV distribution differs markedly from the Extreme Value Type 2 distribution (EV2)

$$F(x) = \exp-\{1-k(x-\theta\ln\lambda)/\theta\}^{1/k} \qquad (6)$$

which is in current use for UK flood data (NERC, 1975). The gradient of the EV2 distribution increases without limit as exp(-ky). For observed parameter values it exhibits a more gradual curvature than the TCEV frequency curve and in the limit has the heavier tail. However the crossover does not occur within the range of the data so for practical purposes it is the TCEV distribution which appears to have the longer tail.

3.3 Moments and Outlier Probability Properties

The parameters θ and λ control standardised moments such as skewness and kurtosis, and also the ratio, p, of outliers within a large sample of annual maxima.

$$p = \lambda/\theta \sum_{j=0}^{\infty} (-1)^{j+1} \lambda^{j} \Gamma((j+1)/\theta)/j! \qquad (7)$$

Moments of the standardised distribution are obtained from the formulae

$$\mu'_r = (-1)^r \left[\Gamma^{(r)}(1) + r\, S_{r-1}\right]$$

where r is the order of the non central moment,

$\Gamma^{(r)}$ is the rth derivative of the Gamma function, and

$$S_r = \sum_{j=1}^{\infty} (-1)^j \lambda^j \Gamma^{(r-1)}(j/\theta)/j!$$

which, for the case of the mean, reduces to

$$E(y') = \gamma + \sum_{j=1}^{\infty} (-1)^{j-1} \lambda^j \Gamma(j/\theta)/j! \qquad (8)$$

The strong similarity between equations (7) and (8) result in similar trends of p and E(y') with θ and λ; typically the standardised mean is between ten and twenty times the outlier probability. Beran et al (1986) present graphical summaries which reinforce this finding. The hinge point, $y'=y_0$, also strongly controls the behaviour of E(y') and p. For example a TCEV distribution with hinge point at y'=3.9 (corresponding to 25 year return period on the TCEV frequency curve) will have a standardised mean of approximately 0.8 and an outlier probability close to 0.05.

4. TCEV FIT TO REGIONAL FLOOD DATA

4.1 A Regional Interpretation

The TCEV distribution can be fitted to data from a single sample, but is intended primarily for regional use. Rossi et al, (1984) and Arnell and Gabriele (1985) describe a two-level parameter estimation scheme in which the parameters θ and λ are assumed constant throughout the region, implying that skewness, kurtosis and the proportion of outliers are also constant. The first level of regionalisation concerns the estimation of θ and λ and is based as above on the standardised variable

$$y' = (x - \theta_1 \ln\lambda_1)/\theta_1$$

The second level of regionalisation can be attained by identifying subregions within which the at-site parameter λ_1 can be assumed constant. This is equivalent to assuming that the coefficient of variation CV is constant in the subregion - all the subregions have the same skewness - and within such a subregion a frequency curve expressed in y' units can be converted to a form expressing the T-year flood as a multiple of the mean annual flood $\bar{x}$ by

$$\frac{x}{\bar{x}} = \frac{y' + \ln\bar{\lambda}_1}{E(y') + \ln\bar{\lambda}_1} \tag{9}$$

where $\bar{\lambda}_1$ is the mean subregion λ_1.

4.1.1 Application to UK Flood Data. A regional TCEV distribution was fitted to a sample of 57 stations with records greater than 30 years (mean record length 41 years) from England, Scotland and Wales. The fit (Figure 3) yielded $\theta = 4.33$ and $\lambda = 0.032$, with an implied skewness coefficient of 3.5. This compares with the fit to Italian data reported by Rossi et al (1984) which gave $\theta = 3.01$, $\lambda = 0.173$ and a skewness coefficient of 2.8.

Given the climate characteristics of the two areas it appears surprising that the implied skewness of the Italian fit is less than that of the U.K. fit, but in fact random samples drawn from a TCEV distribution with Italian parameters have a higher sample skewness

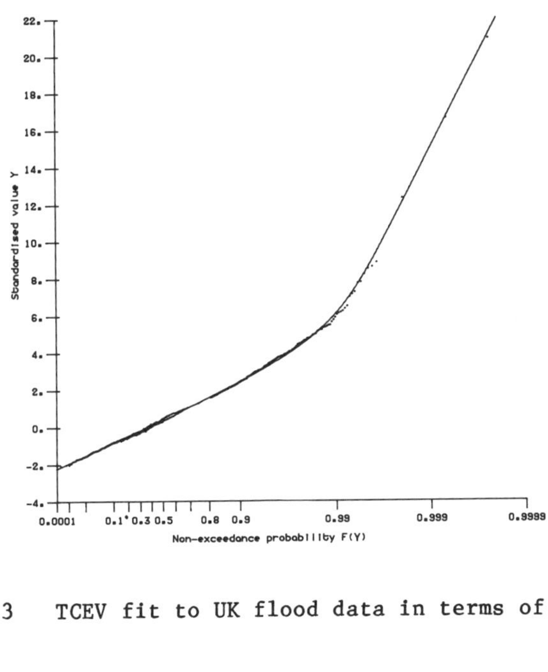

Figure 3 TCEV fit to UK flood data in terms of y'

coefficient than random samples drawn from a TCEV with U.K. parameters. For example, with a record length of 40 the mean sample skewness from 20000 repetitions is 1.70 from the Italian parent and 1.41 from the U.K. parent. This can be attributed to the higher proportion of outliers implied by the Italian fit (about one in seven compared with one in 33 for Britain); for the TCEV distribution the relationship between population and sample skewness depends on the outlier probability. This important result is worthy of further study for the insight it gives into both the Kirby bound and the separation phenomenon.

The fit to the 57 long record stations from the first level of regionalisation yields a dimensionless frequency curve which needs to be rescaled by at-site estimates of θ_1 and λ_1. To compare the TCEV procedure with other regionalisation procedures, a frequency curve is needed which is scaled by the sample mean, $\bar{x}$. This is attained in the second level of regionalisation in which a TCEV distribution constrained by the global parameters θ and λ is fitted to records longer than 20 years in subregions. After computing a subregional average $\bar{\lambda}_1$ equation (9) can be applied to obtain such diagrams as Figure 4 which shows the frequency curve for one of the Flood Studies Report (NERC, 1975) sub-regions. Comparative curves estimated using an algorithm based on the FSR procedure and regional probability weighted moments estimators for the GEV (Hosking et al, 1985) and Wakeby (Wallis, 1980) distributions are also shown. Figure 4 shows that the TCEV curve is considerably steeper than those produced by the other algorithms within the range illustrated.

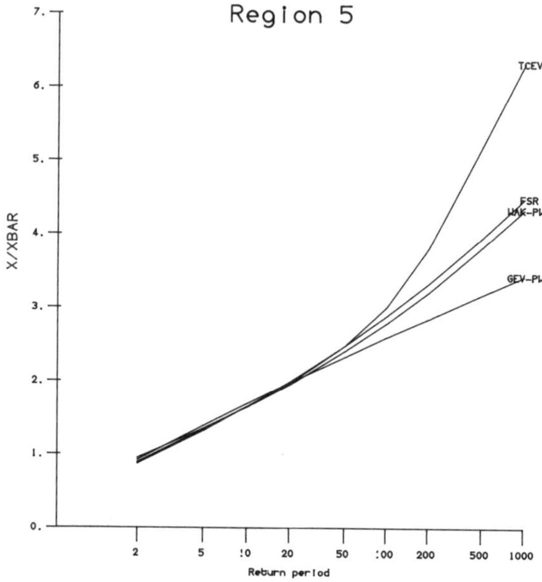

Figure 4 TCEV fit to subregional data in terms of $\bar{x}$

4.2 Replication of Recorded Skewness Characteristics

When estimating the regional frequency curve from the 57 long record samples it was assumed that skewness was constant throughout the UK, and that the considerable variation from sample to sample could be ascribed solely to sampling. One way to test this hypothesis is to draw random samples from a TCEV parent defined by the fitted parameters and compare sample skewnesses and observed skewness coefficients. A close correspondence would suggest that the assumption of constant skewness could be upheld. Accordingly, 1000 samples of size 40 were generated from a TCEV distribution with $\theta = 4.33$ and $\lambda = 0.032$, and skewness was computed for each sample. The skewness values were ranked and plotted on Figure 5 (comparable with Figure 7 of Rossi et al (1984) showing Italian data) together with the 57 observed sample skewnesses. Both visually and through a chi-square test (Table I) it is clear that the sample skewness values are indistinguishable from those drawn from the TCEV parent. It can therefore be concluded that the TCEV distribution with constant θ and λ is capable of producing flood-like skewness values.

TABLE I. Observed and simulated skewness frequencies of samples drawn from a TCEV population

Class interval of skewness	Observed	Simulated from TCEV
0 - 0.5	6	7.35
0.5 - 1	20	17.10
1 - 1.5	13	14.42
1.5 - 2	8	6.21
2 - 2.5	3	3.65
2.5 - 3	2	2.68
3 - 3.5	3	2.17
3.5 - 4	0	0.74
4 - 4.5	2	1.37
4.5 - 5	0	0.40

chi-squared = 1.6969; 95% significance level 7.81 (d.f.= 3).

Figure 5 can be seen as one way of demonstrating the separation phenomenon, which concerns the relative variability of sample skewness from observed and synthetic data. Matalas et al (1975) showed that synthetic data drawn from commonly used flood frequency distributions yielded less variable estimates of skewness than observed data, but Figure 5 indicates that unlike most other distributions, the TCEV can produce realistically variable estimates of sample skewness. This is shown more conventionally by Beran et al (1986).

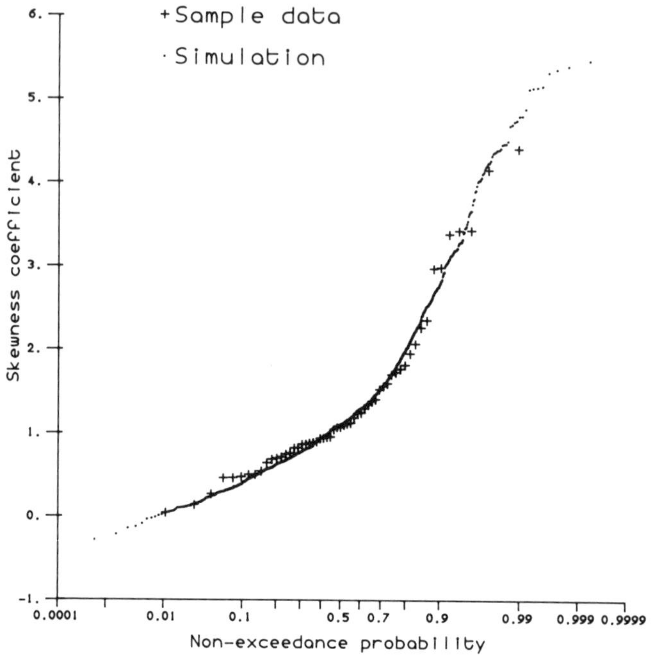

Figure 5 Normal plot of skewness coefficients computed from real and synthetic samples

5. TESTING ROBUSTNESS AGAINST POPULATION AND DATA UNCERTAINTY

5.1 Experimental Design

A series of computer simulation experiments has been conducted to evaluate the robustness of the regional TCEV estimation procedure and to compare it with that of three other index-flood type models. Full results are presented in Arnell and Gabriele (1985), and the experiments will only be summarised here.

The robustness test simulation experiments involve defining a flood-like parent distribution (termed a 'world'), generating many sets of synthetic data from this world, fitting the models to be tested to the generated data, and comparing estimated characteristics (such as the T-year flood) with the known 'world' characteristics. Successful models will, of course, be those which are capable of reconstructing the characteristics of the original worlds. In addition to freedom from bias one seeks also low variance estimators. Here we limit the experiments to two sets of homogeneous flood-like worlds based on TCEV, GEV and Wakeby distributions as parents. Parameters and characteristics of the six parents are shown in Table II. These were selected as far as possible to imitate the properties of the TCEV fit to UK and Italian data. 500 synthetic regions were generated from each parent with 40

TABLE II. Parameters and characteristics of the synthetic worlds

a) Characteristics

	mean	standard deviation	skew	kurtosis
U.K. fit				
TCEV-1	10.0	4.0	3.5	34.923
GEV-1	10.0	4.0	3.5	46.526
WAK-1	10.0	4.0	3.5	62.924
Italian fit				
TCEV-2	33.355	19.764	2.8	17.910
GEV-2	33.355	19.764	2.8	24.572
WAK-2	33.355	19.764	2.8	25.569

b) Parameters

	θ_1	λ_1	θ_2	λ_2	
TCEV-1	2.469	28.423	10.693	0.0693	
TCEV-2	10.0	10.0	30.67	0.3676	
	u	a	k		
GEV-1	8.2026	2.1942	-0.1987		
GEV-2	24.3532	11.6555	-0.1664		
	a	b	c	d	m
WAK-1	5.0924	10.0	10.1847	0.2101	2.6618
WAK-2	14.9002	5.0	74.5012	0.1634	6.3837

stations with 40 years of record.

A regional TCEV distribution was fitted to each set of synthetic regional data. Dimensionless frequency curves expressing the T-year flood as a multiple of the sample mean were computed by rescaling the y'-space values using equation 9.

Three other index-flood type regional flood frequency estimation procedures were applied to the synthetic data. The first two used regional probability weighted moments estimators to fit the GEV (GEV-PWM) and the Wakeby (WAK-PWM) distributions (Hosking et al, 1985; Wallis, 1980) and the third was a computerised version of the graphical regional procedure presented in the UK Flood Studies Report (NERC, 1975). In essence, this FSR procedure fits an average GEV curve through the plotted standardised data points from all the stations in a region.

All four regional procedures yield dimensionless regional frequency curves expressed in terms of the mean annual flood. These curves can be used at ungauged sites but can also be used in conjunction with local data to improve flood frequency estimates at gauged sites. In the U.K. greater interest is shown in the application of regional frequency curves at ungauged sites, and attention in this paper therefore focuses on the ability of the TCEV and other procedures to estimate dimensionless frequency curves.

5.2 Simulation Results on Bias and Error

The bias and root mean square error (RMSE) of each estimator is shown by quantile and parent in Table III, together with the maximum and average absolute bias and RMSE for each estimator. All the indices have been divided by the true dimensionless T-year flood to ensure comparability between parents and quantiles.

Considering bias first, it can be seen from Table III that the TCEV estimation procedure performs best with TCEV and Wakeby worlds, but it considerably underestimates quantiles for the GEV worlds. One index of the robustness of an estimation procedure - its ability to estimate extreme quantiles regardless of the true form of the data without disastrous loss of performance - is the largest absolute bias for a particular quantile over all parents, and the most robust estimator at the quantile is taken to be the estimator with the smallest maximum bias. This 'minimax' rule shows that the TCEV procedure is most robust at lower return periods, but at higher quantiles the FSR procedure produces the least extreme bias.

However, slight changes in the implementation of the FSR procedure can lead to major changes in this performance (as shown by the comparison with Hosking et al's (1985) results, where a different version was used), and if it is excluded from consideration the TCEV procedure is seen to have the smallest maximum absolute bias for all except the 100 year quantile. Another index of robustness is the average absolute bias over all the parents, and Table III shows that the WAK-PWM algorithm has the smallest average bias over most quantiles.

It is clear that evaluation of the alternative procedures depends on the decision rule employed. The 'smallest average' rule takes into account performance over all parents, but does not distinguish well between estimators which perform consistently and those which are variable but with the same mean bias. Conclusions using the minimax rule, on the other hand, can be conditioned by the performance of each estimator with one particular parent; if this parent is not flood-like the conclusions will be of little value. In this study the maximum absolute bias can be seen to occur when fitting to the TCEV worlds - the three alternative estimators (especially GEV-PWM) do not in general perform well when data are drawn from a TCEV parent - but it has been shown (Figure 5) that the TCEV parents used are realistic. Consequently greater weight can be placed on the conclusions reached using the minimax rule, and it can be concluded that the TCEV performs very well in terms of bias.

Overall the quantiles estimated by the TCEV procedure are characterised by greater variability than estimates from the other procedures, and consequently the TCEV performs relatively less well with regard to RMSE. Nevertheless, it is clear from Table III that the TCEV

TABLE III Bias and RMSE in quantile estimates from dimensionless flood frequency curves

	WORLD 1			WORLD 2			absolute bias	
	TCEV	GEV	Wakeby	TCEV	GEV	Wakeby	maximum	average
GEV-PWM								
0.95	0.033	0.002	0.007	0.004	-0.004	-0.001	0.033	0.009
0.98	0.023	-0.003	-0.050	-0.040	-0.006	-0.033	0.050	0.026
0.99	-0.018	-0.006	-0.097	-0.068	-0.008	-0.057	0.097	0.042
0.998	-0.216	-0.016	-0.214	-0.067	-0.012	-0.114	0.216	0.107
0.999	-0.272	-0.020	-0.264	-0.043	-0.012	-0.136	0.272	0.125
WAK-PWM								
0.95	0.033	0.011	0.002	0.008	0.006	-0.004	0.033	0.011
0.98	0.046	0.010	0.003	-0.023	0.002	-0.007	0.046	0.015
0.99	0.022	0.004	0.006	-0.044	-0.006	-0.010	0.044	0.015
0.998	-0.147	-0.022	0.013	-0.030	-0.037	-0.015	0.147	0.044
0.999	-0.191	-0.038	0.017	-0.003	-0.054	-0.017	0.191	0.053
FSR								
0.95	0.046	0.025	0.040	0.054	0.035	0.041	0.054	0.040
0.98	0.078	0.029	0.038	0.027	0.037	0.035	0.078	0.041
0.99	0.075	0.034	0.035	0.014	0.041	0.033	0.075	0.039
0.998	-0.044	0.047	0.028	0.058	0.049	0.031	0.058	0.043
0.999	-0.060	0.056	0.028	0.107	0.056	0.034	0.107	0.057
TCEV								
0.95	-0.005	0.014	0.005	-0.002	0.009	-0.018	0.018	0.008
0.98	-0.007	0.009	-0.040	-0.012	-0.006	-0.023	0.040	0.016
0.99	-0.010	-0.011	-0.062	-0.019	-0.029	-0.012	0.062	0.024
0.998	-0.037	-0.098	0.021	-0.030	-0.109	0.000	0.109	0.049
0.999	-0.040	-0.146	0.085	-0.033	-0.151	-0.014	0.151	0.078

a) normalised bias in quantile estimates

	WORLD 1			WORLD 2			RMSE	
	TCEV	GEV	Wakeby	TCEV	GEV	Wakeby	maximum	average
GEV-PWM								
0.95	0.036	0.012	0.016	0.015	0.015	0.014	0.036	0.018
0.98	0.034	0.022	0.055	0.047	0.025	0.041	0.055	0.037
0.99	0.038	0.033	0.101	0.075	0.034	0.065	0.101	0.058
0.998	0.220	0.056	0.218	0.087	0.055	0.123	0.220	0.127
0.999	0.276	0.068	0.268	0.080	0.066	0.146	0.276	0.151
WAK-PWM								
0.95	0.035	0.017	0.013	0.017	0.015	0.015	0.035	0.019
0.98	0.053	0.026	0.022	0.035	0.026	0.025	0.053	0.031
0.99	0.045	0.034	0.033	0.057	0.037	0.035	0.057	0.040
0.998	0.161	0.067	0.066	0.074	0.076	0.067	0.161	0.085
0.999	0.206	0.085	0.084	0.085	0.096	0.082	0.206	0.106
FSR								
0.95	0.050	0.031	0.044	0.058	0.040	0.046	0.058	0.045
0.98	0.090	0.047	0.053	0.045	0.053	0.050	0.090	0.056
0.99	0.102	0.066	0.069	0.053	0.071	0.063	0.102	0.071
0.998	0.134	0.121	0.130	0.113	0.122	0.110	0.134	0.122
0.999	0.169	0.151	0.167	0.164	0.151	0.136	0.169	0.156
TCEV								
0.95	0.013	0.023	0.010	0.022	0.021	0.025	0.025	0.019
0.98	0.028	0.032	0.044	0.037	0.036	0.039	0.044	0.036
0.99	0.053	0.044	0.072	0.054	0.057	0.052	0.072	0.055
0.998	0.146	0.115	0.146	0.090	0.130	0.100	0.146	0.121
0.999	0.171	0.159	0.206	0.101	0.169	0.116	0.206	0.154

b) normalised RMSE of quantile estimates

is a good estimator, especially at low return periods. Again, the decision rule employed conditions the final conclusions.

5.3 Effect of Record Length

Experiments have been undertaken to explore the effect of reducing sample record lengths (Arnell and Gabriele, 1985). The TCEV estimation procedure is not advocated for short records (Rossi et al, 1984), but the results of the experiments in which sample sizes as small as 10 years are generated shed some light on the characteristics of the TCEV procedure. Data were generated from the UK-based TCEV parent, and it was found that bias and standard deviation of quantile estimates increased significantly. This was most marked for the TCEV estimation procedure, in which RMSE increased by 300% and more compared with the results for 40 year samples. The corresponding increases for the RMSE of the WAK-PWM algorithm were between 120 and 140%, more in line with sample size reduction. These results indicate that the performance of the TCEV declines at a greater rate than that of other procedures as record length decreases.

5.4 Heterogeneous and Error-prone Worlds

The experiments completed so far have been based on homogeneous parents, where all the data in a region are generated from the same distribution. More realistic experiments are planned based on heterogeneous worlds where parent CV varies across sites. This structure will enable investigations into ways of identifying the optimum number of homogeneous subregions within a region with constant skewness: the experiments so far have assumed that there are no distinct subregions.

Real flood data are characterised by measurement error, and it has been suggested that simulation experiments could incorporate this effect (Potter and Walker, 1981). Accordingly, a simple error model was defined representing inaccuracies in stage-discharge rating curves, and was applied to data generated from the UK based TCEV world. The model has the form

$$Q_m = Q_g^c$$

where Q_g is the generated true 'discharge', Q_m is the modified discharge and c is a random normal deviate with unit mean held constant for each station. Different degrees of rating error were modelled by drawing random numbers with standard deviations of 0.05 and 0.1. The WAK-PWM, GEV-PWM and FSR models have been fitted to 500 sets of error-prone regional data, and it has been found that both the mean dimensionless frequency curves and the standard deviation of quantile estimates are increased when rating error is added, and that the effect of rating error increases with return period. The TCEV model has been fitted to only 20 sets of simulated regional data, but an indication of the effect of rating error can be obtained by comparing the TCEV fits with fits from the other models to the same 20 sets of data.

Figure 6 shows the variation in normalised bias and standard deviation with rating error for the 100 and 1000 year quantiles, and it can be seen that the TCEV procedure appears to be less affected by rating error than the GEV-PWM and WAK-PWM at 100 year return period, but more affected at longer return periods. Comparisons between the results for 20 and 500 repetitions for the WAK-PWM and GEV-PWM procedures suggest that the 20 repetitions provide a good indication of performance after 500 repetitions, and the trends apparent in the TCEV performance when rating error is added will perhaps also still be seen after many more repetitions. Further experiments are necessary both to check whether the identified trends continue and to see what happens with different parent distributions or error models.

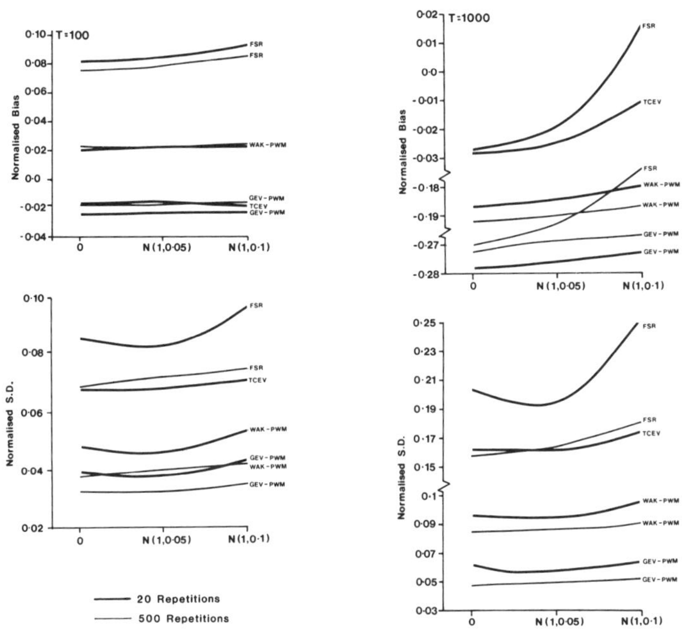

Figure 6 Effect of rating error on bias and standard deviation of quantile estimates.

6. CONCLUSIONS

Our examination of the TCEV distribution reveals some very positive aspects both in terms of its original conception as a maximum of contrasting flood mechanisms and in terms of the hierarchical parameter estimation scheme, in which parameters which can be expected to vary more slowly are estimated from a more global data set than those which are known to exhibit more local variation. The ability to employ a national value for θ and λ reduces the distribution to a two parameter case and will enable design engineers to use the distribution at sites with some gauged data.

Its fit to UK regional and national flood data displays other desirable features; it has good ability to encompass outlying floods and to replicate observed sample skewnesses. UK data exhibit strong separation with respect to alternative distributions such as General Extreme Value and Wakeby but not to the TCEV distribution.

In robustness tests there have been no problems in reproducing the tail shape of TCEV worlds but the TCEV estimation algorithm is less successful with non-TCEV worlds, particularly the GEV world. The introduction of rating error seems not to affect the TCEV distribution to any greater extent than the other candidate algorithms, and from the few numerical results thus far available it appears that most methods are robust against data error. However the detrimental effect of short records is clearly much more severe for the TCEV estimator than for the other cases, although it seems that the UK data base is sufficiently large for successful calibration of the model.

It was hoped to draw clear preferences from the robustness tests. The conclusion drawn, however, is heavily conditioned by the criterion used to judge the results. In the UK context bias is more important than variance and this assists the TCEV distribution. On a minimax criterion the TCEV appears best; but looking overall, and especially if the importance of variance is highlighted, the Wakeby based algorithm is more successful. The final choice of an estimation procedure intended for practical application rests, however, not only on statistical performance but also on experience with real data.

The studies expose a number of areas where further work is required. The contrast between population and sample skewness from TCEV parents deserves attention to explain how the outlier probability interacts with population skewness to determine sample skewness. More work is needed to determine the spatial scales on which the parameters θ, λ, and λ_1 (controlling the coefficient of variation) can be considered constant. More extended numerical experiments are needed to explore further the influence of heterogeneity and data error on the TCEV estimation algorithm.

Overall the Two Component Extreme Value distribution is a useful and conceptually important addition to the armoury of the flood analyst and could be used as the basis of a flood frequency package suited to UK conditions.

7. ACKNOWLEDGEMENTS

Part of the research reported in this paper was funded by the Ministry of Agriculture, Fisheries and Food, and travel funds for collaboration with CNR-IRPI at Cosenza, Italy, were provided by the British Council.

8. REFERENCES

Arnell, N.W. and Gabriele, S. 1985. 'Regional flood frequency analysis with the two-component extreme value distribution: an assessment using computer simulation experiments.' Proceedings of the Workshop on Combined Efficiency of Point and Regional Flood Prediction, December 1985, Perugia, Italy.

Beran, M.A. 1986. 'The UK Flood Studies Report: Continuing responsibilities and research needs.' International Symposium on Flood Frequency and Risk Analysis, Baton Rouge. Reidel.

Beran, M.A., Hosking, J.R.M. and Arnell, N.W. 1986. 'Comment on 'Two-component extreme value distribution for flood frequency analysis'.' Water Resources Research, 22 263-266.

Hosking, J.R.M., Wallis, J.R. and Wood, E.F. 1985. 'An appraisal of the regional flood frequency procedure in the U.K. Flood Studies Report.' Hydrological Sciences Journal, **30** 85-89.

Matalas, N.C., Slack, J.R. and Wallis, J.R. 1975. 'Regional skew in search of a parent.' Water Resources Research. 11 815-826.

Natural Environment Research Council 1975. Flood Studies Report. 5 vols. HMSO: London.

Potter, K.W. and Walker, J.F. 1981. 'A model of discontinuous measurement error and its effect on the probability distribution of flood discharge measurements.' Water Resources Research, **17** 1505-1509.

Rossi, F., Fiorentino, M. and Versace, P. 1984 'Two-component extreme value distribution for flood frequency analysis.' Water Resources Research, 20 847-856.

Wallis, J.R. 1980 'Risks and uncertainties in the evaluation of flood events for the design of hydrologic structures.' In Guggino, A., Rossi, G. and Todini, E. (eds) Eventi hydrologici extremi: piene e siccita. CLUP: Milan.

A METHOD FOR ESTIMATING THE PEAK DISCHARGE THROUGH THE RECORDS OF MEAN DAILY DISCHARGE

P. Canuti
Earth Sciences Department
University of Florence
Via La Pira 4
50121 Firenze
Italy

U. Moisello
Hydraulic and Environmental Engineering Department
University of Pavia
Via Abbiategrasso 213
27100 Pavia
Italy

ABSTRACT. The paper deals with the problem of estimating the peak discharge when only daily discharge data is available. To this purpose the ratio between the yearly maximum of peak and daily discharge is treated as a random variable, whose distribution is lognormal. The main result of the study, conducted for a large number of Italian catchments, consists in giving the values of the two parameters (mean and standard deviation) characterizing the distribution in function of some parameters related to the physical features of the catchment.

Based on the above result, a method is given to estimate the peak discharge from daily discharge data and the corresponding estimation error is studied. Finally, the values of a correction factor, needed to avoid the estimation bias due to some regional differences, have been estimated.

1. INTRODUCTION

Given a maximum daily discharge q with a preset T return period which is obtained by a suitable number of observations, a frequent problem in hydrologic technique is to infer the maximum peak discharge Q with an equal return period. Indeed, for most streams the observations available concern the maximum daily discharge q rather than peak discharge Q. The reasons are that, first, peak discharge values are not always published, in Italy at least, and, secondly, in several cases they do not exist at all. Actually, in many stations, lacking a water-level recorder, a sensible rough value of the mean daily discharge can be in-

V. P. Singh (ed.), Regional Flood Frequency Analysis, 177–184.

ferred from the mean value of the few daily readings (sometimes only one, occuring at fixed times).

Generally this problem can be solved by regional analyses. For further reference, a study by the same authors on some Tuscany catchments outlines the results of these analyses, carried out by different researchers on variously sized regions of Italy (Canuti and Moisello, 1981). The study reports also two relations expressing the statistical parameters (mean and standard deviation) of ratio r. The latter is the ratio of the maximum yearly peak discharge Q and the corresponding maximum daily mean discharge q, treated as a random variable and given in function of the major parameters of the catchment. The estimate of Q distribution parameters is based on the values of r parameters derived from these relations and the observed values of q parameters. Based on the same approach as for Tuscany, this analysis was applied to the whole national territory with the results reported in what follows.

2. BASIC APPROACH

If Q is the peak discharge, q the mean daily discharge and r the Q-to-q ratio (all being considered random variables), the distribution of variable Q can be defined, provided that the joint distribution of r and q is known. The relations between the typical parameters of Q variable and those of r and q are particularly simple when r and q are mutually independent: in fact, the mean of the product of two independent variables equals the product of their means. If m(Q), m(r), m(q) are the means of the above variables, there follows

$$m(Q) = m(r)\ m(q). \qquad (1)$$

The variance of Q, var(Q), is given by the expression:

$$var(Q) = m(Q^2) - m^2(Q).$$

Obviously, the above relation representing the mean of the r-by-q product is valid also for the product of their squares. Hence,

$$var(Q) = m(r^2)\ m(q^2) - m^2(r)\ m^2(q).$$

The expressions of $m(r^2)$ and $m(q^2)$ can be derived from the relations that define the variances of q and r. We obtain

$$var(Q) = m^2(r)\ var(q) + var(r) \left\{ var(q) + m^2(q) \right\}. \qquad (2)$$

Both expressions (1) and (2) are valid on the only condition that r and q variables are regarded as independent, while r and q distributions may be of any type whatever or even be unknown.

The assumption that r and q are independent was proved by the Tuscany study: in actual fact, there is a negative correlation, but this is so small that for pratical purposes r and q can definitely be treated as independent variables. A similar check on a national scale confirmed the validity of this assumption. It is worth pointing out straight that this assumption concerns only the r and q value for a single river and not a group of different streams. In the case of several different rivers the connection between r and q can be seen quite clearly. In fact, for small catchments with low q discharge values, r values are high (i.e., flood hydrographs show particularly elevated peaks). However, in the case of a single stream, the peak discharge-to-mean daily discharge ratio can vary to such a degree, compared to the variability of the maximum yearly value of q, that it is almost impossible to notice weaker peaks in the flood hydrographs with higher daily discharge values.

By introducing in equations (1) and (2) the estimated m(q), var(q), drawn from observation, and the estimated m(r), var(r), derived from catchment characteristics (as explained further on), m(Q), var(Q) can be estimated. If variable Q is distributed according to a preset type of probability law based on previous experience, the estimates of m(Q) and var(Q) enable to estimate also the typical distribution parameters of Q.

As far as the choice of probability law is concerned, a study on Italian rivers by Cicioni et al. (1974) shows that lognormal distribution gives the best results in the analysis of the maximum values of daily discharge q. Beside, a two-parameter lognormal distribution allows to estimate the distribution of ratio r (Canuti and Moisello, 1981), although the lower limit of the variable equals one and not zero. Since the product of two lognormal variables is also lognormal, these results concur with the pratical assumption that also the maximum yearly value of peak discharge is lognormal.

The next step is to define the procedure for estimating the parameters of r. This paper focuses on this point, suggesting a simple approach based on the known values of some typical catchment variables, of which there are two main sorts, i.e. geometrical and hydrological.

The first group includes surface and altitude parameters. The former are perimeter P, surface S, length L (measured along the main channel) plus a form parameter, i.e. the elongation ratio R_e (the ratio of a S surface circle diameter to length L). Height characteristics are expressed by basin relief H_{max} (basin maximum height above outflow point), mean height above sea level H_m and mean channel slope i_m (basin relief-to-length ratio). The choice of R_e elongation ratio as form parameter and i_m mean channel slope as a slope parameter is due to the results of the Tuscany study, supported by a further analysis, which we need not mention here.

To the second group belong rainfall or runoff regime parameters. In order to introduce a rainfall variable for the whole basin the maximum value h_{max}, picked out of the mean values of basin monthly rainfalls, has been chosen as rainfall regime index. Of course, a month is longer than the spell of a rainfall that is the direct cause of floods. However, floods are connected to the monthly maximum rainfall, since both yearly maximum flood and monthly maximum rainfall usually occur in the same month. In point of fact, the maximum areal rainfall values corresponding to short durations (e.g., in the range of hours), which result in floods, are only rarely published. In the attempt to gather significant data by a simple method, the mean value h_{max} has been selected and derived from the Italian Hydrographic Office's publications. Characteristic runoff variables are the mean $m(\overline{q})$ and the standard deviation $s(\overline{q})$ of the yearly maximum $\overline{q}$ of mean daily discharge per unit area. The latter's values are assumed equal to those computed from the available records. Finally, also two indexes related to the presence of lake and glacier areas have been considered relevant to catchment features. Index i_l is the complement to unit of lake area-to-total catchment area ratio; index i_g is the complement of the ratio of glacier area to total catchment area.

3. DEPENDENCE OF r PARAMETERS ON CATCHMENT CHARACTERISTICS

The estimate of r parameters has been solved by working out the expression of mean m(r) and standard deviation s(r) in function of certain variables representing the geometric and hydrologic characteristics of the catchment. Linear regression and logarithmic regression - the two commonest regression procedures - have helped to work out m(r) and s(r) relations. Logarithmic regression is directly connected to linear regression through the logarithmic transformation of the original variable. The ratio f of the dependent variable's observed value to the estimated value is a useful error index for a logarithmic regression. It cannot take negative values and equals 1 when the regression error is null. On the basis of observations recorded for over three hundred catchments scattered throughout Italy and check on the final results, logarithmic regression seems to be the best precedure (whereas the Tuscany study uses linear regression).

The observations cited in this study are taken from the n° 17 publication of the Italian Hydrographic Office (1934, 1939, 1953, 1963, 1980), updated to 1970. Except for catchment length and perimeter, which have been measured on 1:250,000 maps, discharge observations as well as all the other variables have been drawn from the above publication. A first sample of basins was selected depending on the number of years (10 at least) for which both Q and q have been published. Later on, a few items were put aside due to missing values of characteristic variables or to wrong or inconsistent data. On the whole, the catchments which were taken into account were 310 and for nearly all of them discharge observations available for analysis come from a period of 10 years at least.

As a first step it seemed useful to sort out the catchments with areas over 100 km^2. Comparison of the results related to the whole group of 310 basins and those related to the other group (237 basins with an over 100 km^2 area) shows that m(r) and s(r) regressions on the other typical geometric and hydrologic variables are better in the latter case. In fact, especially in penisular Italy, m(r) increases as S decreases only down to values roughly about 100 km^2. In case of lower S values, m(r) is pratically constant or even drops.

To study the dependence of mean m(r) and standard deviation s(r) on the catchment's characteristics, backward, forward and stepwise regression procedures have been used jointly. Each of the above methods provides a feasible alternative leading to different results. With a view to choosing the final formulas, we have taken into account mostly the physical value of the relations and their pratical application.

This paper reports and comments on the final results only, skipping over the intermediate steps. The m(r) and s(r) formulas derived from logarithmic regressions are as follows:

$$s(r) = 14.7\ S^{-0.103}\ H_m^{-0.172}\ s(\overline{q})^{0.114}\ i_1^{\ 0.779},$$

$$s(r) = 111.9\ S^{-0.314}\ H_m^{-0.400}\ s(\overline{q})^{0.333}\ i_1^{2.076},$$

where surface S is given in km², mean height H_m in m and standard deviation $s(\overline{q})$ in m³/s km².

These results vary from those worked out for Tuscany first of all due to the presence of a hydrologic variable among the independent va- variables, i.e. standard deviation $s(\overline{q})$. As this value increases, both m(r) and s(r) increase too. The Tuscany study overlooked the characteristic distribution parameters of the maximum yearly values of mean daily discharges, because of the particular kind of data under consideratio (even more than one flood hydrograph for the same year) and, furthermore, because of the special attention paid to the influence of geomorphic characteristics. It is worth noting that in the regression procedure the $s(\overline{q})$ variable is more significant than variable $m(\overline{q})$. In fact, $s(\overline{q})$ can be regarded as representing the influence of two factors on m(r): unit maximum discharges (measured by $s(\overline{q})$ beside $m(\overline{q})$) and their variability. For equal conditions then the average value of ratio r rises together with $\overline{q}$ and its variability.

As the index i_1 for the lake areas (that comes closer to 1 the lower the lake surface is) increases, also r ratio records an average increase, while its variability increases at even faster rate. This can be fully accounted for by the flood routing effect due to the lake.

As the surface S increases, both m(r) and s(r) decrease. The reason for this is that flood duration rises together with the area and the larger it is the more flood duration exceeds 24 hours (i.e., the duration for calculating mean daily discharge). The difference between peak discharge and the highest observed mean daily discharge during the flood (usually occuring in the same day) decreases, while the duration of flood hydrograph increases. This explains why the ratio between the yearly maximum of peak discharge and the yearly maximum of mean daily discharge becomes lower as the area increases, even though both these maximum values do not always refer to the same flood. In the Tuscany study surface S was not included among the significant variables. In fact, in that case most catchments were quite small, so that their related flood hydrographs were generally contained in a single day.

If the other factors are the same, m(r) and s(r) decrease also when catchment mean height H_m increases. This result tallies with that worked out in the Tuscany study, although it cannot be easily explained. It is possible that, given equal conditions, in catchments sited at a higher altitude floods should be characterized by longer duration due to longer rainfall and the larger areas in which they occur. In other words, heavier rainfall at high altitudes may affect the characteristics of flood hydrographs just like a larger catchment area. This hypothesis, which seems absolutely reasonable, should be checked by a special analysis. However, for all the doubts about the causes of this phenomenon, the result is out of question.

The choice of the independent variables to use in the m(r) and s(r) formulas has taken into account their statistical meaning as well as the pratical application of the formulas. The regressions improve slightly by introducing the rainfall variable h_{max}. However, for pratical purposes, since this variable demands complex calculations, the proposed formulas do not take it into account. It is very interesting to remark that the exponent of h_{max} is negative if this variable is introduced. The decrease of the ratio as the mean monthly rainfall of the rainiest month increases can also be explained by a longer flood duration in a catchment subject not only to heavier but also more frequent rainfalls.

4. RESULT RELIABILITY ANALYSIS

As explained above, the proposed m(r) and s(r) formulas can be applied to the estimates of m(Q) and s(Q) parameters that are linked to m(r), s(r) and m(q), s(q). Thus, once a certain kind of probability distribution has been adopted, the distribution of Q can be defined and the value with a preset return period calculated. Compared to the direct estimate procedure (deriving m(Q) and s(Q) values from observation of peak discharge), the indirect estimate of m(Q) and s(Q) involves an additional error, on which attention has focused in the present analysis.

Q(T) is the value of Q derived from the yearly maximum values of peak discharge and Q*(T) the value derived, through the proposed method, from the yearly maximum values of mean daily discharge q. Then the ratio f between Q*(T) and Q(T) can be assumed as the error index of this method. In logarithm form then:

$$\ln f = \ln Q^*(T) - \ln Q(T) .$$

In the quite reasonable hypothesis that the distribution of random variable Q should be lognormal, the logarithm of f is supplied by the following expression:

$$\ln f = \left\{ m^*(\ln Q) - m(\ln Q) \right\} + z(T) \left\{ s^*(\ln Q) - s(\ln Q) \right\},$$

where m(lnQ), s(lnQ) are the estimates of the mean and the standard deviation of Q logarithm derived from observations; m*(lnQ), s*(lnQ) are the estimates obtained by the suggested method (derived from the sample of q values); z(T) is the standardized gaussian variable corresponding to T return period. Setting $e\{m(\ln Q)\}$, $e\{s(\ln Q)\}$ as the errors affecting m(lnQ), s(lnQ), the expression is as follows:

$$\ln f = e\left\{ m(\ln Q) \right\} + z(T)\, e\left\{ s(\ln Q) \right\} .$$

Given a preset return time, the mean and variance af the logarithm can be inferred from the mean and variance of $e\{m(\ln Q)\}$, $e\{s(\ln Q)\}$ and their covariance.

The above formula implies that the mean and standard deviation of f logarithm increase with T. For the 237 catchments under consideration the mean of f logarithm rises from 0.065 to 0.159 with T increasing from 10 to 1000 years, while standard deviation rises from 0.225 to 0.357. Since lognormal distribution can well approximate the observed f distribution, the mode $\tilde{f}$ is expressed by

$$\tilde{f} = \exp\left\{ m(\ln f) - var(\ln f)\right\}.$$

The mode is always slightly above 1. If we assume that f distribution for a preset return period depends neither on basin nor on Q(T), the obtained estimate of Q*(T) is a random variable, its distribution is lognormal like that of f and its mode is the product $\tilde{f}Q(T)$. In order to obtain Q(T) estimates with a mode equal to Q(T), Q*(T) must be devided by $\tilde{f}$. Since the value of $\tilde{f}$ pratically equals one, it would not be worth correcting the Q*(T) estimates if f distribution could really be taken as independent of the river in question. In actual fact, however, f distribution is largely affected by regional characteristics. This point has been made clear by individual analyses for different areas.

In order to obtain Q*(T) estimates with a mode equal to Q(T), the obtained estimates must be multiplied by a correction coefficient c, equal to the inverse of $\tilde{f}$. The value of c (worked out from observations) varies regularly with T, in function of which can be expressed by the approximate formula:

$$c = aT^b,$$

where T is given in years. The values of coefficients a and b are reported in Table I hereinafter.

TABLE I

Parameters of the correction coefficient formula

District	a	b	District	a	b
Venice	1.038	-0.0009	Pescara	1.044	0.0003
Parma	0.938	-0.0102	Naples	0.881	-0.0176
Bologna	0.955	-0.0169	Bari	1.308	-0.0369
Genoa	0.944	0.0185	Catanzaro	1.017	0.0148
Pisa	0.979	-0.0372	Palermo	1.364	-0.0222
Rome	0.895	0.0040	Cagliari	0.931	0.0496

A simple check shows that the values of c vary noticeably according to the zones. Pisa and Naples have the smallest value, a little over 0.8. In the Parma, Bologna and Rome areas these values are about 0.9, whereas in those of Venice, Genoa and Pescara they are slightly higher than 1. The peak values, from roughly 1.10 to little more than 1.20, are recorded in the Bari, Catanzaro, Palermo and Cagliari districts. Of course also the interval in which Q*(T) estimate is expected to fall with a preset probability varies depending on the region.For instance, in the Cagliari district, for a 10-year return period, the ratio of the corrected cQ*(T) estimate and Q(T), derived from direct observations of peak discharge, can be espected to fall between 0.89 and 1.16 with a 2-to-3 probability. By contrast, for the Venice area the same ratio is expected to be in the interval 0.82-1.43. Obviously, these estimates of error variability are very uncertain, since there is a limited number of observations available (about 10 for some districts). However, given a 2-to-3 probability, with a good approximation the average variability range of the above ratio can be defined as follows: 0.85-1.30 for a 10-year return period, 0.82-1.40 for a 100-year return period, 0.80-1.60 for a 1000-year return period.

The results obtained for the correction coefficients have mainly a pratical purpose. However, they point out a different basin behaviour depending on the area. In the Po valley and in central Italy the correction coefficients do not exceed 1, while they are over 1 in southern Italy and the islands. Aside from the uncertainties due to the few observations available for the various districts, these differences cannot be explained by the geological characteristics of the catchments and therefore the different rainfall regimes should be taken into account. On the other hand, some analyses have proved that the introduction of the variable h_{max} cannot explain these differences, either. Therefore, they must be connected to short duration rainfalls (the cause of floods), whose characteristics actually change depending on the climatic zones.

ACKNOWLEDGEMENTS

The authors thank the Italian Ministry of Public Education for supporting the research.

REFERENCES

Canuti, P., Moisello, U. 1981. 'Relationship between the Yearly Maxima of Peak and Daily Discharge for Some Basins in Tuscany'. Hydrological Sciences - Journal - des Sciences Hydrologiques, 27, pp. 111-128.

Cicioni, G.B., Giuliano, G., Spaziani, F.M., 1974. 'Indagine sulla Distribuzione di Probabilità delle Portate di Piena nei Corsi d'Acqua Italiani' in: XIV Convegno di Idraulica e Costruzioni Idrauliche, Atti, 2, pp. 61-79, Liguori Editore, Napoli.

OBSERVATION AND ANALYSIS OF FLOOD DISCHARGES – EXPERIENCES FROM SWITZERLAND

M. Spreafico
Swiss National Hydrological and Geological Survey
CH-3003 Berne, Switzerland

and

F. Naef
Laboratory of Hydraulics, Hydrology and Glaciology
ETH-Zentrum, CH-8092 Zurich, Switzerland

ABSTRACT. Basic hydrological data is one of the essential requirements for planning and implementing flood protection measures. Unfortunately even today there are many areas where such basic data is still not available. This article is intended to present a number of selected activities and results connected with the resolution of these deficiencies.

A detailed analysis of most flood data from Switzerland has shown, that the existing operational networks are insufficient and additional measurement stations have to be built. There is a need in quantifying the accuracy of the observed data and in improving data quality for model applications. The flood frequency analyses carried out during this investigation is briefly described.

In flood frequency diagrams, the largest events of series of discharge measurements often do not coincide well with the line representing the fitted distribution.

In about 100 small Swiss catchments, event parameters such as rainfall amount and duration, runoff coefficients and unit hydrographs were evaluated for the largest measured events. It was found, that the range of these parameters was much larger than the range of the peak flows and that the runoff intensity, that ist the volume of water available for discharge in the time of concentration of the catchment, often coincided better with a fitted standard frequency distribution than did the peak flow with a similarly fitted distribution.

It was concluded, that out-of-channel retention introduces nonstationarity, as it attenuates flood peaks during large floods, until the retention capacity is exhausted. Then, a much larger flow is suddenly experienced. It is therefore recommended that the influence of the out-of-channel retention is assessed, when the runoff intensity fits a standard frequency distribution well but the peak flow does not.

1. INTRODUCTION

Geologically, topographically and hydrologically, Switzerland's 41'000 sq. km. display many different characteristics. There are great diffe-

V. P. Singh (ed.), Regional Flood Frequency Analysis, 185–195.

rences in the soil types, the land use and the climate within a small area; so there are considerable variations in the quantity and distribution of precipitation, some of which even falls as snow. The mean annual values of the water balance components of Switzerland from 1901 to 1980 are: Precipitation 1456 mm, Runoff 978 mm, Evaporation 484 mm and the change in water reserves (lakes, reservoirs, glaciers) - 6 mm. The 94 reservoirs, the dense urbanization of the country and the intensive cultivation of the soil have a great influence on the natural discharge in numerous rivers. The flood discharges vary very widely in neighbouring hydrological basins. In the nothern part of Switzerland the maximum values are significantly lower than in the southern part. Switzerland's observed specific maximum peak discharges are illustrated in Figure 1 in relation to the world's major drainage areas (Vischer, 1980).

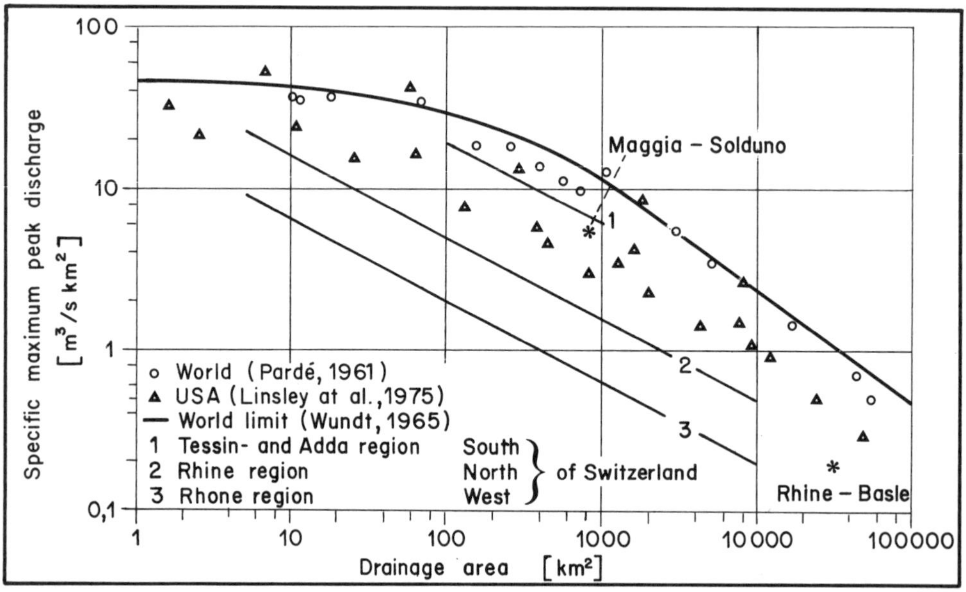

Figure 1 Specific maximum peak discharge in relation to drainage area (Vischer, 1980)

Even though there are higher specific peak discharges in other countries, there are still considerable flood protection problems in Switzerland. The flood danger is countered by controlling the rivers and streams, by creating artificial flood retention reservoirs, by channel improvements, by planning measures in flood-plain land, by flood forecasting and efficient alarm organisations.

2. ENLARGEMENT OF THE DATA BASE

Flood data has been collected in Switerland since the beginning of the 19th century. The observation network established by the Swiss National

Hydrological Survey and some cantons is very dense (1.4 stations per 100 sq. km). The data collection, processing and analysis is done in an operational way and sophisticated measurement techniques are used. Long time series of runoff data are available for most large Swiss streams. However practical experience indicates that such data is often lacking for small streams with catchment areas of 1-150 sq. km. Basically there are 2 ways of improving the situation:

1) Developing new methods or adapting known international procedures for the on-site assessment of flows without direct measurement

2) Construction and operation of additional measuring stations

A research study entitled "Assessment of discharges in rivers and streams at locations without direct measurement" was carried out from 1982 to 1985 as part of the first alternative. During this project,

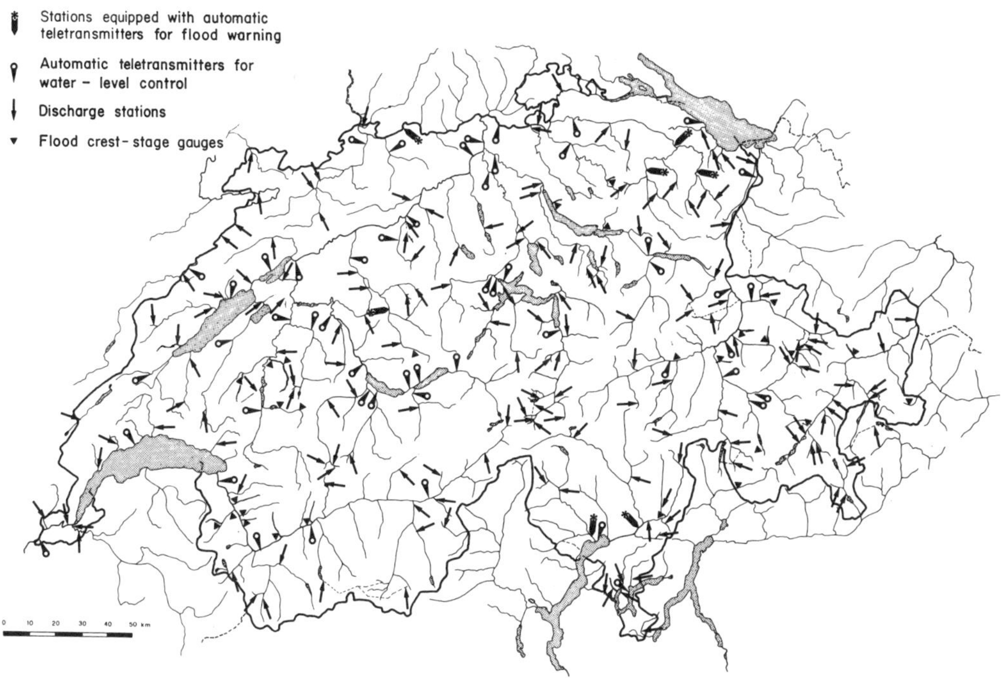

Figure 2 Selected observation stations of the Swiss National Hydrological Survey

known rainfall-runoff models were tested. A new system of estimating peak discharges, specially designed for Swiss conditions, was developed. And process investigations and peak flood analyses were carried out in the field. It became clear during this research study that the number of measuring stations would have to be increased if there was to be any substantial improvement in the accuracy of the information from models used for estimating peak floods. Accurate and reliable information on hydrological observations is an essential precondition for comparison of hydrological models. This experience led to the enlargement of the basic network of the National Hydrological Survey which began some years ago. See Figure 2. For financial reasons in this additional network only the elevation of the flood crest is recorded. The discharges are then calculated from these water levels. Flood crest-stage gauges are fitted in places with natural or artificial overfalls and the discharges are determined employing the formula.

$$Q = 2/3\ b\ \mu\ (2g)^{1/2} H^{3/2}$$

Q = Discharge
b = Width of cross-section
μ = Coefficient of discharge
H = Total energy head
g = Acceleration of gravity

The coefficients of discharge μ are fixed by calibration. Several flood crest stage gauges are installed in series in suitable river reaches without overfalls and the discharges are calculated employing hydraulic calculation procedures. Experience and calculations so far show that the flood crest stage gauges combined with overfalls produce results of sufficient accuracy.

3. PROBLEMS OF DATA QUALITY

Flood discharges in stable channels with optimal flow conditions and minimal sediment load can be determined with a relatively high degree of accuracy. However these conditions only apply to a few of the discharge measuring stations in Switzerland because of natural factors (naturally steep channels with sediment transport, the high cost of constructing artificial channels, hydraulic engineering limitations and environmental considerations). Accurate discharge measurement with hydrometric vanes is often impossible because of the high flow velocity, the enormous sediment and floating material load, the changes in the bed during flood discharge, the rapidly changing water levels, etc. The introduction of new technologies such as cableway equipment and ultrasonic methods made it possible to increase the accuracy, but there are still major difficulties. This is why we are currently looking into the question of measuring instruments without moving parts and into the determination of flood discharges with the help of natural and artificial tracers. However, since discharge observations are needed to standardise assessment procedures and for the implementation of complex rainfall-runoff models, there is a consequent combination of errors from the input data and errors in the model production. The quantification of model and measurement errors thus becomes a central problem of model application.

4. FLOOD FREQUENCY ANALYSES

A new analysis of peak discharges at about 135 stations of the National Hydrological Survey was begun in 1984 using measurement series extending from 30 to 115 years. The aim of this work is to prepare material which will provide the basis for decision-making in hydraulic planning. The people employed in the water resources planning should be provided with as much information as possible on the measuring stations, flow regime, flood observation methods and flood probabilities (Spreafico, Stadler, 1986). The following gives a brief description of some of the work and results:

4.1 Collection and Preparation of the Data

As a first step the station's most important descriptive data is compiled. Data on channel stability, on the highest recorded flow speed and on the highest measured discharge are noted, together with the identification number, location coordinates, drainage area size, observation period, type of measuring equipment and measuring methods employed. Thanks to the fact that all the daily mean discharges and data of the floods were already stored in a data bank it was relatively easy to obtain and process this material and present it in a suitable form. The duration curve for every measurement station were calculated and presented. The highest annual, summer and winter peak discharges during the measuring period were presented in graphic and tabular form. The flood stage graph of the most important stations were digitized and the flood discharge volume was determined and plotted. Analysis of precipitation data was generally avoided since storm precipitation statistics are already available for a great part of Switzerland.

4.2 Control of Consistency and Homogeneity

Consistency and homogeneity analysis of long series proved to be very difficult and time consuming, even though the National Hydrological Survey has systematically and operationally obtained, compiled and documented flood discharges since 1863. Tracing inconsistencies in particular led to major problems since, naturally enough over such long periods of measurement, there has been many changes in the measuring methods and instruments, in data processing procedures and in fixing the relationship between stage and discharge. The data series were statistically evaluated using different procedures (Kruskall-Wallis-Test, Mann-Whitney-Test, Kendall Rank Correlation Coefficients, Spearman Rank Order Correlation Test For Trend, etc.). However it soon became apparent that the accuracy of these tests was merely sufficient to bring out major inconsistencies or trends. An attempt was therefore made to establish the type and time of human and natural influences on the data series employing documents on the mesurement stations and their drainage areas. Where individual measurements or series of measurements appeared unreliable, attempts were made to check and correct them using old measurements, stage and discharge rating curves and hydraulic calculations.

4.3 Flood Frequency Probability Distribution

It was possible to fall back on experiences gained on an old investigation in the 1970's (ASF, 1974) when choosing the theoretical distribution function. Finally Pearson III, log-Pearson III and Gamma III were used. In view of the slight differences, these distribution functions were eventually employed to give uniform results even when another distribution function could have given a still better approximation of measured values.

Overall it was possible to establish the following:

1) The choice of distribution function (log-normal, Pearson III, log-Pearson III, Gamma III) and the choice of method for determining the plotting positions is not normally of crucial importance for small recurring periods (Q_{50}, Q_{100}).

Calculations of Q_{50} and Q_{100} using different distribution functions showed only minor divergences. Generally the mistakes based on errors of measurement, incorrect stage-discharge relationship, inconsistencies and inhomogeneities were substantially greater. The elimination of these mistakes is especially difficult when dealing with long series.

2) It does not appear sensible to us to forecast rare events (Q_{100}, Q_{10000}) just on the basis of peak discharge statistics.

5. ON THE INFLUENCE OF OUT-OF-CHANNEL RETENTION DURING LARGE FLOODS ON THE ESTIMATION OF EXTREME FLOODS

To estimate floods of high periods, extreme value distributions together with measured yearly maximum flows at the point of interest are used. With the measured flows, the parameters of the distribution are estimated and the flows are then extrapolated to the desired return period.

In a river like the Glatt (fig. 3a), this procedure can be used with some confidence, as the points for the largest events approximate on the probability paper more or less a straight line, as proposed by the theory.

However, in many, mainly smaller catchments, the largest flood discharges lie below the fitted line, as in the example from the Murg (fig. 3b).

In the Langete, the largest floods of the 50 years of measurements lie quite close together (fig. 3c).

The 100-year flood was estimated on the basis of this 50 years of measurement to be 52 cumec. In 1975, a flood occured with an estimated peak of 90 cumce. According to the fitted distribution, this event would have a very high return period. But the rainfall, which caused

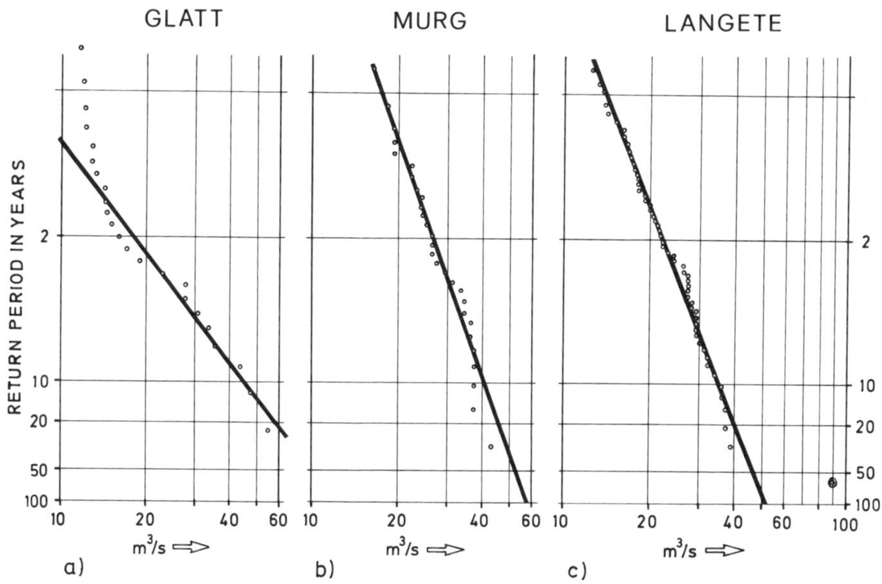

Figure 3 Flood frequency diagrams for 3 small Swiss rivers (Glatt, 16 sq.km, 21 years; Murg, 79 sq.km, 28 years; Langete, 115 sq.km, 60 years)

the flood had a return period of only 50 to 100 years. So what is the real return period of the 1975 flood? How reliable are these methods, if such discrepancies occur even when using long and reliable records? What 100-year flood discharge should be expected in the Murg catchment, which shows a similar behavior in the flood frequency diagram as the Langete does?

Some answers to these questions were found in a research project, which aimed to develop or adapt procedures to estimate floods of given return periods in small catchments in Switzerland without discharge records. Results of the project are reported elsewhere (Naef, 1985; Naef et al., 1986).

In this project, data from 600 flood events in nearly 100 small catchments with areas of up to 100 sq. km and a mean period of discharge measurement of 20 years have been collected. They comprised rainfall amount and duration, rainfall volume, runoff coefficient and peak flow, time-to-peak an two recession constants, as well as antecedent rainfall and the date of occurence for each event.

These parameters differed considerably for different events. It was neither possible to explain these differences with event related parameters, like season of the year, antecedent rain or rainfall amount, nor was it possible to relate them meaningfully to catchment characteristics like slope, soil parameters and vegetation. In a second step, these data were to be used to calibrate rainfall-runoff relationships on the basis of catchment parameters.

The results were disappointing. The scatter of the evaluated parameters was so large, that their mean values could not be used to estimate peak discharge with reasonable accuracy. The means over all catchments of the coefficients of variation, estimated for the different event parameters are:

	duration of rainfall	amount of rainfall	runoff coeff.	UHmax	Qmax
Coeff. of variation	.65	.48	.37	.57	.25

The coefficients of variation of the parameters needed in a rainfall-runoff relationship to estimate the peak flow, namely rainfall duration, rainfall amount, runoff coefficient and maximum ordinate of the unit hydrograph are larger than that of the variable to be predicted, the peak flow. As the variations of the parameters could not be explained with the data available and the present knowledge of the runoff process, it seems more reasonable to estimate the peak discharge directly with a simple statistical formula from catchment characteristics than with a more detailed model.

In the hope of explaining some of the scatter of the parameters, the following graphs were drawn (fig. 4, 6, 7), where the deviations of

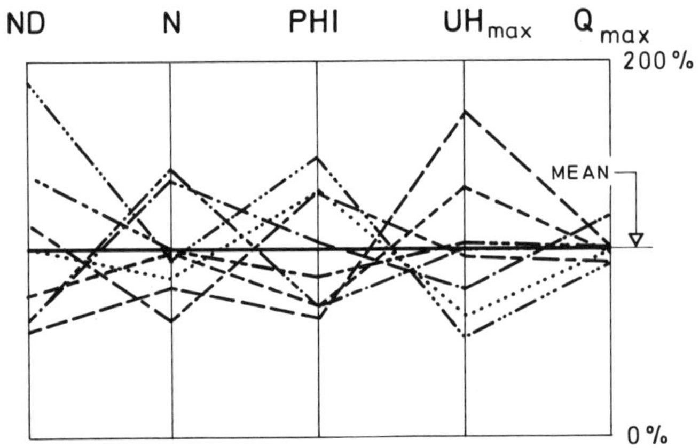

Figure 4 Deviations of the event parameters from the mean for the 8 largest events measured in the Murg catchment. Parameter values of the same event lie on the same type of line

the parameters of all events in one catchment form the mean values are plotted. All the parameter values for the same event lie on the same type of line.

Figure 4 shows this graph for the Murg river. The parameters rainfall duration ND, rainfall amount N, runoff coefficient PHI and the maximum ordinate of the unit hydrograph UHmax of the 8 largest events of the 28-year long period of measurement show a large scatter, whereas the range of the peak flow Qmax is quite small. This fact explains some of the scatter of the parameters: If Qmax, which is the product of N, PHI and UHmax changes little, a large N*PHI must combine with a small

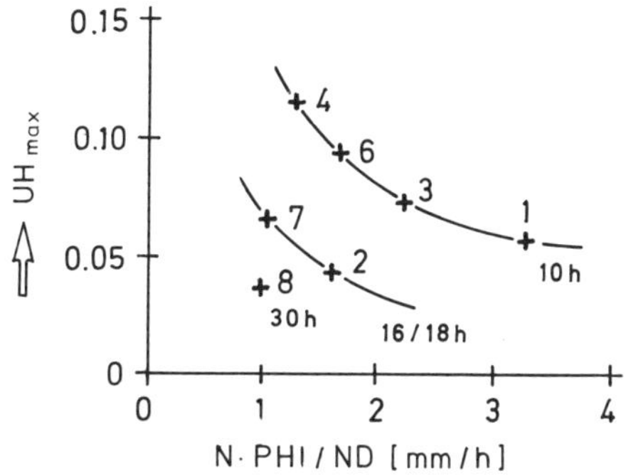

Figure 5 Relation between the maximum ordinate of the standardized hydrograph, the runoff intensity and the duration of the rainfall for the Murg

UHmax to result in a more or less constant Qmax or vice versa. Figure 5 shows this effect even more strongly. On the x-axis, the runoff intensity is plotted, that is the amount of water available for discharge per unit of time (N*PHI/ND); on the y-axis, the maximum ordinate of the unit hydrograph is plotted. The relationship between the two variables can be explained with the increased retention due to flooding during large discharges. If the runoff intensity is high, the capacity of the

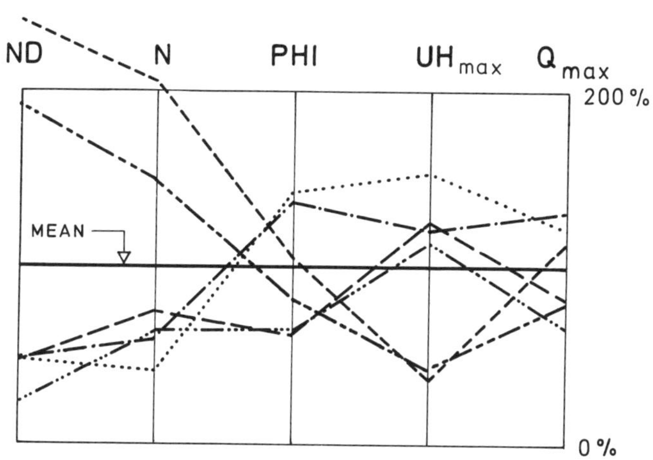

Figure 6 Deviation of the event parameters from the mean for the Glatt

river channel is exceeded, flooding occurs, the retention is increased and the peak discharge further down-stream corresponds to the capacity of the channel and is more or less constant. This cause the higher discharges to lie below the expected frequency distribution (fig. 3b).

Figure 6 shows the situation for the Glatt river, where the channel capacity is not so much a limiting factor, therefore the scatter of Qmax is larger than in the Murg. This can also be noted in the frequency distribution and extrapolation can be made with more confidence, as already mentioned.

Figure 6 also shows, that events of quite different duration cause flood peaks of similar size (see parameter ND). The time of concentration for the Glatt is around two hours, but two of the largest events have a much longer duration. Furthermore, it was not the event with the largest rainfall amount which caused the highest peak, but the one which was combined with a high runoff coefficient. Thus the assumption, that the n-year return period rainfall causes a flood with the same return period, which is made in most of the flood estimation procedures used today, does not hold for the Glatt catchment. Indeed, this assumption was valid in only 8% of the catchments analysed.

Figure 7 shows the scatter graph for the Langete, which looks quite similar to that for the Murg, except for the 1975 event. Obviously,

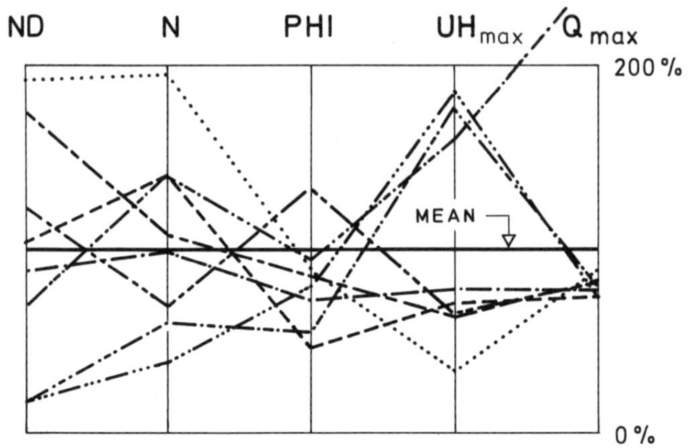

Figure 7 Deviation of the event parameters from the mean for the Langete

the increased retention due to flooding makes the peak flow nearly constant, although the event parameters vary considerably. The 1975 event parameters are not exceptionel, why then is the discharge so much higher?

The combination of the parameters of the 1975 event is such, that it delivers by far the largest runoff volume during 8 hours. The water volume delivered in this time can fill the retention volume available along the river channel, so that the retention capacity is exhausted and a much higher peak flow results.

The flood frequency statistics for the Murg and the Langete based on the discharge alone give a rather too optimistic picture for flood discharges with high return periods. The scatter graphs and the frequency statistic of the runoff intensity during the critical duration (fig. 8) are more disturbing. To estimate peak flows of higher return periods for the Murg with some confidence, one has to decide, when the

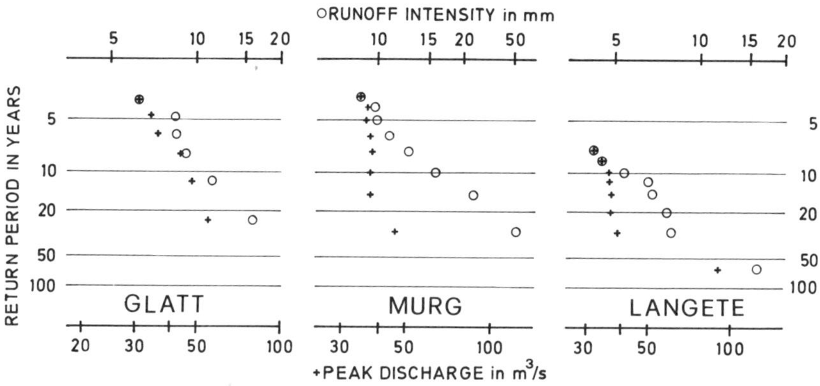

Figure 8 Frequency diagrams for the runoff intensity (N*PHI/D*tc) and the peak flow for the largest events

retention capacity along the river channel is exhausted, so that a breakthrough occurs. Although it does not appear so at first. extrapolation to floods of high return periods is easier in the Glatt catchment, which is steeper, has a higher specific discharge and a larger scatter in Qmax than the seemingly tame Murg or Langete.

In such catchments, a flood frequency analysis based on the yearly maximum flows alone, can often not give a reliable estimate of a design flood. A frequency diagram of the runoff intensity can help to show the influence of the out-of-chan-nel retention on large flows and can indicate catchments, where more detailed studies of this retention capacity are required to improve estimates of the design flood.

6. REFERENCES

ASF, 1974. The maximum flood discharges of Swiss rivers up to 1969. Mitteilung des Eidg. Amtes für Strassen- und Flussbau.

Naef, F. 1985. "What parameters control the flood peaks in small Swiss catchments?" Proceeding of Fourth International Hydrology Symposium on multiviariate analysis of hydrologic processes, Fort Collins

Naef, F., Zuidema, P. and Kölla, E. 1986. Abschätzung von Hochwassern in kleinen Einzugsgebieten. Beiträge zur Geologie der Schweiz Nr. 33

Spreafico. M., Stadler, K. 1986. Flood discharges in Swiss rivers. Mitteilung Nr. 7, Swiss National Hydrological and Geological Survey.

Vischer, D. 1980. "The greatest possible flood and the empirical upper limit discharge" (in German). Schweiz. Ingenieur und Architekt, Heft 40.

ESTIMATING THE REGULATORY FLOOD ON A DEGRADING RIVER

Brian M. Reich, Visiting Professor
School of Renewable Natural Resources
Donald R. Davis, Professor
Department of Hydrology and Water Resources
University of Arizona

ABSTRACT. Two floods that occurred in the last seven years of a 70-year gage-record on a catchment crossing the Mexican/United States border have caused concern about the earlier estimate of the 100-year flood. These recent events exceed by two and three times the previous maxima which had been relatively constant for this semi-desert catchment of about six thousand square kilometers. The various flood frequency computations that have been made by statistical and deterministic methods are presented. Historic descriptions of a pristine catchment and progressive channel incisement of large floodplains suggest hydrologic mechanisms that may account for the worsening flood condition. As a result, the flood regime is not stationary and is best represented by the last 25 years of the record.

INTRODUCTION

Seventy years of measured floods on the Santa Cruz River at Congress Street, Tucson, Arizona from a 5,755 square kilometer desert watershed in the southwestern U.S. illustrate one case where computation with a long flood series yields unreliable estimates of future floods for a rapidly urbanizing environment. Testing for nonstationarity, which should always precede flood frequency computations, shows that for this Arizona community, the latest third of the data better represent future floods than does the entire series.

A flood of 1,500 m^3/S in the Santa Cruz River inflicted itself on the city of Tucson in 1983. Some may wish this away by calling it a 100-year, or even a rarer event! One can hardly blame them in light of the Federal Emergency Management Administration 1982 floodplain map's use of 850 m^3/S for Q100, a flood defined as having only 0.01 exceedance probability each year. On the other hand, recent flooding seems to be getting worse on this degrading alluvial river in the last quarter century. Its latest peak was 2.2 times as great as the previous flood of record that occurred in 1977. The 1977 peak was 1.4 times the size of the previous 63-year maximum set in 1965. Two other floods since 1960 also exceeded the previous flood of record, 425 m^3/S,

V. P. Singh (ed.), Regional Flood Frequency Analysis, 197–212.

that occurred in 1914.

Various statistical tests indicate that a significant increase in flood magnitude has occurred in annual maximum floods throughout the 70 years of streamgauging. Inspection of a 35 Km reach upstream of the gauge, a series of aerial photographs, and historical reports have suggested that recent increased channel capacity and reduced overbank flooding onto previously wide, well vegetated, floodplains may be responsible for the nonstationary flood series.

Log Extreme Value analysis of the last 25 years of data yield 2,800 m^3/S for the present Q100. This is independently substantiated by analysis of flood series and floods of record on similar sized Arizona watersheds. These do not consider possible future flood growth from further deterioration of the Santa Cruz watershed. Underestimating the regulatory flood by a factor of three would result from cranking the entire 70-year data series through an LPIII computer program. This would lead to serious community losses in the growing city. Examination for non-stationarity, recommended in the Bulletin 17B text, is seldom performed. Even a visual examination of the chronologically plotted series of floods could warn engineers and administrators of changing flood regime. However, administrative inertia against increasing earlier underestimates of the regulatory flood persists.

Floodplain degradation followed landuse changes, groundwater overdraft, erosion at and between many bridge crossings, and other impacts of man. Under semi-desert conditions, vegetation cannot exert the restorative function observed in other riverine environments. The common government reluctance to admit change for fear of liability suits appears inappropriate here because a broad mosaic of contributory causes has progressed over about one century.

PREVIOUS FLOODS AND OFFICIAL 0.01-PROBABILITY ESTIMATES

Twelve estimates of the 0.01-probability flood (Q100) are listed in Table I along with the five largest events since 1914. Tucson's October 1983 flood of 1,500 m^3/S inflicted considerable damage because little protection was available beyond the 850 M^3/S, which was the 0.01-probability flood (Q100) used to prepare the 1982 federal Flood Insurance Administration (FIA) maps. The 1983 event was twice the magnitude of Q500, the 0.002 probability flood, estimated by Log Pearson III (LPIII) analysis of records from 1914 through 1976 (Roeske, 1978). In fact, the 1978 water year flood gave 670 M^3/S at Congress Street in Tucson, which was 1.43 times as large as a previous record set in 1961. Fortunately, the openings and footings of two downtown bridges had been rebuilt, for 1,275 M^3/S, after the October 1977 flood. Otherwise, significant damage would have occurred to the city's motel district and transportation system. Two bridges in southern suburbs and more in the county's northern neighborhoods of this river were rendered useless for many months by the 1983 event. The City also gained from the State of Arizona imposing higher standards than the Federal Government's Q100 of 850 M^3/S, and thereby obviated destruction

TABLE I

Summary of Observed Floods and Q100 Estimates for Santa Cruz River at Congress Street, Tucson, AZ, 5,755 Square Kilometers

Month	Year	Cubic Meters/Second Observed	Q100	Source	Notes
Dec.	1914	425		USGS	
Aug.	1961	470		USGS	
Dec.	1967	456		USGS	
-	1972		1,300	Corps of Eng.	Based on Floods in S.E. Arizona
Oct.	1977	671		USGS	
	1978		1,080	USGS	Area Basal Equation
	1978		1,275	Reich	Pima Co.→State→City
	1981		625/657	USGS	LPIII Station and Weighted Estimate
	1982		850	FIA	Flood Insurance Maps (pre-1978 decision)
Oct.	1983	1,492		USGS	Flooded People Outside FEMA's Q500
Dec.	1983		2,720	Reich	LEV Using 1960-1984
July	1984		2,180	Boughton	Regional Data Without 1984 Water Year
Oct.	1984		1,900	Hydro software	Computer Model Assuming Rainstorm
Feb.	1985		1,416	Pima Co.	Board of Supervisors
July	1985		1,700	" " FCD	Regulatory Flood??
July	1985		2,000	" " "	Design Flood??
Nov.	1985		830	USGS	1915-1984 Using WRC

of a 1982 housing project by requiring lowest finished floors to be one foot above the 1,275 M^3/S water surface. The state was responding to the Pima County Flood Control District (Reich, 1978) which reported that the 64-year flood record was nonstationary even before 1983's flood.

A routine application of LPIII analysis to the entire data record from 1914 including October 1983's flood calculates Q100 as 906 M^3/S. According to that flood frequency curve, the 1,500 M^3/S, which occurred that month, has a 500-year return period. Following Water Resources Council 1981 treatment of "outliers", one gets 830 M^3/s for Q100. However, its inclusion into the 70-year record does not drastically affect the LPIII analysis. Some might even choose to further downplay its effect upon the frequency calculations, because a 500-year event has as much as a 13% chance of occurring during a 70-year period. The importance of the 1983 event becomes far greater, however, if it is conceded that only data since 1960 reflect the present flood regime.

TABLE II

0.01 Probability Flood Estimates Assigned by FIA to Four Drainage Areas Affecting the Tucson Metropolitan Community

River	Location	Drainage Area Km^2	Q100, M^3/S
Santa Cruz	Congress Street	5,755	850
Rillito Creek	First Avenue	2,310	960
Pantano Wash	Houghton Road	2,271	877
Tanque Verde Creek	Above Aqua Caliente Tributary	1,684	650

How should engineers and planners in Tucson, which will house 1,000,000 people by the year 2000, react to ever increasing floods? In January 1985, the Pima County Board of Supervisors adopted a new Flood Plain Ordinance which decreed, among many other items, that the Santa Cruz River at Tucson (Congress Street) had Q100 of 1,416 M^3/S. In July 1985 the Pima County Flood Control District proclaimed "regulatory" and "design" discharges of 1,700 and 2,000 M^3/S. The latter decision was based upon a deterministic runoff model (Hydrosoftware, 1984, Ponce et al., 1985). Frustration at changing estimates from flood frequency analysis can be appreciated. However, deterministic runoff models are only approximations to the runoff process themselves and depend on estimating the 100-year storm. Added complexity arises from uncertainty concerning duration, aerial coverage, and travel direction, time distribution of the storm, the joint probability of antecedent watershed moisture, etc. (Ely and Peters, 1984). Different deterministic runoff models will yield different peaks. All have serious calibration problems in this data-deficient region.

The remainder of this paper reports statistical and physical evidence to show that the Santa Cruz flood record is not randomly "stationary," and that the flood regime is altering due to watershed changes. Bulletin 17B states that "Only records which represent relatively constant watershed conditions should be used for frequency analysis." This paper addresses the question of how much early data must be separated from the "after" change floods before the computation of Q100 for a hypothetically stationary future.

INCLUDE ONLY FLOODS FROM A POST-CHANGED WATERSHED FOR PROGNOSIS

Bulletin 17B discusses the philosophy of needed stationarity of annual flood series. Although not succinctly presented in one place, it appears in its introduction and in the section on nonconforming special situations. Some details are added in its refinements and reliability sections. It states: "Man's activities which can change flow conditions include urbanization, channelization, levees, construction of reservoirs, diversions, and alteration of cover conditions.... Special efforts should be made to identify those records which are not homogeneous." Including earlier data from a more pristine watershed into a frequency analysis violates this. Future deployment of private or public funds along the Santa Cruz deserves an updated design parameter, which can hopefully be used for about a 30-year planning horizon. Evidence has been reported by Betancourt and Turner (1985) of various works of man over the past 100 years which have combined to produce far higher velocities of floods that approach Tucson from tributaries and the mainstream in Mexico (Reich, 1985). Hydrologic change was not sudden, as would follow the removal of a flood control dam, but grew progressively.

The population in Pima County, which is largely concentrated in Tucson, showed an accelerated growth in the 1950's. The chronological series of annual maximum floods through Tucson (Fig. 1) suggests that big floods have become larger and more frequent after that decade. Growth of a modern city increases the demand for sand and gravel. Laursen and Carmody (1980) wrote that the elevated highway built upstream of town in the early 1960s on a fill of 5 meters high and 40 meters wide required about 700,000 cubic meters of dirt which was readily available from the dry riverbed that paralleled it for about 20 kilometers. They explained how riverine borrow pits produce erosion both upstream and downstream. Evidence of other channel straightening and widening which could increase downstream flood peaks was reported (U.S. Soil Conservation Service, 1937). Decreased soil cover on the watershed over almost a century has been photographically documented by Cox et al. (1982).

San Xavier Wetlands

Twelve kilometers upstream of Tucson's Congress Street this highway crosses the Santa Cruz between two hills near the San Xavier church that was built 286 years ago by a Jesuit priest. Prehistorically natural waterspreading at this site stimulated riverine vegetation. In fact Kupel (1984) wrote that aboriginals were irrigating 600 hectars at San Xavier 500 years ago. In 1699 the Spanish explorer, Manje, described it as "sufficient lands watered by means of good ditches" to support 2,000 people. It was only after Mexican control in 1822 that water resources were abused along the Santa Cruz, culminating in a copper smelter in 1856. "Tucson was using groundwater heavily by 1870. Railroad construction and wagon roads also contributed to erosion." Within two years after the 1877 Bavispa earthquake, channel entrenchment started. Overpumping has lowered the groundwater during the last 50 years. Tall trees still occupied the floodplain in a 1940 photo (Reich and Davis, 1985). By 1978, huge conveyance had been created by the devegetation and erosion. In describing riverine change over the last 50 years in the seven miles upstream of the bridge site, Lowe (1984) reported that channel degradation was somewhat controlled half a mile upstream of the hill by a concrete dam until 1977. Progressive degradation since 1940, however, further transformed a shallow, small stream that had previously braided through a mile-wide floodplain into a 30-foot deep gully with more than enough capacity to carry the largest record flood. Today's dry wasteland with 10-meter deep, vertical, sand banks that have now eroded to as much as 200 meters apart contrast with Commander Emory's (1857) observation: "Approaching San Xavier ... we again came upon running water ... The settlement of Tucson occupies the lowest line of constant running water and, consequently, the last fertile basin lying in the course of this valley." Another portrayal by Bartlett (1856) of his travels south (upstream) of Tucson in July 1852 refers to:

> "a fertile (San Xavier) valley, a very small portion of which is now tilled ... Leaving the village we rode on a mile (1.6 km.) further and stopped in a fine grove of large mesquite near the river, where there was plenty of grass ... we resumed our journey along the valley as before, through a forest of mesquite trees ... The rain having continued all night, we were much delayed in getting off this morning. The whole country was drenched with water and the road almost impassable ... After 30 kilometers upstream we stopped on the banks of the river and strange as it may appear, notwithstanding all the rain that had fallen, the river, such is the uncertainty of the streams in this country, was quite dry."

This appeared to have been a case where runoff from a tributary had spread across the two or three kilometer wide floodplain of the Santa Cruz; and was traversing it at extremely low velocity with much associated infiltration losses. Today such tributary inflows are being transmitted downstream to Tucson at high velocity by a channel network of headcutting gulleys. This is replacing previous vegetative retardance along the lateral fringes which contained the alluvial fans onto which tributaries as large as 250-square kilometers had previously dissipated their floods.

Changes Further Upstream in the Second Half of the Twentieth Century

Another reach between 15 and 30 kilometers south of San Xavier has experienced considerable change over the past 50-years. Overpumping of groundwater was an early precursor of hydraulic change. Since 1960 erosional deepening and widening of the river through the Green Valley reach followed man's attempts to keep floods out of floodplain orchards and fringe areas of a new residential community. In the first half of this century, floods traveling down the Santa Cruz valley had spread out onto wide grassy floodplains. Originally, the full peak intensity of mainstream floods from the south was not transmitted downstream, but was attenuated by temporary overbank storage among dense vegetation and by infiltration. The pre-1960 upper-limit of Tucson floods at about 300 M^3/S as seen in the chronological series of floods (Fig. 1) may have represented the throttling of larger inflows which exceeded the then limited channel capacity. By 1969 the channels probably carried 600 to 700 M^3/S without significant overbank losses in this 15 Km reach. Fig. 1 shows that six Tucson floods exceeded 400 M^3/S since 1960. Aerial photographs substantiate man-made cutoffs of meander bends and other changes in these flood plains which must have reduced natural flood attenuation.

Independent validation of this hypothesis can be obtained by expressing winter Tucson flood peaks, which are not confounded by an isolated summer storm on a downstream tributary, as a ratio to the corresponding peak above Green Valley. Hydrographs observed at both locations in the late 1940's show Tucson peaks were 0.6 times the size of those at Green Valley. From the 1960's through 1980, they averaged 0.85 of upstream peak magnitudes. In October 1983, Tucson's peak was 1.2 times that at Green Valley.

STATISTICAL ANALYSIS AT CONGRESS STREET

A visual means (Reich and de Roulhac, 1985) of examing whether two different periods of a record were produced by the same hydrologic mechanism is to plot both segments on one sheet of probability paper (Fig. 2). The floods from 1915-1959 are shown as x's. The record from 1960-1984, indicated by dots, lies systematically far above the earlier data set. This is a graphical algorithm of a non-parametric test. One such numerical test is the Kruskal-Wallis test, which showed that there is only a 7.2% chance that the disproportionate number of recent bigger floods occurred simply by chance. It is most unlikely that these widely differing flood frequency arrays were generated by the same hydrologic mechanism.

Flood Statistics of Progressive Thirds of Data

Dividing the series into virtually equal segments allowed the calculation of averages and coefficients of skewness and variation of the logarithms of the floods. Two common flood prediction techniques are added in the last two columns. The first three lines in Table III highlight the progressive growth in flood indices. The second three illustrate the effect of eliminating outliers.

The Geometric Means

This statistic is a relatively stable measure of the average flood size for each segment of the record. Stability is illustrated by the fact that adding October 1983's 1,492 M^3/S to the previous 24 years' floods only raised the geometric mean by 9%. A large single value within a sub-series will not affect the geometric mean as substantially as will the same total increase in annual floods spread throughout many years. Even without the October 1983 event the geometric mean of Congress Street flood peaks has been continually increasing, as shown in the first three lines.

TABLE III

Statistics from Segments of Flood Peak Series; M^3/S

Segment of Record	Geometric Mean	Coefficient of Skewness of Logs of Floods (CSL)	L.P. III 100-year Prediction	Log E.V. 100-year Prediction
1915/1937	123	0.093	623	1,189
1938/1961	155	-0.056	566	1,048
1962/1984	184	0.264	1,416	2,775
1960/1983	176	-0.269	792	1,880
1960/1984	192	0.288	1,529	2,718
Same without 1965 low-outlier	206	0.677	1,671	2,350

Omitting 33.7 M^3/S, which was a low-outlier compared to the second smallest flood of 70.2 M^3/S, from the 1960 through 1984 caused the geometric mean to rise 7%. Addition of 1984's 1,492 M^3/S to the 1960/1983 series which had been led by October 1977's maximum of 670 M^3/S, further increased the geometric mean by 6%.

Variance Changes

The variance gives the steepness to a flood frequency line and is higher in the last 25 years. An F-test suggests that there is only a 6% chance of randomness accounting for the increases.

Log Extreme Value Estimates

Another flood "statistic" which shows that Tucson's flood potential increased about 2 1/3 times in the last 25 years is the 0.01 probability estimate (LEV Q100) according to Log Extreme Value theory.

This corresponds to mathematically fitting a straight line on LEV paper. The solid line on Fig. 3 presents an example of the goodness-of-fit through the data, particularly the largest observations. The last column of Table I shows about a 9% decrease in LEV Q100 between the first and second thirds of the record, traceable to the coefficient of variation of the logs of the floods being 0.0814 and 0.0626 for the sub-histories. A 9% decrease in LPIII Q100 was experienced in the same time frame. Such variability is the expected outcome of sampling in time.

Drastic changes occurred in the record that started in the 1960's, even without the inclusion of October 1983. Recent increases in the coefficient of variation as well as the geometric mean and standard deviation of recent years brought LEV Q100 up to 2,720 M^3/S and 1,880 M^3/S with and without the October 1983 event. These increases above an average LEV Q100 of about 1,188 M^3/S for years between 1915 and 1960 are hard to dismiss as random variability caused by sampling at different times in an unchanging river record.

Bulletin 17B explains that special situations may require other approaches than Log Pearson Type III (LPIII). While the solid line in Fig. 3 shows how well the LEV fits the 19 largest floods, and gives Q100 as 2,720 M^3/S, the dashed LPIII passes 370 M^3/S below the largest event.

Log Pearson Type III Estimates

This 0.01 probability estimate (LPIII Q100), which has been popularized by Bulletin 17B, is listed in Table I to the right of CSL upon which it is heavily dependant. Whether any LPIII Q100 is an acceptable estimate of the 100 year flood can only be established after studying how the LPIII curve fits the plotted data. An example of how this method is controlled by a CSL, determined largely by small outliers, is given in Fig. 4. The curve is positioned across the required prediction range, from 25 to 100-year return periods, by curvature established by the numerous floods which are not generally of concern in flood control work.

The dashed LPIII curve in Fig. 3 passes 368 M^3/S below the largest event because of curvature established from the 12 smallest floods. As a result, LPIII predicts Q100 as 1,510 M^3/S, compared to 1,492 M^3/S which was the maximum observed in these 25 recent years. Another property of observed floods having larger and less random scatter about the LPIII than about the LEV line favors the latter. In this case, LPIII Q100 is only the size of an event already observed in 25 years. However, even this statistic shows in Table I that the post-1960 flood potential at Congress Street had increased by 33%, even without the inclusion of October 1983's 1,492 M^3/S.

When Should the "Present" Flood Series Commence?

The progressive nature of the flood changes has been discussed with some history. It was not sudden, as would follow a dam collapse. A series of Kruskal-Wallis tests (Conover, 1971) were run on alternative

native "early" and "after" floods separated by five omitted years (Reich and Davis, 1985). These tests indicated that transition in the "nonstationarity" occurred somewhere between 1958 and 1961. Q100 increases as the record shortens below 27 years. The smallest LPIII estimate for "after" change (1958-1984) gave Q100 = 1,400 M^3/S.

Risk and NonStationarity

The most dramatic change produced by the nonstationarity of the flood regime on the Santa Cruz River is the increase in the risk of flooding as illustrated in Table III. Structures that were considered to be in the 10,000-year flood plain based on the 1915-1959 data record are in the 50-year flood plain based on the 1960-1984 data.

TABLE IV

Flood Exceedence (LPIII) for the Santa Cruz River at Congress Street Based on Different Subhistories

Flow	Systematic Record	
M^3/S	1915-1959	1960-1984
283	0.1	0.3
566	.01	.1
850	.001	.04
1133	.0001	.02
1416	.00002	.01

Management of the flood hazard where the flood regime in nonstationarity must take into account the changing risk of flooding. Flood protection measures that reduce flood risk to a satisfactory level now will not have the same effectiveness in the future. Management of nonstationary flood hazards must specifically consider the changing risk over time. If a structure is to be protected, the expected life of the structure must be considered so that protective measures can be designed to be effective throughout the life of the structure. Flood risk analysis for the nonstationary case will be more difficult than the stationary case as additional information about the floodplain and the flood regime will be needed. Commonly accepted relations between flood risk descriptors may no longer be valid. When the flood regime is stationary, the expected recurrence interval of a flood of a specified magnitude is the reciprocal of the probability of that flood occurring next year. This relation between the probability of a flood next year and the expected time to recurrence of the flood does not hold when the flood regime is not stationary.

DISCUSSION

Surprisingly large flood changes occurred in about half a century in this semi-arid valley. Aboriginal irrigation over five centuries had no significant influence upon riverine environment. For almost two more centuries, Hispanic civilization in this valley harmonized with natural controls. After this, the introduction of well pumps, channels, roads, trains, large farms, mines and air-conditioned housing progressively over the last century has irrepairably damaged the river. A tenfold increase in floods may ultimately destroy the outpost of civilization which could not foresee dangers of floodplain encroachment. Local government has not recognized the impact on floods themselves. Even if the truly large floods were acceptable to the engineering profession, vast areas of land would need to be purchased along this river and its major tributaries. Sediment transport and discontinuity pose great technical problems. Huge capital costs and maintenance fees are associated with necessary engineering works. While floods continue to be passed off as natural disasters, other causes will continue to receive the funds.

Great care should be taken before using flood records even as long as 40 or 50 years, in case they contain nonstationarity. Modern societies can impose drastic changes on the environment. It may be better to predict future floods from shorter post-change periods and to temper the estimates with safety factors.

Present flood regulatory mechanisms give no indication of how nonstationarities in the flood regime should be handled. If the flooding potential is increasing, a flood whose probability of occurrence next year is one in a hundred will be expected to next occur in a period less than 100 years. A comprehensive approach to flood risk management will have to consider the expected time to the occurrence of a flood of specified magnitude as well as its annual probability of occurrence.

If the flood regime of the Santa Cruz continues to change as indicated, the risk of flooding will be underestimated even if the most recent record is used.

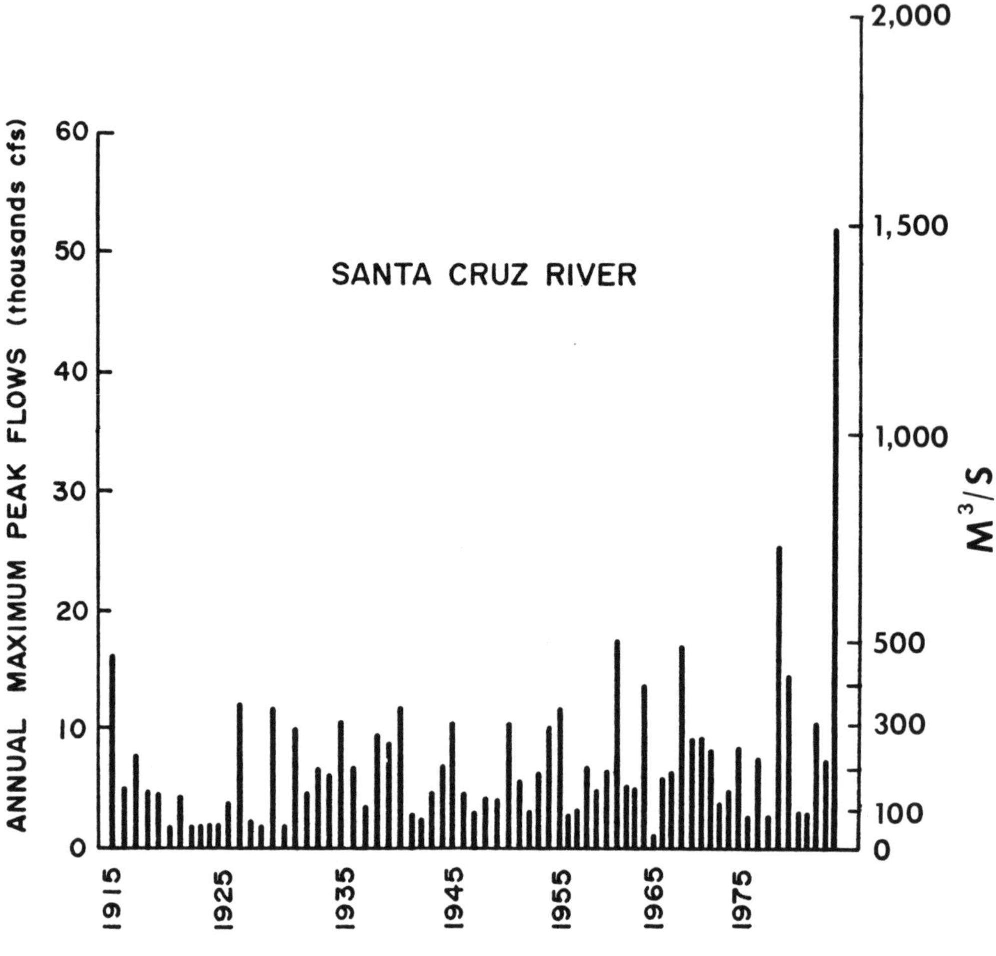

Fig. 1 Chronological Series of Annual Maximum Floods Measured at Congress Street, Tucson.

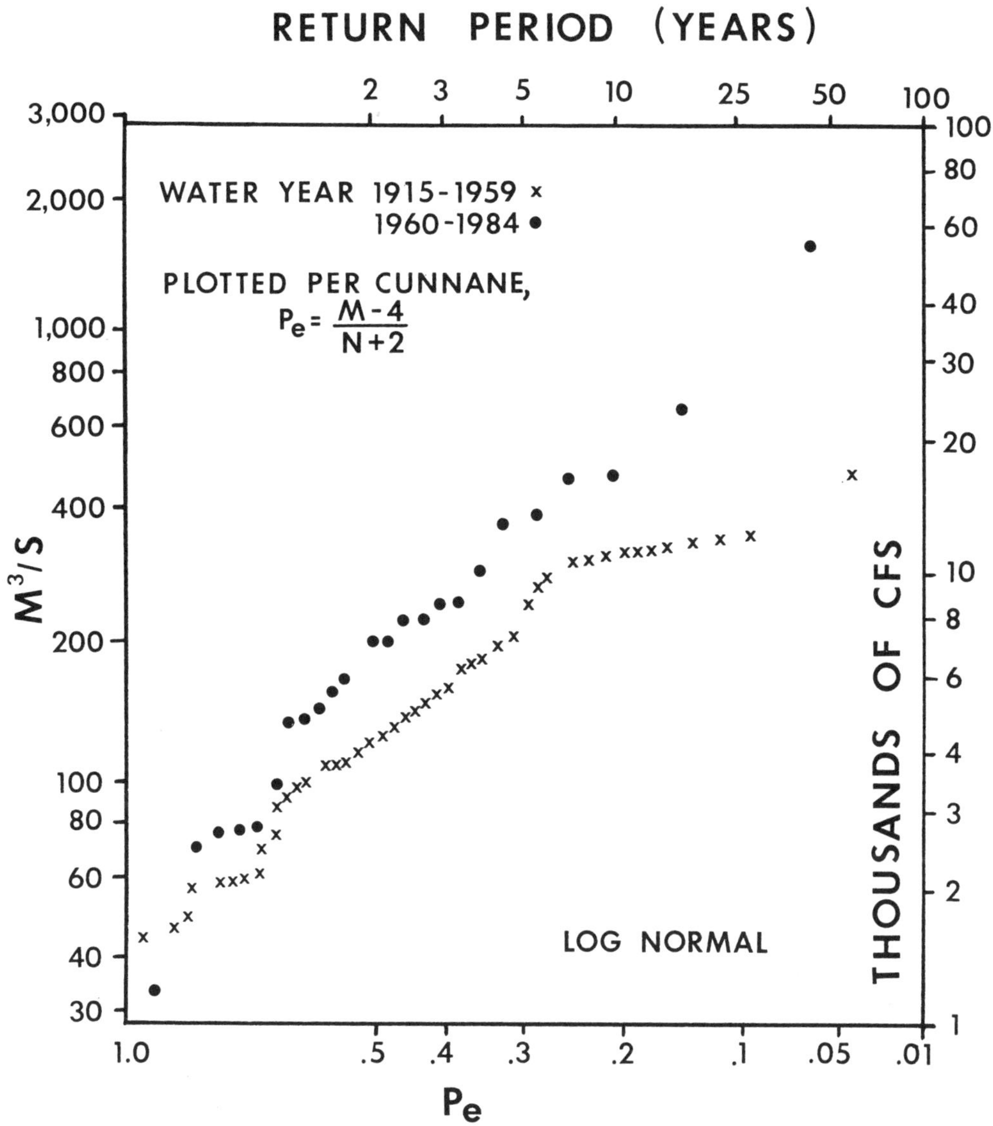

Fig. 2 Graphical Non-parametric Test Shows Post-1959 Floods were Larger than Earlier Ones.

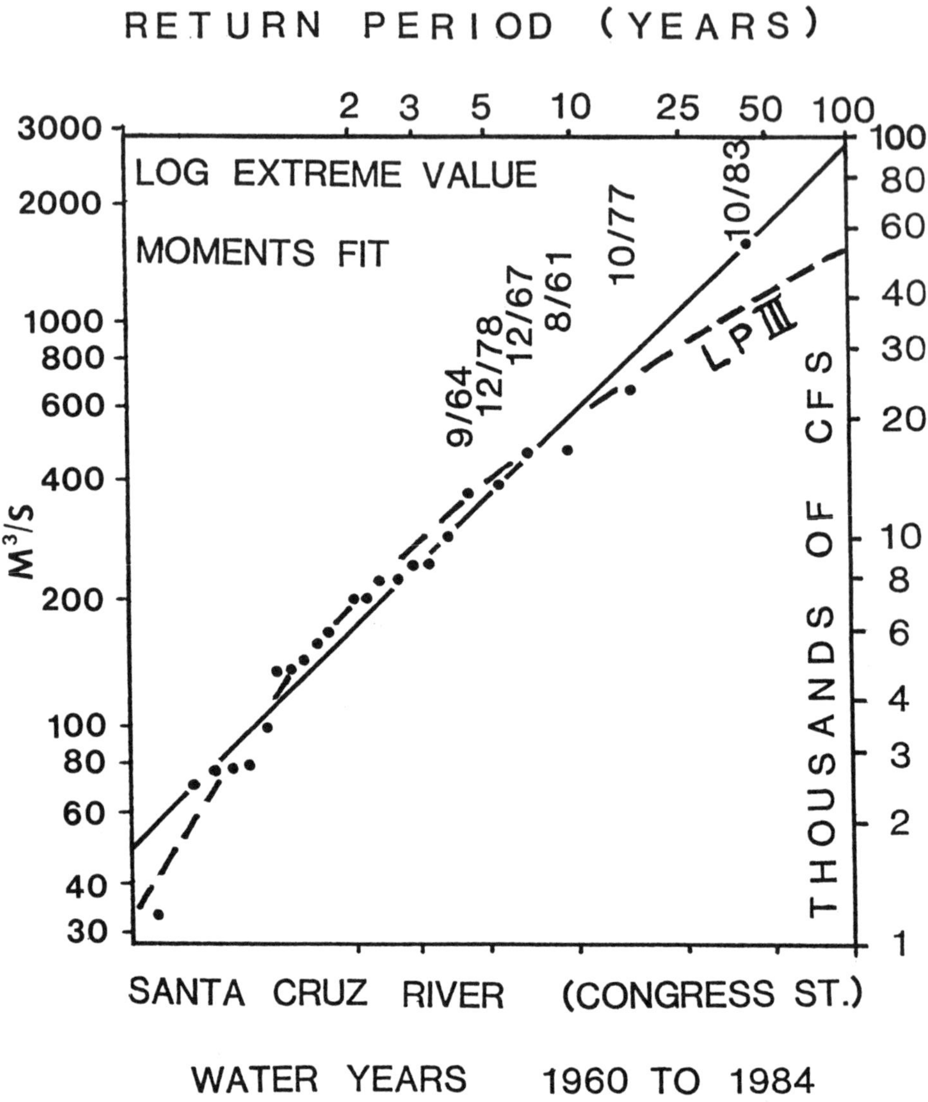

Fig. 3 Log Extreme Value Line is 20 m^3/S Above the Smallest Flood; LPIII Curves 368 m^3/S is Below the Largest Flood.

REFERENCES

Aldridge, B.N. and J.H. Eychaner, 1984. Floods of October 1977 in Southern Arizona and March 1978 in Central Arizona. U.S. Geological Survey Water Supply Paper 2223, 143 pp.

Bartlett, J.R., 1856. Personal Narrative of Explorations and Incidents in Texas, New Mexico, California and Chihuahuan. D. Appleton & Co., New York.

Betancourt, J.L. and R. M. Turner, 1985. Historic Arroyo-cutting and Subsequent Channel Changes at the Congress Street Crossing, Santa Cruz River, Tucson, Arizona. Proc. of the "Arid Lands: Today and Tomorrow" Conference, Univ. of Arizona, Tucson.

Cox, J.R., H.L. Morton, T.N. Johnsen, G.L. Gilbert, S.C. Clark, and L.C. Fierro, 1982. Vegetative Restoration in the Chihuan and Sonoran Deserts of North America. USDA, ARS-Research Manual No. 28, 37 pp.

Conover, W.J., 1971. Practical Nonparametric Statistics. pp. 256-263, John Wiley and Sons, Inc., New York.

Ely, P.B. and J.C. Peters, 1984. Probable Maximum Flood Estimation--Eastern United States.

Emory, W.H., 1857. Report on the United States and Mexican Boundary Survey (in two volumes). U.S. 34th Congress, 1st Session, Senate Document 108.

Eychner, J.H., 1984. Estimation of Magnitude and Frequency of Floods in Pima County, Arizona, With Comparisons of Alternative Methods. U.S. Geological Survey, Water-Resources Investigations Report 84-4142, 69 pp.

Federal Emergency Management Agency, 1982. Flood Insurance Study: City of Tucson, Arizona, Pima County. 45 pp. plus 34 profile sheets.

Hydrosoftware, Inc., 1984. Hydrologic Evaluation of Santa Cruz Basin, Arizona. Report to Pima County Department of Transportation and Flood Control District, 235 pp.

Kupel, D.E., 1984. Diversity in the Face of Adversity: Water Control at San Xavier. Manuscript Submitted to: Terramara International Services, Tucson, 83 pp.

Laursen, E.M. and T. Carmody, 1980. Hydrology and Hydraulic Report for Irvington Road. Report prepared for Pima County. By Ruiz Engineering Co., 141 pp.

Lowe, P.O., 1984. Santa Cruz River Changes South of Tucson. Hydrology and Water Resources in Arizona and the Southwest, Vol. 14, 10 pp.

Ponce, V.M., Z. Osmolski, and D. Smutzer, 1985. 'Large Basin Deterministic Hydrology: A Case Study.' Jour. of Hydraulic Eng., ASCE, Vol. 111, No. 9: 1227-1245.

Reich, B.M., 1976. 'Magnitude and Frequency of Floods.' CRC Critical Reviews in Environmental Control, 6 (4), pp. 297-348.

Reich, B.M., 1978. Analysis of Santa Cruz River Flood in Tucson, Including the Largest of Record That Occurred in October 1977. Report from Pima County Flood Control District to: FIA; USGS; Corps of Engineers; and City of Tucson, 12 pp.

Reich, B.M., '1984 Recent Changes in a Flood Series.' American Water Resources Association, Hydrology and Water Resources in Arizona and the Southwest, Vol. 14, pp. 231-238.

Reich, B.M., 1985. 'Effect of Floodplain Erosion and Flood Probability.' Paper to Bilateral Symposium on the Analysis of Extraordinary Flood Events. Nanjing, Peoples Republic of China, 22 pp.

Reich, B.M., and D.R. Davis, 1985. 'The 1983 Santa Cruz Flood: How should Highway Engineers Respond?' Transportation Research Record, No. 1017, National Academy of Science, Washington D.C., pp. 1-7.

Reich, B.M., and D.G. de Roulhac, 1985. 'Microcomputer Graphics Flood Stationarity Test.' Proc. of Speciality Conference, Hydraulics/Hydrology in the Small Computer Age, Hy. Div., ASCE, pp. 1401-1406.

Roeske, R.H., 1978. Methods for Estimating the Magnitude and Frequency of Floods in Arizona. Arizona Dept. of Transportation, ADOT-RS-15-121, 82 pp.

U.S. Army Corps of Engineers, L.A. District, 1972. Interim Report on Survey for Flood Control, Santa Cruz River Basin, Arizona, 43 pp.

U.S. Soil Conservation Service, 1937. Report on Santa Cruz Watershed. 36 pp.

U.S. Water Resources Council, 1981. Guidelines for Determining Flood Flow Frequency. Bulletin 17B of the Hydrology Committee, U.S. Water Resources Council, 138 pp.

COMPARISON OF REGIONAL FLOOD FREQUENCY METHODS IN SOUTHERN ONTARIO USING ANALYSIS OF VARIANCE TECHNIQUES

R. Condie[1] Head, Statistical Hydrology Section
P.J. Pilon[1] Senior Hydrologist
K.D. Harvey[1] Operational Hydrologist
H. Goertz[2] Operational Hydrologist

1) Water Resources Branch, Environment Canada, Ottawa, Ontario, Canada K1A 0E7
2) Water Resources Branch, Environment Canada, Guelph, Ontario, Canada N1H 3N4

ABSTRACT. There are many methods of estimating design floods at locations where no site-specific hydrometric data exist. Broadly, the methods fall into two categories: rainfall-runoff models and statistical models. Only the latter category is considered here, the candidate models being the Index Flood Method using the three-parameter lognormal distribution, the Direct Regression Method where T-year floods are related to physiographic and climatic parameters, and the Regional Wakeby Method. Eighty hydrometric stations were used with a pseudo-random split of 40 calibration stations and 40 verification stations. After the three models were developed using the calibration stations, they were applied at the verification stations, and the deviations in the flood estimates of calibration from verification were expressed as percentages. These percentage deviations then were the data required for the Analysis of Variance (ANOVA). A three factor ANOVA was performed, the first factor being method, with three levels, the second factor being drainage basin size with four ranges, and the third factor being the return period of the estimated flood with three return periods. There were no significant interactions nor differences between methods. The drainage basin size was a very highly significant factor, and the return period was very nearly significant at the five percent level. It is emphasized that these findings apply to Southern Ontario only, and probably different conclusions would be reached if the methods were applied in different parts of Canada or the world.

V. P. Singh (ed.), Regional Flood Frequency Analysis, 213–222.

require in the design stage a knowledge of the flood frequency regime at the proposed site. This knowledge is also needed for non-structural means of flood damage reduction such as flood plain zoning. Since hydrometric records from which an acceptable flood frequency analysis can be made are seldom available at the location of interest, the floods have to be estimated using indirect methods which can be roughly categorized as either rainfall-runoff models or statistical models. Only a limited selection from three of the latter category are considered here, and the intent of this study is not to come up with any firm recommendation. Rather, it is to set a fairly simple procedure which can be followed by hydrologists in other parts of Canada to determine a method suitable for their own regional needs.

The U.S. Water Resources Council (1981) published a pilot study which set forth a proposed nationwide test for determining a methodology for estimating flood frequencies at ungauged sites. In that study, procedures representative of eight broad categories of method were tested. This presentation is a simplified version of the aforementioned publication and only three methods of regional flood frequency analysis are considered: Index Flood, Direct Regression, and the Regional Wakeby.

SITE SELECTION

Defining Southern Ontario roughly as that part of the province south of the 46th parallel of latitude and bounded by the Great Lakes on the south, west and northwest, and the Ottawa and St. Lawrence rivers to the east, hydrometric stations were selected which met the following criteria: the station had an automatic recorder gauge, the record length exceeded 10 years, and the sites were on unregulated streams. Over 80 stations met the foregoing criteria. Four of these stations were in extremely small drainage basins and were ruled out since floods could easily be estimated by the Rational Method. One or two others on the extreme boundaries were also removed, leaving 80 stations on which the study was based.

SINGLE STATION ANALYSES

In the first instance, one member of the study team independently carried out the single station analysis of all 80 stations, and he was charged with providing estimates of the 2, 10, and 100-year floods using any of the following distributions: the generalized extreme value distribution, the three-parameter lognormal distribution, 3LN, the log Pearson type 3 distribution and the Wakeby distribution. Appropriate treatment was given to low outliers and historic information and, with the exception of the Wakeby, the distributions were fitted by maximum likelihood. The Wakeby was fitted using probability weighted moments as described by Landwehr, Matalas and Wallis (1979). The analyses performed were accepted as final by the other three team members. For the Direct Regression method, the best subjective estimates of the 2, 10 and 100-year floods were available.

REGIONAL ANALYSES

The eighty selected stations were then pseudo-randomly split into 40 stations from which the regional analyses were to be made, and 40 stations for verification against which regional estimates could be tested. Because of Canada's alphanumeric system of allocating gauging station numbers, after all 80 stations were arranged in alphanumeric order, calibration and verification stations were allocated alternately down the list and a good geographical spread of calibration and verification stations was obtained.

Rather than carry out the regionalization with one team member, each using one method, it was decided that the three remaining team members would together produce mutually acceptable Index Flood, Direct Regression and Regional Wakeby studies. This approach avoided unwittingly building into the experiment the undesirable factor of hydrologists' skills.

Index Flood Method

The Index Flood Method was first documented by Dalrymple (1960), and the method used in this study was presented by Harvey, Condie and Pilon (1985). The method substantially follows that of Dalrymple but with substantial improvements. Single station analyses were performed by fitting a 3LN distribution by maximum likelihood to the data sample, with appropriate treatment given to any low outliers and/or historic floods. A homogeneity test appropriate to the 3LN distribution is also included. Using this technique and an interactive program, the study area was easily subdivided into three regions, in each of which all calibration stations were shown to be homogeneous. Dimensionless 3LN regional flood frequency curves were thus developed. The index chosen was the 2-year flood, and defined by regression equations to be next described.

Direct Regression Method

This method of regionalization is currently extremely popular, and appears to have been first documented by Benson (1962) in an application to determine floods of recurrence interval T years. For a first attempt, the following structure was used:

$$Y = a_0 X_1^{a_1} X_2^{a_2} \dots X_v^{a_v} \tag{1}$$

where Y is the T-year flood, X_1 etc. are physiographic or climatic characteristics of the drainage basin, and a_1 etc. are parameters to be determined. Logarithmic transformation then gives the linear equation shown below which is easily dealt with in any standard computer package.

$$\ln Y = \ln a_0 + a_1 \ln X_1 + a_2 \ln X_2 \ldots a_v \ln X_v \tag{2}$$

By a trial and error process, using an interactive regression program, Southern Ontario was subdivided into three regions with an acceptably small standard error of estimate of the 2-year flood. Stepwise regression was primarily used, with verification by backward elimination. In this fashion, two-thirds of the area of Southern Ontario could be covered with two equations using drainage area only, with a standard error of about 23 percent. The pattern of the residuals was satisfactory in these two regions, but in the third region another variable, area controlled by lakes and swamps, was significant and the residual pattern indicated the need for a power transform of the dependent variable, the 2-year flood. A maximum likelihood estimate showed a square root transform was needed. The standard error of estimate of the root transform was 0.69, and the percentage standard error of the 2-year flood can then be computed from:

$$200 Q_2^{-1/2} \text{ (s.e. of e of } Q_2^{1/2}) \tag{3}$$

The average value of the 2-year flood for the 16 stations in the region was about 50 cubic metres per second, so the standard error for floods of this magnitude is about 20 percent. The required regressions were then developed for the Index Flood (the 2-year flood estimated from a 3LN distribution); for the Direct Regression (the 2-year flood estimated from the subjectively chosen best fitting distribution); and for the Regional Wakeby (the arithmetic mean annual flood). Additionally, regressions were developed for the 10 and 100-year floods which are required for the Direct Regression. For interest, the standard errors of estimate of the 100-year flood in the two regions where equation (1) held and drainage area only was used were 23 percent and 40 percent. For the third region where a square root transform was needed, for a typical 100-year flood of 140 cubic metres per second, the standard error of estimate was about 50%.

The Regional Wakeby

This is a comparatively new method of regionalization and is fully described by Wallis (1980). It is essentially an index flood type of method, and the required T-year floods are developed as multiples of the mean annual flood. A regression equation is derived for the mean annual flood, thence using the multipliers, the T-year flood at an ungauged site can be estimated. There is no homogeneity test in the strict statistical sense, but to quote Wallis, "It would appear that using common sense and a modicum of hydrologic expertise should lead to regions that are no worse than quasi-homogeneous". In this study, it was assumed that an experienced hydrologist would arrive at approximately the same regions as were derived in this study using a 3LN distribution and the appropriate homogeneity test.

ANALYSIS OF VARIANCE

This statistical technique was developed by Sir Ronald Fisher and essentially resolves the total variation in a data set into component variations due to independent factors. Each of the components provides an independent estimate of the population variance and can be compared for homogeneity using the "F" test. If non-homogeneity is indicated then it can be concluded at a chosen level of significance that different factors have different effects on the variable under consideration. A very easily read explanation of the topic, and several hand-worked examples from simple to complex, are given in Moroney (1964).

The Variate

The variate, upon which we wish to assess the effects of the various factors is the percentage deviation of the T-year flood at a particular site, as estimated by the model, from the T-year flood at the same site, as estimated by the single station analysis at that verification site. Thus, the variate are given in:

$$100(Q_c - Q_v)/Q_v \tag{4}$$

where the subscripts c and v indicate, respectively, estimates made from the regional analysis and from the single station flood frequency analysis.

The Method Factor

Three methods of preparation of regional flood frequency analysis were considered; the Direct Regression, the Index Flood, and the Regional Wakeby methods. Thus, there are three levels of the method factor.

The Return Period Factor

It was suspected that return period may have an influence on the variate; therefore, return periods of 2, 10, and 100-years were considered. The range of return period covers the usual range for which regional estimates are made, and the 10-year return period was chosen simply because on a probability scale it lies approximately midw'y between the 2 and the 100-year recurrence intervals.

The Range Factor

Four different ranges were chosen by ranking the verification stations in ascending order of drainage basin area. Then, from the 40 stations, four area bands were chosen such that 10 stations were allotted to each range.

To summarize so far, there are three levels of the method factor, three levels of the return period factor, and four levels of the area range factor. Since 40 verification stations were used, there are then 10 replicates in each band of drainage basin area. Thus, the body of the table of variates for analysis has 360 entries. Rather than show the entire table, it is presented in the form shown in Table I, where the entries in each cell are the totals of the 10 replicates per cell. Presenting it in this fashion makes the overall picture much clearer to the reader. The ANOVA was of course carried out on the 360 variates. The program used in the ANOVA is included in the SAS (Statistical Analysis System) package, and the table of analysis of variance is shown in Table II.

DISCUSSION AND CONCLUSIONS

Discussion

From the ANOVA shown in Table II, it can be seen that at the 5% significance level there are no interaction, method, or return period effects; but, before dismissing these factors it is as well to note that the return period factor is significant at the 10% level. By far the most significant factor is the drainage basin size, and in fact the probability of exceeding the computed F value of 18.65 shown on Table II with 3 and 324 degrees of freedom is an exceedingly remote one in ten thousand. It can be asserted that there is almost no probability of making a Type 1 error, whereby a null nypothesis is rejected as false when it is really true. As can be seen from Table I, the cell totals in the smallest range of basin areas, row 1, are all negative regardless of method or return period, except for the Wakeby model at the 100-year return period, and for the largest range of basin areas, row 4, the cell totals are all positive. These observations seem to indicate bias in the regression equations. It will also be noted from the column (method) totals on Table 1 that the Wakeby model at the 100-year flood level is rather high compared to the other eight, and it might have been expected that some model effect is present. As a side investigation and using deviations at the 100-year flood level, a two-factor ANOVA was made; the factors used were method with three levels, drainage basin size with four levels, and 10 replicates per cell. Again there was no significant method effect, and there still was a significant range effect, although this effect was then only significant at the 5 percent level.

To return to the most vexing question of the significant effect of drainage basin size, Region 1 was chosen for a more detailed investigation. Regions are purposely not identified since this is a comparative study of regionalization methods, and not a regional flood frequency study as such. This will avoid having a casual user using the material herein for estimating floods at ungauged sites in Southern Ontario. The regression equation for the 2-year flood in Region 1 was first re-examined but could not be faulted on the pattern of the residuals. Fifteen stations were used to calibrate the model, giving the following regression equation in drainage area A only, with a quite acceptable standard error of estimate of 23 percent.

$$\log Q_2 = 0.91886 \log A - 0.6109 \qquad (5)$$

When this calibration model was applied to the 11 verification sites, a very biassed pattern of residuals resulted. If the verification sites are aligned in ascending order of drainage area, the pattern of the residuals (sign only) is:

- - - - - - - - - + + -

This pattern is of course reflected in Table 1, and was also present, though to a lesser degree, in the other two regions. As a first attempt to explain this pattern, a regression equation was then derived for the 2-year flood in Region 1, but this time using the 11 verification sites. Again an acceptable pattern of residuals was obtained and a standard error of estimate of about 19 percent.

$$\log Q_2 = 0.81563 \log A - 0.32301 \qquad (6)$$

We can now compare estimates made by each of the regression equations at drainage basins roughly in the middle of each of the four ranges, to get:

A	50	140	260	1400	
Q_2	11.6	26.8	44.3	175	from calibration model
Q_2	8.9	23.0	40.6	191	from verification model

Thus, the calibration model when applied to the verification sites will in general give a preponderance of negative residuals in the lower three ranges of basin size, with a preponderance of positives in the highest range. This then raised the question of whether the regressions at (5) and (6) are significantly different. To answer this requires first pooling both calibration and verification stations, as would have been done anyway had the intention been simply to produce a regional regression equation for the 2-year flood in Region 1. Again, a good regression equation was obtained with an acceptable pattern of residuals and a standard error of estimate of 23%. The regression equation for the pooled stations is:

$$\log Q_2 = 0.86250 \log A - 0.45868 \qquad (7)$$

Table I. CELL TOTALS OF PERCENT DEVIATIONS.

	M_1			M_2			M_3		
	T_1	T_2	T_3	T_1	T_2	T_3	T_1	T_2	T_3
R_1	-130	-10	-28	-131	-158	-168	-121	-146	+188
R_2	-98	-125	-36	-97	-46	-26	-96	-7	+121
R_3	-64	-24	+92	-65	-5	+52	-44	+39	+241
R_4	+238	+253	+255	+239	+255	+299	+242	+346	+463
Column Totals	-54	94	283	-52	46	157	-19	232	1013

M_1 = Direct Regression M_2 = Index Flood M_3 = Regional Wakeby

$T_1 = Q_2 = T_2 = Q_{10}$ $T_3 = Q_{100}$

R_1 = less than 90 km^2 R_2 = from 90 to 185 km^2

R_3 = from 186 to 330 km^2 R_4 = from 331 to 2500 km^2

Table II. ANALYSIS OF VARIANCE.

Source of Variation	Sums of Squares	Degrees of Freedom	Mean Square	F	F (5%)
M, (method)	3 480	2	1 740	1.21	3.00
R, (range)	80 244	3	26 748	18.65***	2.60
T, (return period)	7 951	2	3 795	2.77	3.00
M * R	2 375	6	396	0.28	-
M * T	2 811	4	711	0.49	-
R * T	1 470	6	245	0.17	-
M * T * R	1 482	12	123	0.09	-
Error	464 669	324	1 434		

*** very highly significant indeed

There was no significant difference between regressions (5) and (6). When the regression at (7) was applied to the verification sites in Region 1 and the residuals calculated, the pattern of signs was:

- - - - - - - - - + + -

This is exactly the same pattern of residuals as obtained previously when the regression equation as derived from the calibration stations was applied to the verification stations.

Conclusions

A method has been presented whereby the statistical technique of Analysis of Variance can be applied to determine whether several different methods of preparing regional flood frequency studies are significantly different from each other. The technique also allows an assessment of the influences of drainage basin area and return period. From this study, there are no apparent interaction effects, nor do the methods or return periods have any significance, but the drainage basin size has a very highly significant effect. Paradoxically, because this effect is so extremely high, and because of the peculiar patterns of the residuals, the authors are reluctant to conclude that drainage basin size is in fact a hydrologically significant factor. There remains an irritating suspicion that there may have been a fault in the pseudo-random process of allotting the available stations to calibration and verification, or faults in the regression equations.

Finally, it is emphasized that the tentative conclusions reached are only applicable in Southern Ontario, and quite different conclusions could be reached in other parts of Canada or the world. This study merely presents a method by which hydrologists may assess the various methods of preparing regional flood frequency studies, and the effects thereon of drainage area and return period.

ACKNOWLEDGEMENTS

Acknowledgements are due to J.R. Wallis of the IBM Thomas J. Watson Research Center, Yorktown Heights, New York, who provided the program for the Regional Wakeby; to W.O. Thomas Jr., Surface Water Division, U.S. Geological Survey, who reviewed all the regression equations and dimensionless flood frequency curves; and to D.A. MacLeod of Systems Information Directorate, Environment Canada, who ran and interpreted the three-factor ANOVA.

REFERENCES

Dalrymple, T. 1960. Flood Frequency Analyses, U.S. Geological Survey Water-Supply Paper 1543-A.

Harvey, K.D., Condie, R. and Pilon, P.J. 1984. 'Regional Flood Frequency Analysis with the Three-Parameter Lognormal Distribution.' in: Proceedings, 7th Canadian Hydrotechnical Conference of the Canadian Society for Civil Engineering, Saskatoon, Saskatchewan.

Landwehr, J., Matalas, N.C. and Wallis, J.R. 1979. 'Estimation of Parameters and Quantiles of Wakeby Distributions', Part 2. Water Resources Research, Vol. 15:6, pp. 1373-1379.

Moroney, M.J. 1964. Facts from Figures, Penguin Books Ltd., Harmondsworth, Middlesex, England, pp. 371-457.

U.S. Water Resources Council. 1981. 'Estimating Peak Flow Frequencies for Natural Ungauged Watersheds. A Proposed Nationwide Test'. 2120 L Street, N.W., Washington, D.C. 20037.

Wallis, J.R. 1980. 'Risk and Uncertainties in the Evaluation of Flood Events for the Design of Hydraulic Structures'. Seminar on Extreme Hydrologic Events: Floods and Droughts, Centro di Cultura Scientifica, "E. Majorana", Erice, Italy, March 22-25.

COMPARISON OF FLOOD FREQUENCY CURVES FOR MANY DIFFERENT REGIONS OF THE WORLD

F.A.K. Farquharson,
Chief Hydrologist,
Overseas Consultancy Group,
Institute of Hydrology,
Wallingford, UK.
C.S. Green,
Hydrologist,
Institute of Hydrology.
J.R. Meigh,
Hydrologist,
Institute of Hydrology, and
J.V. Sutcliffe,
Hydrological Consultant
Goring on Thames
formerly Institute of Hydrology

ABSTRACT. Annual maximum flood data have been collected from 1121 gauging stations in 70 different countries and have been analysed in a uniform manner to allow comparison of results between stations and between countries and larger regions. The data set comprises about 31,000 station years. For each station, the annual maximum flood series was fitted to the general extreme value distribution using the method of probability weighted moments. The derived station flood frequency curves were made dimensionless by dividing the annual maximum flood series by the mean annual flood. Where data were sufficient, the station growth curves for each country or region were combined to produce a flood frequency curve intended to be representative of floods in that region.

It is suggested that the flood frequency curves produced by this study could be used by engineers and hydrologists during planning and pre-feasibility studies where comparable flood estimates may be required at a large number of sites and in areas where data are scarce. If used with care and with understanding of the underlying assumptions implicit in this study, the regional flood frequency curves may be sufficient for design flood estimation of very small schemes or structures.

The study reported on in this paper does not yet comprise a full flood estimation procedure. To estimate the flood of any specific return period at a site, an estimate of the mean annual flood is required to complement the flood frequency curves presented here. Such an estimate of the mean annual flood must be derived either from local data or from comparison with adjacent gauged catchments.

V. P. Singh (ed.), Regional Flood Frequency Analysis, 223–256.

1. INTRODUCTION

The study upon which this paper is based was undertaken at the Institute of Hydrology, Wallingford, U.K., with the aim of examining and classifying the characteristics of floods in as many countries and from as wide a range of climates as possible. To date, annual maximum flood series and basic catchment data have been collected from 1121 gauging stations in 70 countries, totalling about 31,000 station years. Data from all stations have been analysed in a uniform manner to allow comparison of results between stations and between countries and larger regions.

The annual maximum floods were made non-dimensional by dividing by the mean annual flood and plotted in the form of a flood frequency curve or growth curve for each station. Where data were sufficient, the station growth curves were combined to produce a growth curve intended to be representative of floods in that region.

2. DATA SOURCES AND QUALITY

The data assembled for this study consist of annual maximum flood series and catchment characteristics for each gauging station. The density of coverage varies widely in different countries, but can be broadly classified in two groups:

(1) Relatively sparse coverage. These data have in general been obtained from the two catalogues of very large floods published by UNESCO (1976) and by IAHS (1984). They vary from some countries where there are only one or two stations to others where there is better coverageAn example of the first type is Bolivia which has only one station with 30 years of record, and of the second are Japan which has 15 stations with a total of 572 years and the USSR with 47 stations and 2631 years of data. The data in this category came from all parts of the world, and because of the large amount and our lack of any substantial knowledge of the areas concerned, we have been unable to check the individual records. Therefore the data have been accepted as given, except where obvious typographic errors have been noticed.

(2) More detailed coverage. These data are usually all the available data from a country or from a region within a country. In some cases, for example Denmark and Thailand, the sources of information were hydrological year books published by the relevant authorities in those countries. These records have been subjected to some scrutiny but again thorough checking of the data in collaboration with the gauging authority was not possible.

In other cases, for example Kerala in India, data have been gained through contacts in the country of origin, and doubtful points have been checked by correspondence with the person who supplied the data.

Data have also come from within the Institute of Hydrology and have been collected over the years for various projects such as regional flood studies and water resources studies. These data come from a wide range of developing countries and are usually concentrated in particular regions of those countries. They have been subjected to careful checking during the course of the original projects.

The amounts and sources of data in each country or region are listed in Table I, and the worldwide distribution of data can be seen in Figure 1. Where data are from only part of a country the approximate position and size of the region is indicated by the area of stippling on Figure 1 and sparse coverage is indicated by dots for individual stations. A greater idea of the density of coverage can be obtained by examining the figures for number of stations and number of years of data.

3. METHODOLOGY

For each station the series of annual maximum flood peaks was processed to produce a non-dimensional flood frequency or growth curve using the method of probability weighted moments, PWM, proposed by Hosking et al (1984). The analysis was undertaken on non-dimensional flood series by dividing by the mean annual flood to enable comparison of the derived flood frequency curves for all stations within a region.

Regional flood frequency curves were produced by pooling all individual station curves within a country or region and fitting the data to a general extreme value (GEV) distribution as described in the UK Flood Studies Report (NERC, 1975). This has the form:

$$F(q) = \exp\{-[1 - \frac{k(q-u)}{\alpha}]^{1/k}\} \qquad \text{if } k \neq 0,$$

and if k = 0 this reduces to the Gumbel or EV1 distribution:

$$F(q) = \exp\{- \exp[- \frac{(q-u)}{\alpha}]\}$$

Introducing the Gumbel reduced variate, y, we obtain:

$$q = u + \frac{\alpha}{k}(1-e^{-ky}) \qquad \text{for } k \neq 0$$

or

$$q = u + \alpha y \qquad \text{for } k = 0$$

The GEV distribution was chosen because in practice, it has been found to be of general applicability. It is a three parameter distribution which allows the full range of flood regimes to be adequately modelled.

3.1 Choice of Regions

The flood growth curves presented in this report are for regions which are considered to be reasonably homogeneous. This means that the growth curve of an ungauged catchment within any particular region can be assumed to be reasonably close to the regional curve, provided its catchment characteristics are not markedly different from those of the stations used to derive the curve. The method for deciding on homogeneity was to compare the individual station growth curves and inspect catchment characteristics such as AREA, average annual rainfall (AAR) and climatic regime. It was expected that there would be a moderate spread of station curves within a homogeneous region, especially as many of the records are short. Where the spread was not too great and other conditions were reasonably uniform a region was accepted as homogeneous. One or two station curves may have departed considerably from the general line of the majority. Where there seemed to be no good reason, such as noticeably different AREA or climate, this has been taken to be the result of chance and the station has been accepted as part of the region. The averaging technique of pooling many curves to form a region means that one or two untypical stations will not greatly affect the result.

Where it appeared that conditions were variable and the stations could be divided into two or more groups this was done. The division was done on the basis of AAR or AREA, or on more general characteristics such as climatic regime where the catchments were in a mountainous area or the flood behaviour was dominated by snowmelt. Having divided the stations, the resulting sub-regions were inspected for homogeneity in the same way as before. All this work was done using curves fitted by PWM. The method was subjective, depending on the interpretation given to words such as reasonable and moderate. Although work has been done to develop statistical tests to show whether a group of stations is homogeneous or whether two groups are significantly different (S.E. Wiltshire, personal communication and Wiltshire, 1985), no satisfactory objective techniques of grouping catchments has yet been established. Much of the present data set is composed of short records, which are possibly unreliably observed, and flood response is controlled by numerous variables which are not measured, in addition to the known catchment characteristics. In view of this, it is felt that the rigorous statistical test of regionality is inappropriate for this study and, therefore, that subjective methods must remain.

4. RESULTS

4.1 Pooling catchments by country

Initially work was done treating each country as a unit. Where the available data were concentrated in a particular part of a country, the results should be taken to apply only to this region. The location of all the regions in the study are shown in Figure 2. If the region has been further divided into sub-regions, according to the method of section 3.1, results have been given both for the sub-region so found, and for the whole region. In a number of cases two or more countries have been treated as a unit. For example, this was done for Togo and Benin because they are small adjoining countries of similar characteristics, and for Saudi Arabia and Yemen, because there is only one station in Yemen, close to the border and again of similar characteristics to the rest.

For convenience of presentation the results have been grouped by continent. Results are summarised in Tables II to V and the flood frequency curves are plotted in Figures 3 to 6. The Tables give the fitted GEV parameters and predictions for return periods of 50, 100 and 500 years obtained using the PWM method. The Tables also give, in the final column, an indication of the expected quality of each predicted curve. This is based on the amount of data that is available in each region and has been derived by calculating the number of stations per 1000 square kilometres in each country or region. Only two regions, Denmark and Kerala State in India, have been shown as 'Very Good', having more than 1.4 stations per 1000 km^2. Growth curves in 'Good' regions have been derived with nearly all the possible data in that region and the results are likely to be representative of nearly every catchment in the region. 'Fair' indicates that the stations were more scattered and that there are parts within the region of which the growth curve may not be typical, while 'Poor' regions have only been included to allow comparisons with other regions.

4.2 Discussion of results

It seems that the most important factor in determining the shape of regional growth curves is climate. Generally extreme climates have extreme growth curves. Arid and semi-arid regions have the steepest curves (Saudi Arabia has the steepest of any in this study) and tropical rain forest tends to have the flattest curves (Congo & Zaire and the surrounding area has the flattest curve of any). For arid and semi-arid regions this result was consistent: the growth curves for Iran, Jordan, North Africa, Saudi Arabia and the semi-arid parts of South Africa, the U.S.A. and U.S.S.R. were all similar. However, in the case of tropical rain forest, while the curves were often very flat this was not always so. Congo & Zaire and Guyana have very flat curves, but Indonesia and Papua New Guinea are much less so. It is clear that, while climate is by far the strongest factor in some cases, in others, factors such as topography, soils or vegetation cover may be as important or more important. Other climates tend to have growth curves lying betweeen these extremes above. However, while they are not so easily classified, a few general tendencies are mentioned below.

The second important factor in determining growth curve shape is catchment area (AREA). In any particular region large catchments very often have flatter growth curves than small catchments. This was particularly so in Europe, but was found less often elsewhere. However, in some regions this tendency may not have been discovered because there are few gauged catchments available or because the range of AREA available is insufficiently wide. The opposite tendency: for large catchments to have steeper growth curves than small was found in only one case (Thailand and Malaysia, Malay peninsula). Unfortunately there is only a small data set in this region and little other information was available to substantiate the result.

The regional growth curves found for Europe are spread over a wide range. There is some tendency for southern parts to have steeper growth curves than those in the north. For instance, Bulgaria and Romania are both almost as steep as the curves for semi-arid regions, while Denmark and Sweden & Norway are rather flat. The curves for the sub-arctic parts of the U.S.S.R. and north western U.S.A. and Canada are very similar to that for Sweden and Norway, indicating that there may be a homogeneous northern region where the flood producing mechanism is the annual snowmelt. The available results indicate that flood behaviour is similar in all such northern areas.

In Africa there are a number of results from regions which have a tropical savannah climate. In West Africa the results for Guinea & Mali, Ivory Coast & Ghana and Togo & Benin are fairly similar flat curves. However, other African regions with similar climate, for example Kenya or Malawi, have much steeper curves. Another climate which does not produce consistent growth curves is the tropical monsoon. We have examples of this in Sri Lanka and in Kerala in India, which also both have similar topography but have very different growth curves.

From this general discussion it is clear that, while there are some general tendencies due to climate, each region must be considered on its merits.

4.3 Larger Regions

Thus it appears that two regions could be formulated in which the climate is the dominant control on growth curve shape; these are an arid and semi-arid region and a sub-arctic or tundra region. Similarly there are two geographical regions where there is good areal coverage of data and which it may be possible to represent by more general growth curves. These are eastern Europe and West Africa.

As the main interest of this study was in developing countries, where there tends to be a shortage of data, the derivation of regional growth curves has only been attemped for the arid and semi-arid region and the West African region. The results are summarised in Table VI and Figure 7 in the same manner as for the other regions.

(i) Arid and semi-arid region

The location of stations used in the arid and semi-arid region is marked on Figure 2. As mentioned in Section 4.1, individual semi-arid regions all have similar growth curves; therefore catchments have been combined from suitable areas in all parts of the world. A strict limit on average annual rainfall, AAR<600mm, has been used to select them, and this has been supplemented by other criteria. For instance, temperate and arctic areas often have very low rainfall but these are excluded. Such rivers as the Ob and Lena in northern U.S.S.R. and the Mackenzie in Canada have annual rainfall in the range 200 to 400mm, yet their flood producing mechanism is quite different from that in hot semi-arid areas.

If the distribution of stations shown in Figure 2 is examined it will be observed that a number of catchments are omitted which are surrounded by other catchments which have been included in the region. Such a catchment is the Syr Darya at Tiumen-Aryk in the U.S.S.R, which flows from the Tien Shan mountains on the U.S.S.R/Chinese border westwards to the Aral Sea. The catchment average rainfall is 200mm, but the mountains where the Syr Darya rises have annual rainfall of about 1000mm, while around the Aral Sea rainfall falls to only about 50mm. The AAR of 200mm is misleading because the flood response is likely to be controlled by the higher rainfall, mountainous, upper reaches than by the arid climate of much of the catchment. Other catchments have been excluded for similar reasons.

While an attempt was made to follow the selection criteria rigorously, some catchments near the edge of semi-arid regions may have been included erroneously. Through working from small scale maps, it is not always possible to decide to which climate type a catchment should belong. The selected catchments all have similar steep station growth curves. They were examined to see if sub-regions could be formed on the basis of AREA or AAR. Dividing on AREA, no distinguishable groups were found: even very large catchments with AREA>100,000 km2 did not have markedly less steep growth curves. Some difference was noticed between groups when divided on rainfall. Very dry catchments (AAR<150mm) had a slightly less steep curve than those at the wetter end of the range (AAR 500 to 600mm). However, it was felt this difference was not great enough to justify presenting separate curves, and a single curve has been presented for the region. There are substantial data for this regional growth curve and it is possible that it would be suitable for application to other hot semi-arid regions, provided all the criteria were carefully checked.

(ii) West African region

The West African region has been formed on a geographical basis as opposed to the strict climatic and rainfall definition used for the arid and semi-arid region. The area used extends from Guinea in the west to the Gongola basin in Nigeria and the two catchments in Chad and northern Cameroon in the east. It includes most of West Africa to the south of the semi-arid Sahel region. The region is marked on Figure 2. The climate of the region is tropical savannah, but one or two catchments in the south western part of the region may fall into the tropical monsoon zone. Also a number of gauging stations are to the north of the climate zone, but the great majority of the catchment area of each is within the zone.

The catchments used include all those used in the four smaller regional groupings, as derived in Section 4.1. There are three additional catchments: the Niger river at Niamey and Onitsha and the Senegal river at Bakel. The regional growth curves for the four smaller regions are all fairly similar. In each case there was a moderate spread between the individual station curves, but no significant sub-regions were found, despite the wide ranges of AAR and AREA. Since the larger region has similar climate combining all data within it is considered valid. The larger groupings allowed the search for sub-regional groups to be repeated and because the data set is larger it is more likely to be a useful exercise. The catchments were examined for sub-regions on the basis of AAR and AREA. Using AAR there was some separation into groups, but this was small despite the wide range of AAR in the region (750 to 2630mm). Using AREA the separation of the sub-regional curves was better, but still not large. However, it was observed that while the behaviour of very large and medium catchments was similar and did not depend on rainfall, that of small catchments did. Small catchments with high rainfall had growth curves similar to the larger catchments but those which were drier had steeper curves. On this basis two sub-regions were formed: one having AREA <10,000 km^2 when AAR>1500mm, and the other having AREA<10,000 km^2 when AAR>1500mm and AREA>10,000 km^2 for any AAR. It is notable that the small, dry sub-region has a steeper growth curve than any of the original four regional curves in West Africa. This may indicate that in those four regions, AREA would have produced differences in the growth curves, but these were not found, probably because of the relatively small data sets. The larger region reveals this difference. However, there is still a shortage of data from small catchments in the region, especially those smaller than 1000 km^2, and more data would be required to clarify the flood characteristics of the region. This example is described in some detail to illustrate the sort of approach required when adequate data are available.

5. CONCLUSIONS

The regional flood frequency curves presented in this paper should enable hydrologists and engineers to compare the flood regimes of different countries of the world. This facility should be useful to those engaged in estimating floods at ungauged sites or in areas with insufficient or poor quality data. It is anticipated that the presented results could be used for planning and pre-feasibility studies and possibly for the design of small schemes and small structures.

However, the study presented here is at present incomplete. The Institute of Hydrology is presently undertaking a complementary study to relate the mean annual flood to climatic and physiographic characteristics of the catchments studied. Results of this second study will complement the results presented in this present paper.

The authors would like to encourage individuals and organisations to submit data sets of annual maximum flood series to the Institute of Hydrology for analysis. Full results of the analysis undertaken will be provided for the originators of any data sets.

ACKNOWLEDGEMENTS

The authors of this report would like to thank Dr. J S G McCulloch, Director of the Institute of Hydrology, for permission to publish the paper. They would further like to thank the Overseas Development Administration who funded the bulk of the work. Our colleague, Mr M A

Beran, was a constant source of helpful advice and finally a great debt of gratitude is due to Mrs V J Bronsdon for entering so much of the data on to the computer system and for production of all final report diagrams.

REFERENCES

Hosking, J.R.M., Wallis, J.R., and Wood, E.F., 1984. 'Estimation of the generalized extreme value distribution by the method of probability weighted moments.' *Technometrics,* **vol 27**, pp251-261.

Natural Environment Research Council, 1975. *Flood Studies Report, 5 volumes*

Rodier, J.A. and Roche, M. 1984. 'World Catalogue of Maximum Observed Floods', *International Association of Scientific Hydrology (IAHS publication no.143)* publication no. 143.

UNESCO, 1976. 'World Catalogue of Very Large Floods', *The UNESCO press, Paris.*

Wiltshire, S.E., 1985. 'Grouping basins for regional flood frequency analysis'. *Hydrological Sciences Journal,* **vol 30**, pp 151-159.

TABLE I LIST OF COUNTRIES AND DATA SOURCE

Country	Data source	No. of stations	No. of years	Key to Figure 1
Algeria	IAHS	3	94	ALG.
Australia	IAHS, CORR	28	1039	AUS.
Bangladesh			- see India	BAN.
Benin			- see Togo	BEN.
Bolivia	IAHS	1	30	BOL.
Botswana	IH	4	38	BOTS.
Brazil	IH, IAHS	60	1340	BRA.
Bulgaria	UNESCO	5	272	BUL.
Cameroon	IAHS	2	84	CAM.
Canada	IAHS, UNESCO	20	864	CAN.
Central African Republic	IAHS	1	77	CAR.
Chad	IAHS	1	40	CHAD.
China	IAHS	4	214	CHINA.
Colombia	IAHS	1	40	COL.
Conga & Zaire	IAHS, UNESCO	6	154	CON.& ZAI.
Costa Rica	IAHS	1	26	COS. RIC.
Cuba	IAHS	1	19	CUB.
Czechoslovakia	IAHS, UNESCO	16	774	CZE.
Denmark	CORR	62	1946	DEN.
France	IAHS	3	456	FRA.
French Guyana	IAHS	1	29	FRE.GUY.
Gabon	IAHS	1	49	GAB.
Germany	IAHS	1	40	GER.
Ghana	IAHS, UNESCO	7	185	GHA.
Guatemala	IAHS	1	24	GUA.
Guinea	IH	17	271	GUI.
Guyana	UNESCO, IH	8	203	GUY.
Hungary	IAHS, UNESCO	10	248	HUN.
India & Bangladesh	CORR, IH, IAHS, UNESCO	97	1582	IND.& BAN.
Indonesia	IH	95	1009	INDO.
Iran	IH	25	343	IRAN.
Iraq	IAHS	1	35	IRAQ.
Italy	IAHS, UNESCO	22	744	ITA.
Ivory Coast	IAHS, UNESCO	10	222	IV. COAST
Japan	IAHS, UNESCO	15	572	JAP.
Jordan	UNESCO	6	57	JOR.
Kenya	IH, CORR	19	338	KEN.
Korea	IH, IAHS	24	542	KOR.
Madagascar	IAHS	4	92	MAD.
Malawi	IH	28	509	MAL.
Malaysia	IAHS, UNESCO	4	110	MALAS.
Mali	IAHS	1	73	MALI.
Mexico	IAHS	2	77	MEX.
Morocco	IAHS, UNESCO	5	148	MOR.
Netherlands	IAHS	1	76	NET.
Niger	IAHS	2	57	NIG.
Nigeria	IH, IAHS	8	117	NIGERIA
Norway	IAHS	2	125	NOR.
Pakistan	IAHS	1	111	PAK.
Papua New Guinea	CORR	50	450	PAP.NEW GUIN.
Poland	UNESCO	11	364	POL.
Romania	UNESCO	14	420	ROM.
Saudi Arabia & Yemen	IH	30	378	S.ARA & YEM.
Senegal	IAHS	1	75	SEN.
South Africa	CORR, IAHS	99	3846	S.AFRICA.
Spain	IAHS	5	221	SPA.
Sri Lanka	IH, UNESCOA	13	280	SRI LANKA
Sudan	IH	4	215	SUD.
Sweden	IAHS, UNESCO	20	1134	SWE.
Thailand	CORR	121	2186	THAI.
Togo & Benin	IH, IAHS	19	443	TOG. & BEN.
Tunisia	IAHS	2	103	TUN.
Turkey	IAHS	1	44	TUR.
Upper Volta	IAHS	1	24	U. VOL.
Uruguay	IAHS	1	83	URU.
U.S.A.	IAHS, UNESCO	43	2445	U.S.A.
U.S.S.R.	IAHS, UNESCO	47	2631	USSR.
Yemen		see - Saudi Arabia		
Yugoslavia	IAHS	2	100	YUG.
Zaire			- see Congo	
Total	70 countries	1121	30,937	

Key to data sources
IAHS - IAHS (1976)
UNESCO - UNESCO (1984)
IH - data from IH projects
CORR - data published by relevant authority, or supplied to IH by individuals or organizations through correspondence

TABLE II SUMMARY OF RESULTS - AMERICAS

Country/Region	No. of stations	No. of years	GEV parameters			Predicted Floods			Quality
			u	α	k	q_{50}	q_{100}	q_{500}	
Brazil (Rio Grande do Sul)	57	1209	0.830	0.348	+0.0959	1.96	2.12	2.46	GOOD
Guyana & French Guyana[2]	9	232	0.915	0.222	+0.2369	1.48	1.54	1.64	FAIR
U.S.A. & Canada (North West)[2]	15	684	0.857	0.254	+0.0163	1.82	1.98	2.36	POOR
U.S.A. (South West) & Mexico[2]	14	663	0.478	0.341	-0.4966	4.56	6.53	14.82	POOR
U.S.A. & Canada (Central & East)[2]	31	1776	0.759	0.331	-0.1327	2.45	2.86	3.96	
U.S.A. & Canada (Central & East: Area <30,000 km^2)	21	1068	0.738	0.334	-0.1742	2.61	3.10	4.49	POOR
U.S.A. & Canada (Central & East: Area >30,000 km^2)	10	708	0.792	0.327	-0.0572	2.22	2.51	3.23	

[2] Region for which data is available only covers part of country. See Figure 2

Note that in all the following Tables; $Q_T = \bar{Q} \times q_T$

TABLE III SUMMARY OF RESULTS - EUROPE & U.S.S.R.

Country/Region	No. of stations	No. of years	GEV parameters			Predicted Floods			Quality
			u	α	k	q_{50}	q_{100}	q_{500}	
Bulgaria	5	272	0.667	0.390	-0.2202	3.08	3.78	5.86	FAIR
Czechoslovakia	16	774	0.717	0.356	-0.1821	2.74	3.28	4.82	
Czechoslovakia (Area< 250 km^2)	6	224	0.640	0.352	-0.3156	3.34	4.28	7.45	GOOD
Czechoslovakia (Area> 250 km^2)	10	550	0.751	0.358	-0.1073	2.48	2.88	3.91	
Denmark	62	1946	0.814	0.302	-0.0368	2.08	2.33	2.92	
Denmark (Area <20 km^2)	11	271	0.756	0.356	-0.0997	2.46	2.84	3.82	VERY
Denmark (Area:20-200 km^2	34	1047	0.803	0.314	-0.0476	2.15	2.42	3.07	GOOD
Denmark (Area >200 km^2)	17	628	0.859	0.258	+0.0314	1.81	1.96	2.32	
France, Germany & Netherlands[2]	5	572	0.875	0.279	+0.1487	1.70	1.81	2.01	POOR
Hungary & Yugoslavia[2]	12	348	0.793	0.281	-0.1389	2.25	2.61	3.57	
Hungary & Yugoslavia (Group 1)[1]	5	120	0.657	0.348	-0.2962	3.22	4.07	6.89	GOOD

TABLE III cont.

Country/Region	No. of stations	No. of years	GEV parameters			Predicted Floods			Quality
			u	α	k	q_{50}	q_{100}	q_{500}	
Hungary & Yugoslavia (Group 2)[1]	7	228	0.871	0.245	+0.0544	1.73	1.87	2.17	
Italy	22	744	0.785	0.348	-0.0408	2.26	2.54	3.24	
Italy (Area < 100 km^2)	6	149	0.721	0.321	-0.2318	2.76	3.36	5.18	FAIR
Italy (Area:100-5000 km^2)	10	340	0.770	0.364	-0.0524	2.35	2.66	3.45	
Italy (Area > 5000 km^2)	6	255	0.849	0.339	+0.1511	1.85	1.97	2.21	
Poland	11	364	0.734	0.373	-0.1205	2.59	3.03	4.19	
Poland (Area <10,000 km^2)	6	150	0.660	0.433	-0.1756	3.09	3.73	5.54	FAIR
Poland (Area >10,000 km^2)	5	214	0.787	0.331	-0.0614	2.25	2.55	3.29	
Romania	13	289	0.669	0.336	-0.2962	3.14	3.97	6.68	
Romania (Area <5000 km^2)	7	118	0.585	0.364	-0.3682	3.75	4.97	9.34	FAIR
Romania (Area >5000 km^2)	6	171	0.729	0.318	-0.2216	2.70	3.27	4.98	
Spain[2]	5	221	0.730	0.402	-0.0860	2.60	3.00	4.04	POOR

TABLE III cont.

Country/Region	No. of stations	No. of years	GEV parameters			Predicted Floods			Quality
			u	α_{50}	k	q	q_{100}	q_{500}	
Sweden & Norway[2]	22	1259	0.853	0.258	+0.0069	1.85	2.02	2.42	
Sweden & Norway (Area <5000 km²)	10	485	0.812	0.296	−0.0532	2.10	2.36	3.00	FAIR
Sweden & Norway (Area >5000 km²)	12	774	0.879	0.233	+0.0616	1.69	1.81	2.08	
U.S.S.R.	47	2631	0.783	0.306	−0.1186	2.30	2.65	3.59	
U.S.S.R. (AAR <500mm, Area <200,000 km²)	12	483	0.706	0.374	−0.1771	2.81	3.36	4.94	
U.S.S.R. (Area >200,000 km²)	11	625	0.892	0.228	0.1131	1.61	1.71	1.91	POOR
U.S.S.R. (Latitude >60°N)	12	502	0.849	0.254	−0.0177	1.88	2.07	2.52	
U.S.S.R. (Uncategorised)	0	1339	0.756	0.321	−0.1566	2.49	2.92	4.14	

[1] Hungary & Yugoslavia (Group 1) is: Area <10,000 km², and Area 10,000 − 50,000 km² when AAR <800 mm
Hungary & Yugoslavia (Group 2) is: Area >50,000 km², and Area 10,000 − 50,000 km² when AAR >800 mm

[2] Region for which data are available only covers part of country. See Figure 2

TABLE IV SUMMARY OF RESULTS - AFRICA

Country/Region	No. of stations	No. of years	GEV parameters			Predicted Floods			Quality
			u	α	k	q_{50}	q_{100}	q_{500}	
Algeria, Morocco & Tunisia[2]	10	345	0.544	0.418	−0.3454	4.00	5.27	9.70	POOR
Congo & Zaire, Cameroon (South), C.A.R., Gabon[2]	9	316	0.937	0.161	+0.2224	1.36	1.40	1.48	POOR
Guinea & Mali[2]	18	344	0.906	0.234	+0.2131	1.53	1.59	1.71	FAIR
Ivory Coast, Ghana & Upper Volta[2]	18	431	0.846	0.360	+0.1733	1.87	1.99	2.21	FAIR
Kenya[2]	19	338	0.651	0.459	−0.1577	3.13	3.75	5.50	GOOD
Madagascar[2]	4	92	0.642	0.382	−0.2699	3.28	4.12	6.80	POOR
Malawi	28	509	0.655	0.422	−0.1968	3.13	3.81	5.80	GOOD
Nigeria (Gongola basin), Chad, Cameroon (North)[2]	9	181	0.845	0.328	+0.1155	1.87	2.01	2.30	FAIR
South Africa & Botswana[2]	103	3884	0.485	0.422	−0.3990	4.45	6.06	12.06	
South Africa & Botswana (AAR <1250 mm)	95	3651	0.469	0.428	−0.4065	4.56	6.25	12.58	GOOD
South Africa & Botswana (AAR >1250 mm)	8	233	0.733	0.343	−0.1710	2.64	3.13	4.53	

TABLE IV cont.

Country/Region	No. of stations	No. of years	GEV parameters			Predicted Floods			Quality
			u	α	k	q_{50}	q_{100}	q_{500}	
Sudan[2]	4	215	0.903	0.215	+0.1422	1.55	1.63	1.79	POOR
Togo & Benin	19	443	0.818	0.413	+0.1579	2.02	2.17	2.45	GOOD
			(0.829	0.425	+0.2342	1.92	2.03	2.22)	

[2] Region for which data is available only covers part of country. See Figure 2

TABLE V SUMMARY OF RESULTS - ASIA & AUSTRALASIA

Country/Region	No. of stations	No. of years	GEV parameters			Predicted Floods			Quality
			u	α	k	q_{50}	q_{100}	q_{500}	
Australia[2]	28	1039	0.515	0.487	-0.3013	4.13	5.36	9.41	POOR
Australia (Goulburn basin)	20	516	0.620	0.490	-0.1686	3.32	4.03	6.00	FAIR
Australia (Eastern catchments)	8	523	0.419	0.486	-0.3899	4.88	6.66	13.23	POOR
India & Bangladesh[2]	21	411	0.817	0.311	-0.0098	2.05	2.28	2.81	
India & Bangladesh (Area <100,000 km^2)	14	207	0.788	0.349	-0.0290	2.23	2.50	3.16	POOR
India & Bangladesh (Area >100,000 km^2)	7	204	0.847	0.273	+0.0164	1.88	2.06	2.46	
India (Kerala)	76	1171	0.747	0.370	-0.0991	2.51	2.90	3.92	VERY GOOD
Indonesia (Java & Sumatra)	95	1009	0.837	0.266	-0.0340	1.95	2.16	2.68	
Indonesia (Java & Sumatra): Area <600 km^2)	47	541	0.812	0.290	-0.0671	2.10	2.37	3.05	GOOD

TABLE V cont.

Country/Region	No. of stations	No. of years	GEV parameters			Predicted Floods			Quality
			u	α	k	q_{50}	q_{100}	q_{500}	
Indonesia (Java & Sumatra: Area >600 km^2)	48	468	0.866	0.239	+0.0175	1.77	1.92	2.27	
Iran[2]	25	343	0.590	0.394	−0.3235	3.67	4.76	8.46	
Iran (Area <7500 km^2)	16	198	0.559	0.376	−0.3806	3.94	5.27	10.10	FAIR
Iran (Area >7500 km^2)	9	145	0.636	0.419	−0.2307	3.29	4.07	6.43	
Japan	15	572	0.709	0.366	−0.1822	2.79	3.34	4.93	
Japan (AAR <1950 mm)	9	329	0.650	0.384	−0.2553	3.22	4.01	6.50	FAIR
Japan (AAR >1950 mm)	6	243	0.794	0.341	−0.0267	2.20	2.46	3.10	
Jordan[2]	6	57	0.525	0.474	−0.3039	4.07	5.28	9.28	GOOD
Korea[2]	24	542	0.775	0.373	−0.0256	2.31	2.60	3.29	GOOD
Papua New Guinea	50	450	0.818	0.280	−0.0682	2.07	2.33	2.98	GOOD

TABLE V cont.

Country/Region	No. of stations	No. of years	GEV parameters			Predicted Floods			Quality
			u	α	k	q_{50}	q_{100}	q_{500}	
Saudi Arabia & Yemen[2]	30	378	0.427	0.459	−0.4094	4.84	6.67	13.57	FAIR
Sri Lanka	13	280	0.656	0.328	−0.3275	3.25	4.17	7.32	GOOD
Thailand (excl. S. peninsula)	109	2012	0.762	0.343	−0.1049	2.42	2.79	3.77	
Thailand (alt. <100 m, Area <15,000 km^2)	18	290	0.828	0.310	+0.0233	1.98	2.18	2.62	
Thailand (alt <100 m, Area >15,000 km^2)	13	284	0.919	0.243	+0.3128	1.47	1.51	1.58	GOOD
Thailand (alt. >100 m, Area <5000 km^2)	54	942	0.691	0.388	−0.1835	2.90	3.50	5.19	
Thailand (alt >100 m, Area >5000 km^2)	24	496	0.780	0.330	−0.0829	2.30	2.63	3.47	
Thailand & Malaysia (Malay peninsula)[2]	16	284	0.708	0.352	−0.2050	2.81	3.40	5.13	
Thailand & Mal. (Malay peninsula: Area <2000km^2)	11	165	0.740	0.355	−0.1379	2.57	3.02	4.23	FAIR

TABLE V cont.

Country/Region	No. of stations	No. of years	GEV parameters			Predicted Floods			Quality
			u	α	k	q_{50}	q_{100}	q_{500}	
Thailand & Mal. (Malay peninsula: Area >2000km²)	5	119	0.668	0.348	−0.2804	3.13	3.93	6.51	

[2] Region for which data is available only covers part of country. See Figure 2.

TABLE VI SUMMARY OF RESULTS - LARGER REGIONS

Country/Region	No. of stations	No. of years	GEV parameters			Predicted Floods			Quality
			u	α	k	q_{50}	q_{100}	q_{500}	
Arid & Semi-arid	116	2912	0.475	0.421	-0.4087	4.52	6.19	12.50	
West Africa	67	1537	0.855	0.334	+0.1672	1.81	1.93	2.15	
*West Africa (Group 1)	17	319	0.804	0.409	+0.1074	2.11	2.29	2.66	
*West Africa (Group 2)	50	1218	0.869	0.314	+0.1892	1.74	1.83	2.02	

* West Africa (Group 1) is: Area <10,000 km^2 when AAR <1500 mm
West Africa (Group 2) is: Area >10,000 km^2, and Area < 10,000 km^2 when AAR >1500 mm

Distribution of data

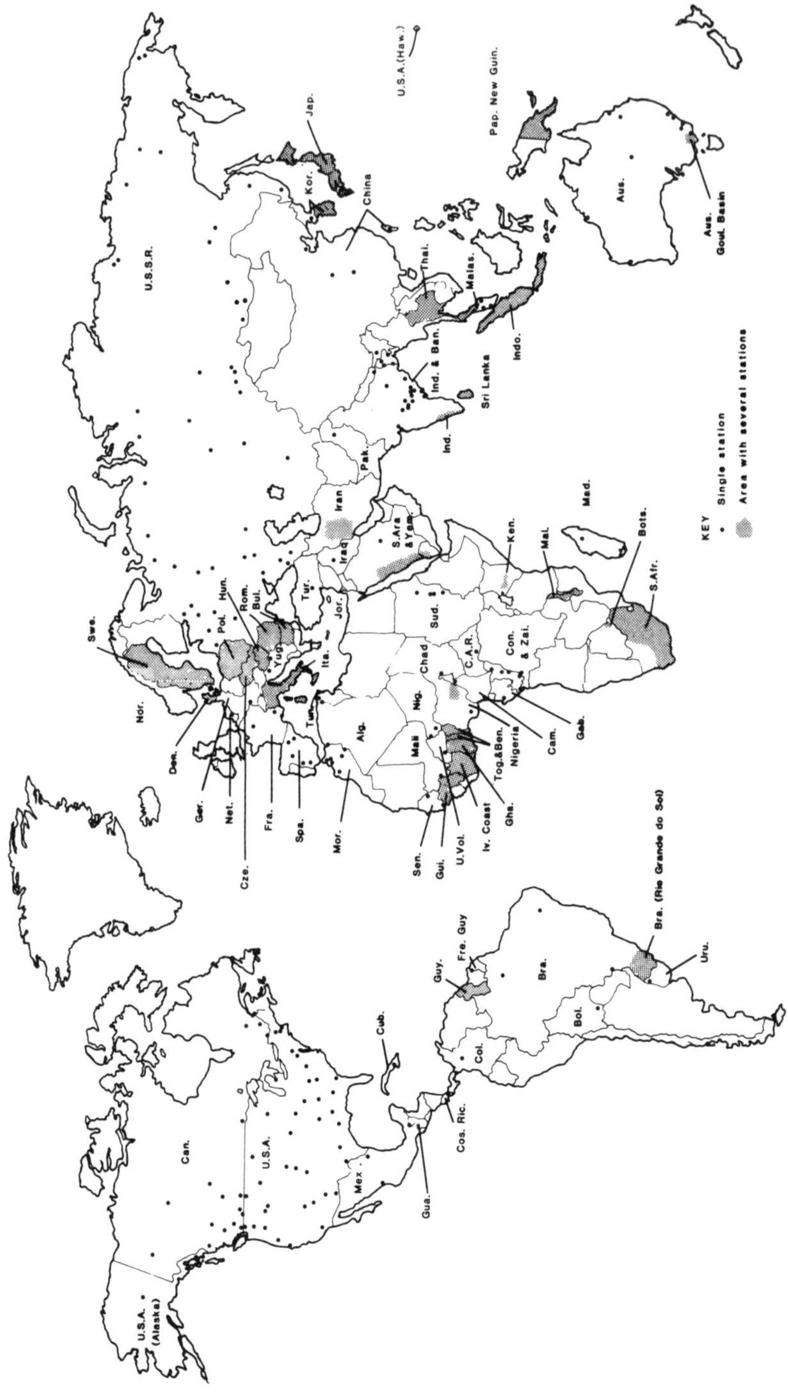

Figure 1

Regional groups studied

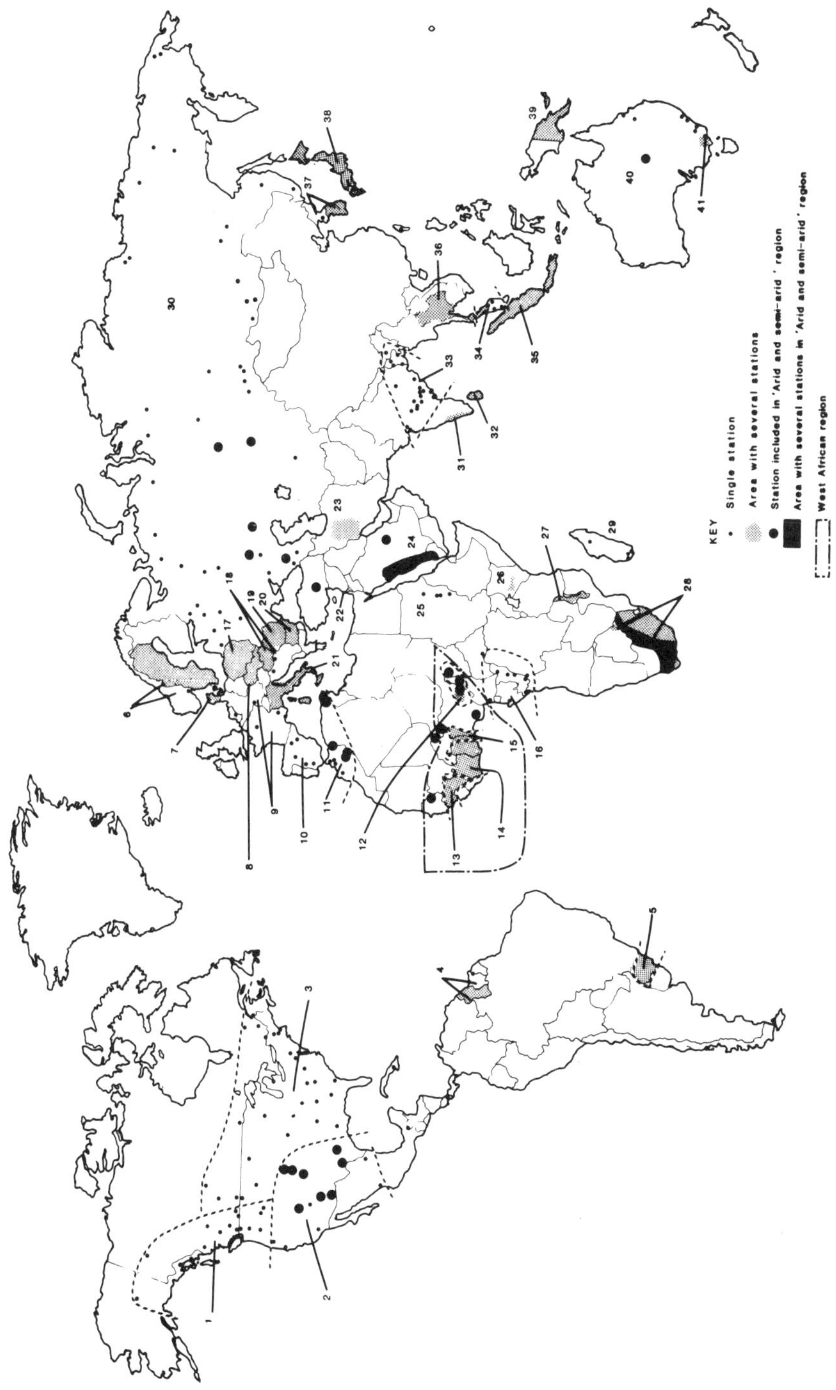

Figure 2

KEY TO FIGURE 2

KEY	REGION	NO.STNS/STATION YEARS
1	U.S.A. & Canada (North West)	15/684
2	U.S.A. (South West) & Mexico	14/663
3	U.S.A. & Canada (Central & South East)	31/1776
4	Guyana & French Guyana	9/232
5	Brazil (Rio Grande Do Sol)	57/1209
6	Sweden & Norway	22/1259
7	Denmark	62/1946
8	Czechoslovakia	16/774
9	France, Germany & Netherlands	5/572
10	Spain	5/221
11	Algeria, Morocco & Tunisia	10/345
12	Nigeria (Gongola Basin), Chad, Cameroon (N)	9/181
13	Guinea & Mali	18/344
14	Ivory Coast, Ghana, Upper Volta	18/431
15	Togo & Benin	19/443
16	Congo & Zaire, Cameroons (S), C.A.R., Gabon	9/316
17	Poland	11/364
18	Hungary & Yugoslavia	12/348
19	Romania	13/289
20	Bulgaria	5/272
21	Italy	22/744
22	Jordan	6/57
23	Iran	25/343
24	Saudi Arabia & Yemen	30/378
25	Sudan	4/215
26	Kenya	19/338
27	Malawi	28/509
28	South Africa & Botswana	103/3884
29	Madagascar	4/92
30	U.S.S.R.	47/2631
31	India (Kerala)	76/1171
32	Sri Lanka	13/280
33	India & Bangladesh	21/411
34	Thailand & Malaysia (Malay Peninsula)	16/284
35	Indonesia	95/1009
36	Thailand	109/2012
37	Korea	24/542
38	Japan	15/572
39	Papua New Guinea	50/450
40	Australia	8/523
41	Australia (Goulburn Basin)	20/516

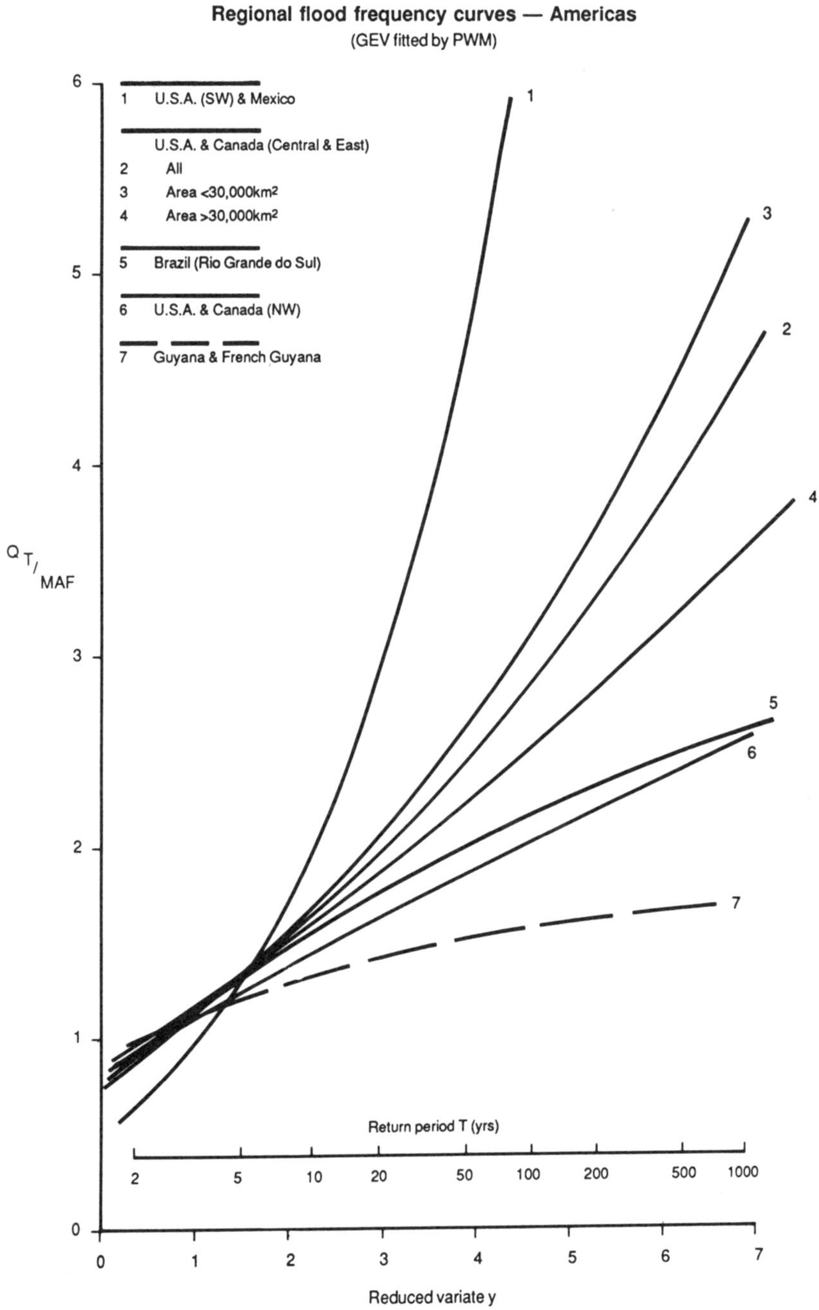
Regional flood frequency curves — Americas
(GEV fitted by PWM)
1 U.S.A. (SW) & Mexico
U.S.A. & Canada (Central & East)
2 All
3 Area <30,000km2
4 Area >30,000km2
5 Brazil (Rio Grande do Sul)
6 U.S.A. & Canada (NW)
7 Guyana & French Guyana
QT/MAF
Return period T (yrs)
2 5 10 20 50 100 200 500 1000
Reduced variate y
0 1 2 3 4 5 6 7

Figure 3

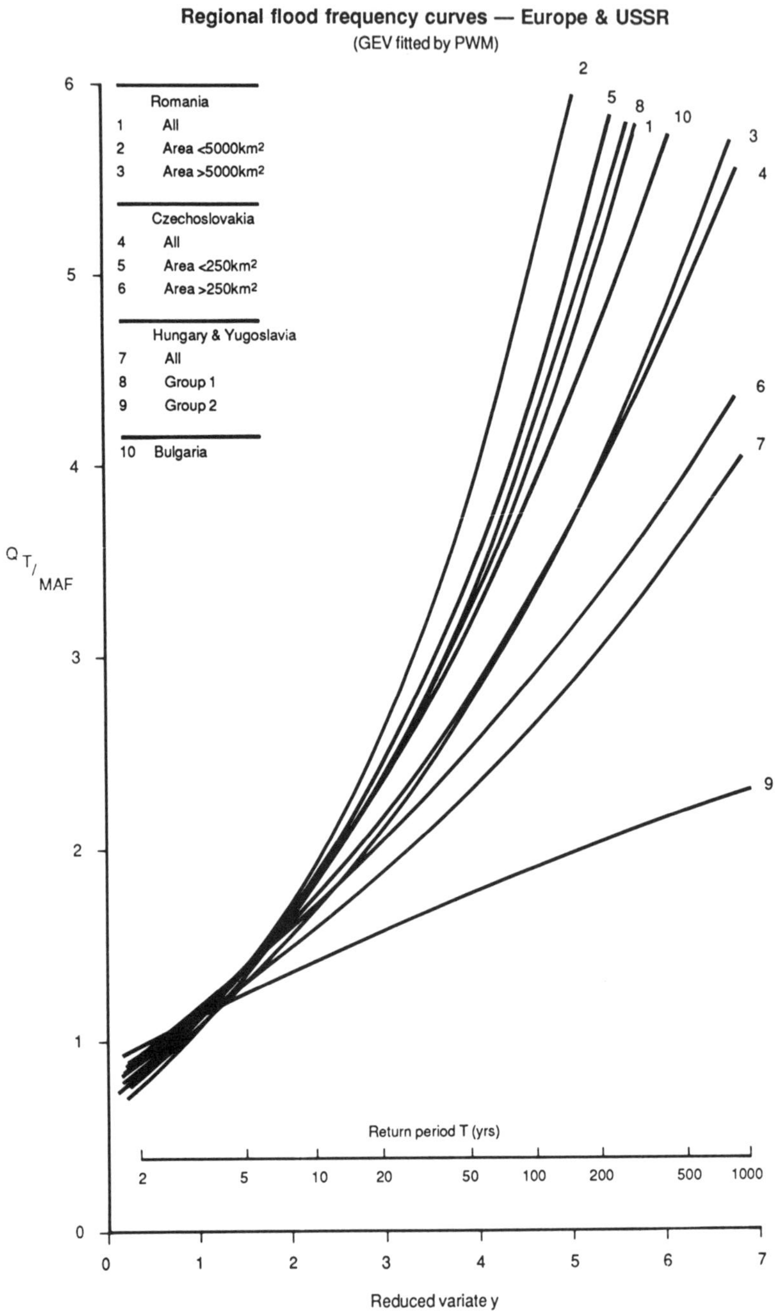
Regional flood frequency curves — Europe & USSR
(GEV fitted by PWM)
Romania
1 All
2 Area <5000km2
3 Area >5000km2
Czechoslovakia
4 All
5 Area <250km2
6 Area >250km2
Hungary & Yugoslavia
7 All
8 Group 1
9 Group 2
10 Bulgaria
QT/MAF
Return period T (yrs)
2 5 10 20 50 100 200 500 1000
Reduced variate y
0 1 2 3 4 5 6 7

Figure 4

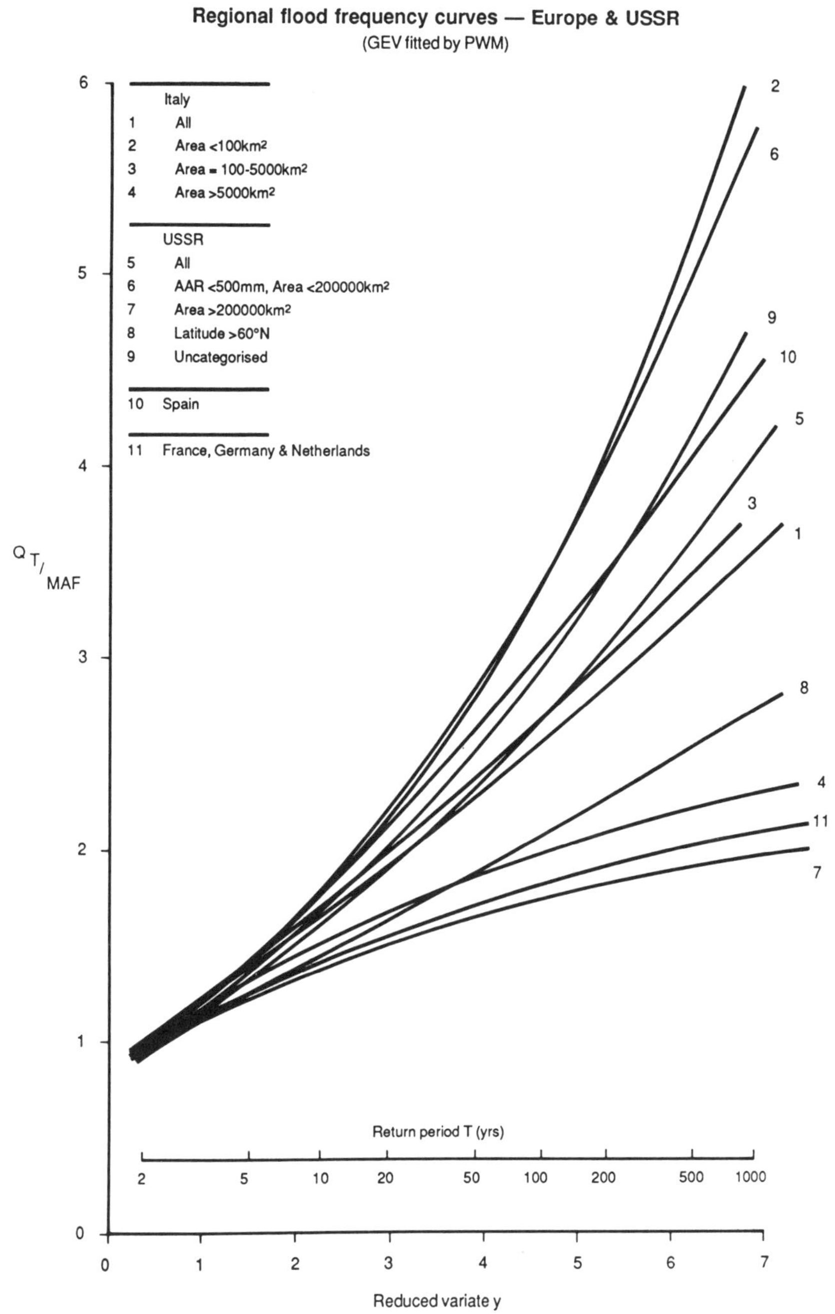

Figure 4 (cont.)

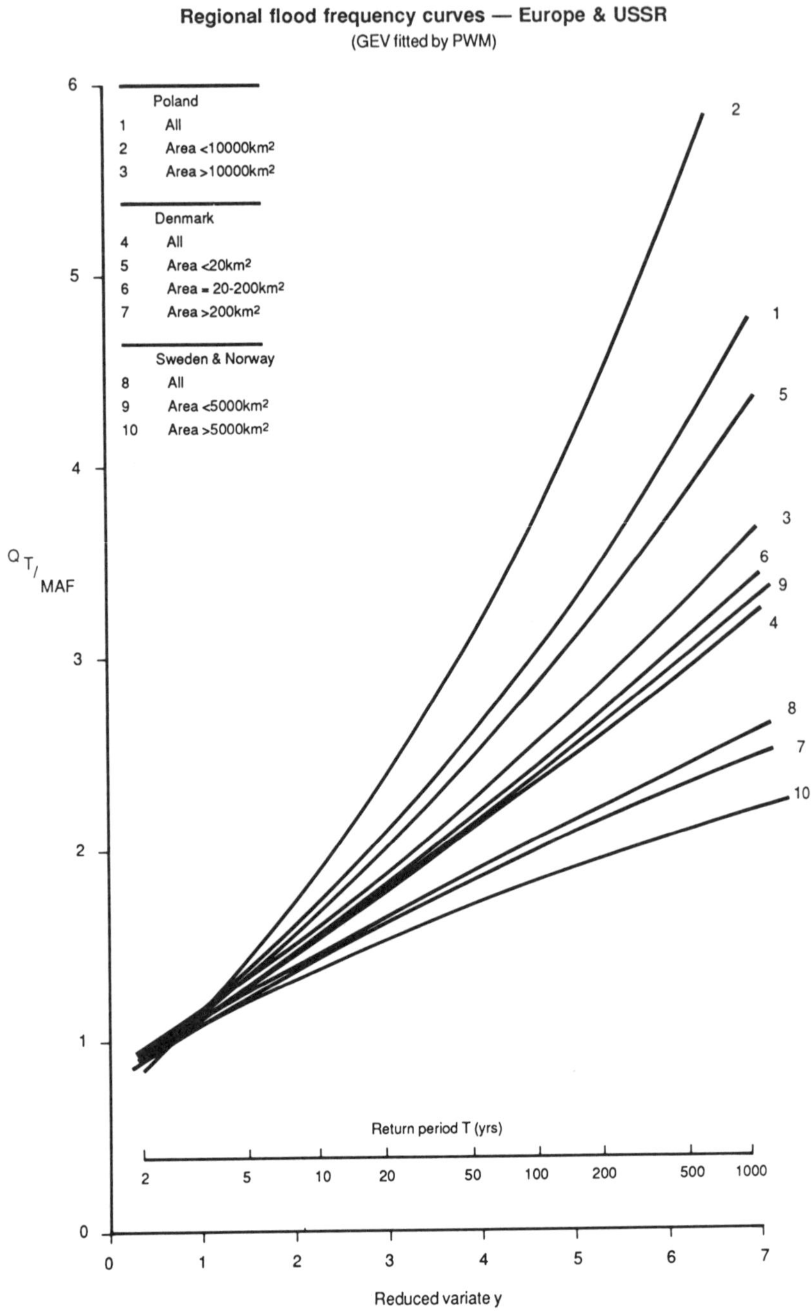

Figure 4 (cont.)

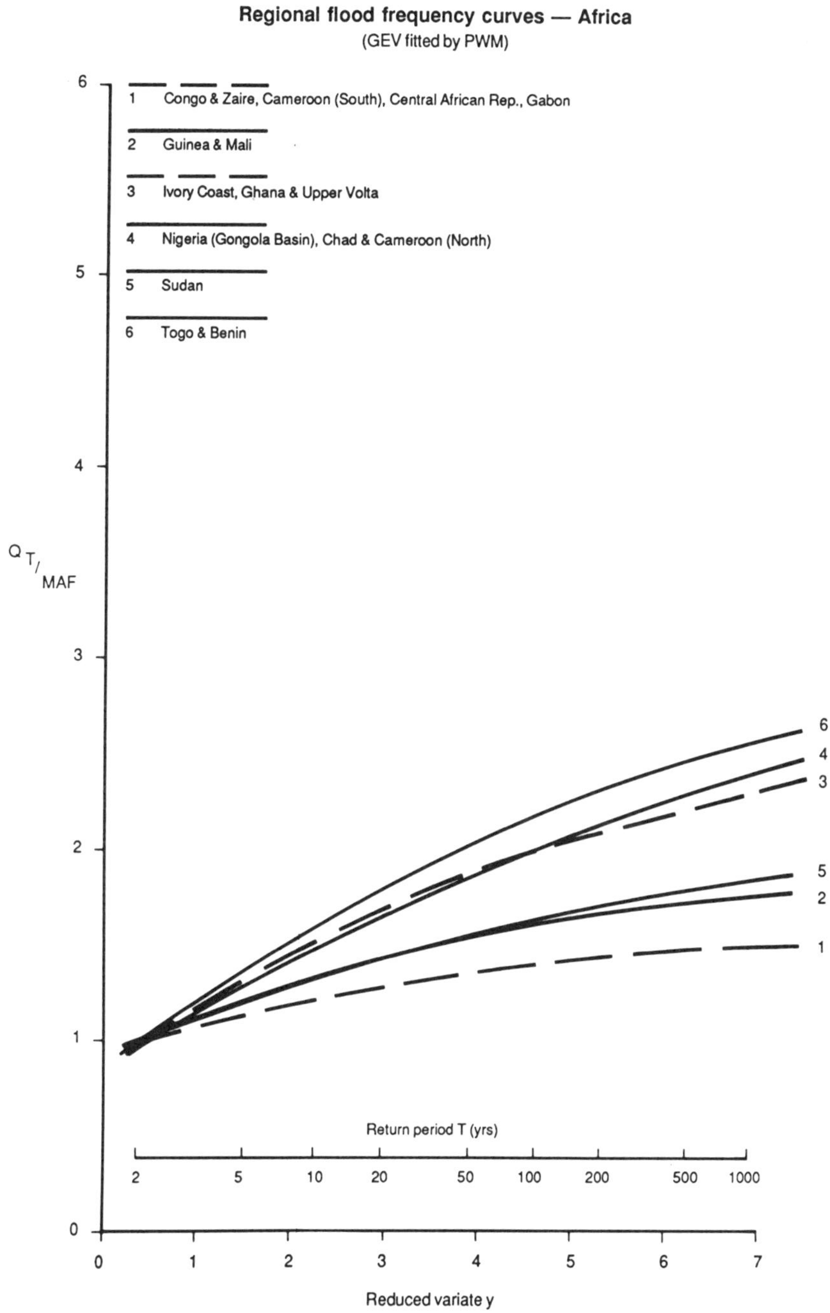
Regional flood frequency curves — Africa
(GEV fitted by PWM)
1 Congo & Zaire, Cameroon (South), Central African Rep., Gabon
2 Guinea & Mali
3 Ivory Coast, Ghana & Upper Volta
4 Nigeria (Gongola Basin), Chad & Cameroon (North)
5 Sudan
6 Togo & Benin
Q T/MAF
Return period T (yrs)
2 5 10 20 50 100 200 500 1000
Reduced variate y
0 1 2 3 4 5 6 7

Figure 5

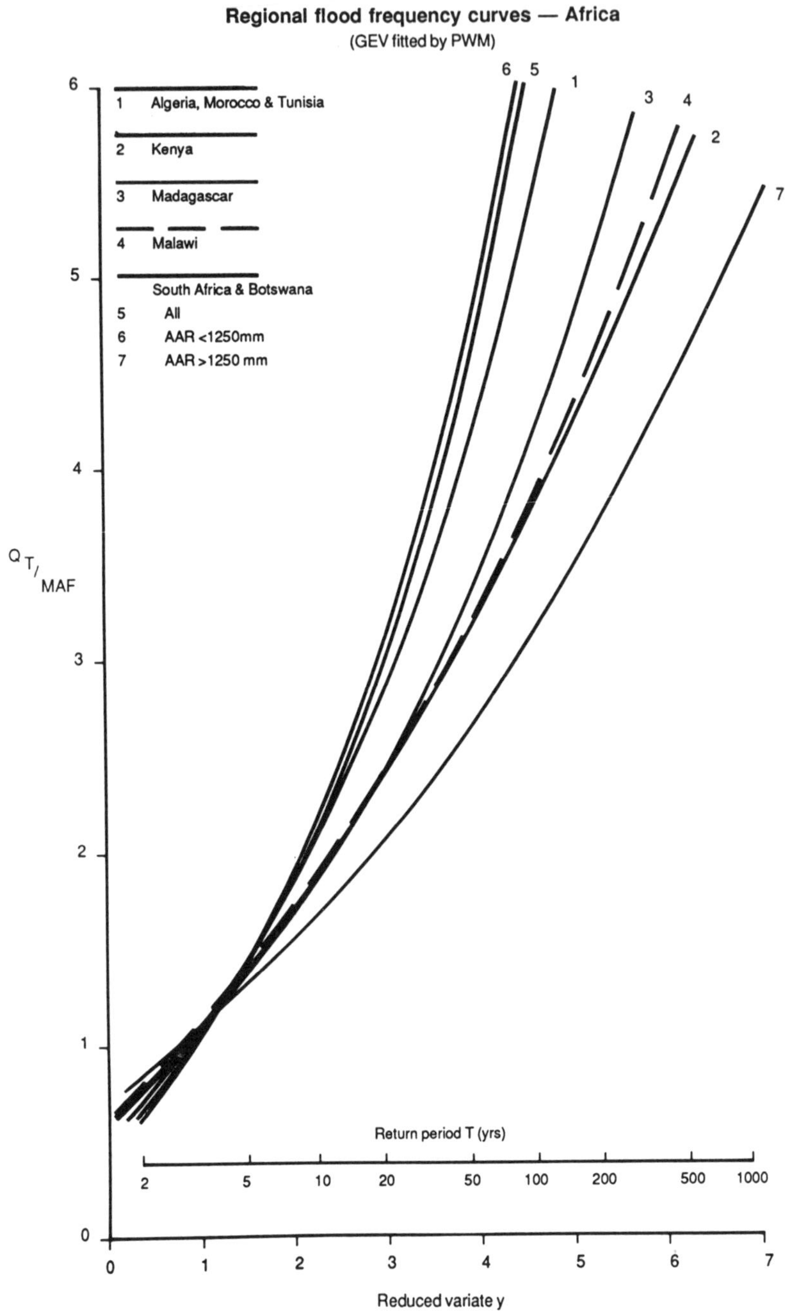

Figure 5 (cont.)

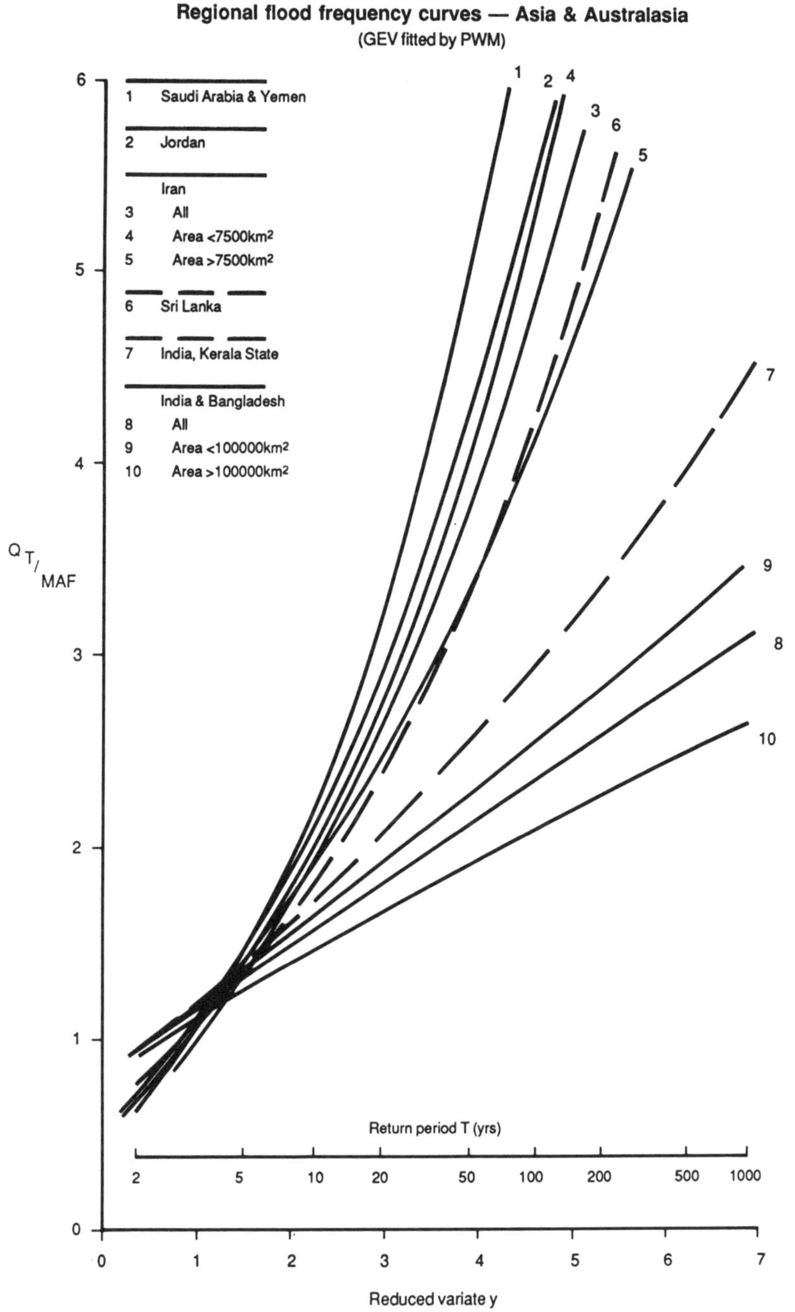
Regional flood frequency curves — Asia & Australasia
(GEV fitted by PWM)
1 Saudi Arabia & Yemen
2 Jordan
Iran
3 All
4 Area <7500km2
5 Area >7500km2
6 Sri Lanka
7 India, Kerala State
India & Bangladesh
8 All
9 Area <100000km2
10 Area >100000km2
QT/MAF
Return period T (yrs)
2 5 10 20 50 100 200 500 1000
Reduced variate y
0 1 2 3 4 5 6 7

Figure 6

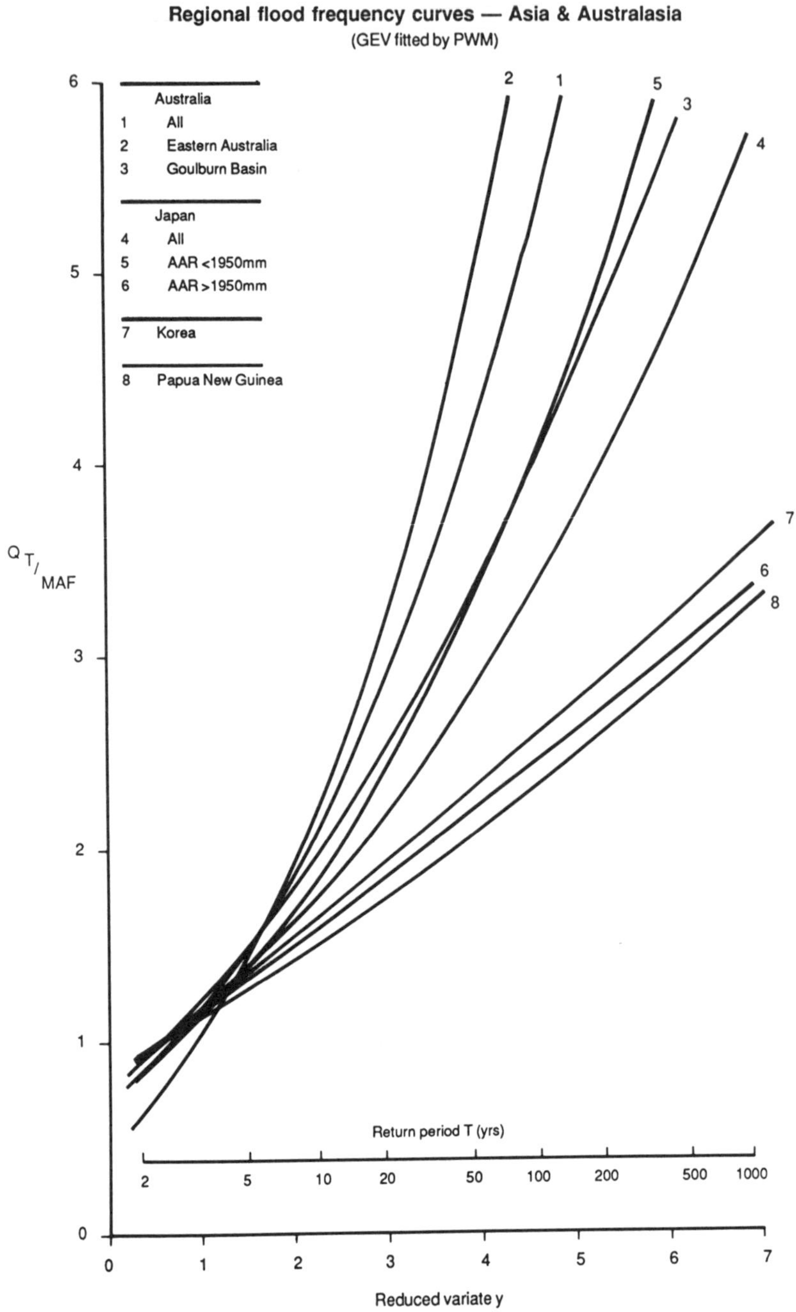

Figure 6 (cont.)

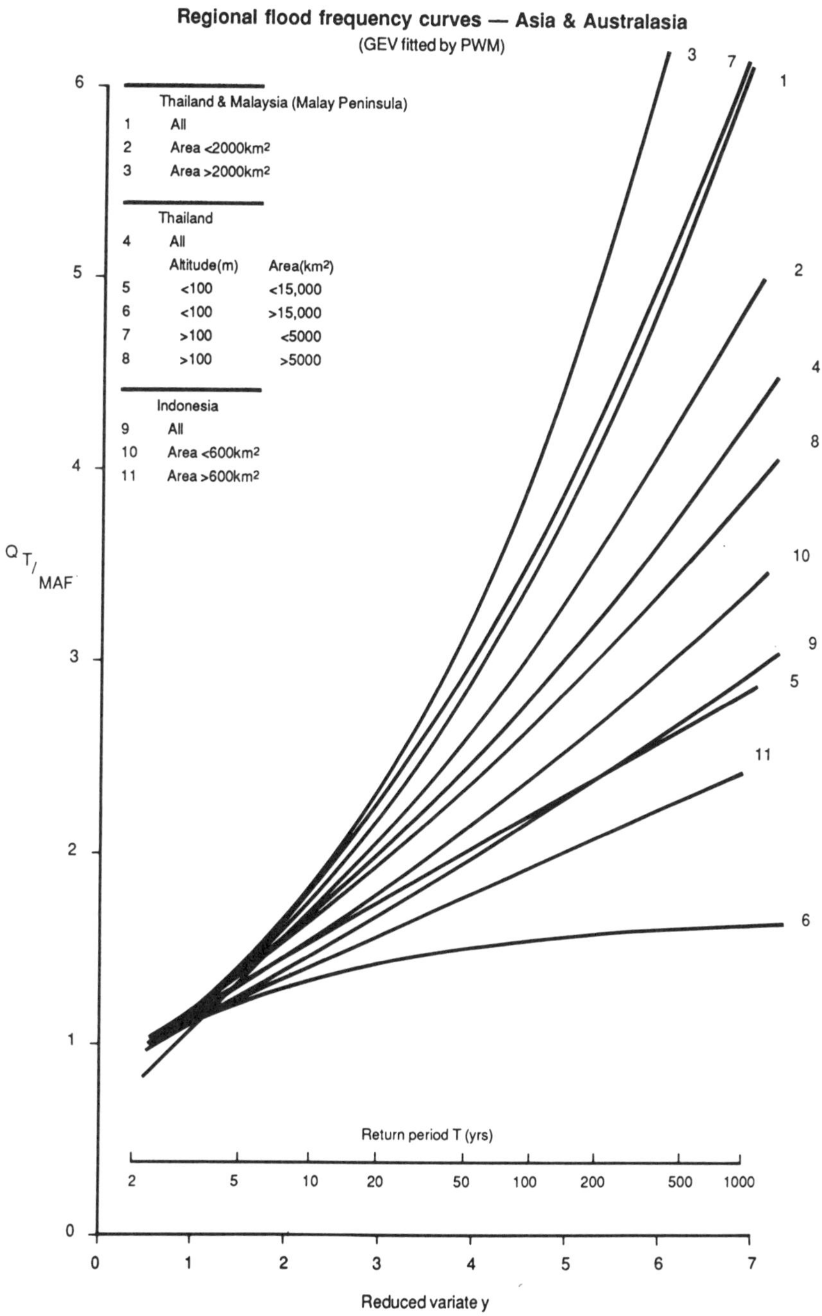

Figure 6 (cont.)

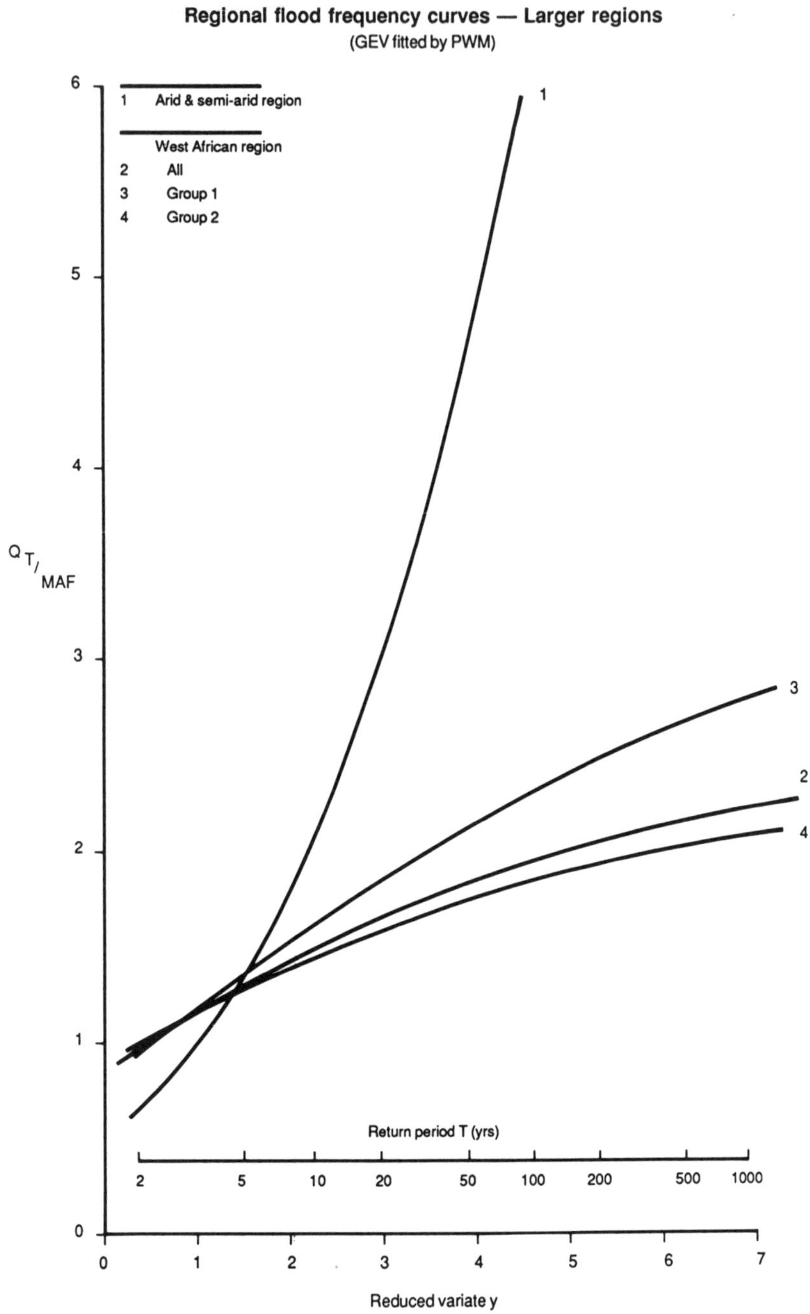
Regional flood frequency curves — Larger regions
(GEV fitted by PWM)
1 Arid & semi-arid region
West African region
2 All
3 Group 1
4 Group 2
Q T/MAF
Return period T (yrs)
2 5 10 20 50 100 200 500 1000
0 1 2 3 4 5 6 7
Reduced variate y
1
2
3
4
0
1
2
3
4
5
6

Figure 7

ON THE TWO-COMPONENT EXTREME VALUE DISTRIBUTION AND ITS POINT AND REGIONAL ESTIMATORS

M. Fiorentino
Research Engineer
Dipartimento di Difesa del Sudo
Universita della Calabria
Cosenza, Italy

Vijay P. Singh and K. Arora
Department of Civil Engineering and
Louisiana Water Resources Research Institute
Louisiana State University
Baton Rouge, Louisiana 70803-6405, USA

ABSTRACT. The principle of maximum entropy (POME) was used to derive the two-component extreme value (TCEV) distribution which has recently been shown to account for most of the characteristics of the real flood experience. The derivation specified the information with respect to which TCEV distribution could be thought to be least-biased and consistent, and yielded equations for estimation of the distribution parameters. This method of parameter estimation is suitable for application in both the site-specific and regional cases. Statistical properties of this regionalized estimation procedure were evaluated using a Monte Carlo approach and compared with those of the maximum likelihood regional estimators.

1. INTRODUCTION

The random variable x is defined to have a two-component extreme value (TCEV) distribution if its probability density function (pdf) is given by

$$f(x) = [\frac{\Lambda_1}{\theta_1} \exp(-x/\theta_1) + \frac{\Lambda_2}{\theta_2} \exp(-x/\theta_2)] \exp[-\Lambda_1 \exp(-x/\theta_1) - \Lambda_2 \exp(-x/\theta_2)];\ x > 0 \quad (1a)$$

$$= \exp(-\Lambda_1 - \Lambda_2);\ x = 0 \quad (1b)$$

where $\Lambda_1 > 0$, $\Lambda_2 \geq 0$, $\theta_2 \geq \theta_1 > 0$ are parameters. Its cumulative distribution function (cdf) is

V. P. Singh (ed.), Regional Flood Frequency Analysis, 257–272.

$$F(x) = \exp[-\Lambda_1 \exp(-x/\theta_1) - \Lambda_2 \exp(-x/\theta_2)];\ x \geq 0 \qquad (2)$$

This cdf has been shown (Versace, et al., 1982; Rossi, et al., 1984) to represent the distribution function of the annual maximum x of a non-negative random variate z whose number of occurrences, k, in a year is a random variate when the following hypotheses hold: (1) z is an independent, identically distributed (iid) variable with probability density function defined by a mixture of two exponential distributions; (2) k is an iid Poisson distributed variate; and (3) z and k are not dependent upon each other.

The two components of the distribution of both z and x are usually referred to as basic component (subscripts of parameters = 1) and outlying component (subscripts of parameters = 2). Theoretical properties of the TCEV distribution have been widely investigated (Rossi, et al., 1984; Beran, et al., 1986; Rossi, et al., 1986). Briefly summarized, the TCEV distribution permits a reasonable interpretation of the physical phenomenon which generates floods and is able to account for most of the characteristics of the real world flood data, important among them being the large variability of the sample skewness coefficient which mostly gives rise to the poor performance of other commonly used flood frequency distributions. The TCEV distribution also offers a practical approach to regional flood frequency estimation.

Thus far, only two parameter estimation methods have been proposed for fitting the TCEV distribution to annual flood series. Canfield (1979) suggested a least squares technique, while Rossi, et al. (1984) presented a procedure based on the maximum likelihood estimation (MLE) method. The latter was further investigated and a regional estimation algorithm based on it was developed (Fiorentino, et al. 1985). Small sample properties of the site-specific and regionalized TCEV-MLE procedure were assessed by Fiorentino and Gabriele (1985), and Arnell and Gabriele (1985). In particular, the latter compared with regionalized algorithm with other regional estimators. Although various features of the TCEV-MLE method exhibited a competitive performance, an improvement of the site-specific estimators was suggested as noted by Fiorentino and Gabriele (1985). Furthermore, Fiorentino, et al. (1986) noted that regional estimates of some parameters could be still improved. In light of this discussion, it may perhaps be desirable to investigate a new method of fitting the TCEV distribution.

The objective of this paper is to derive this distribution using the principle of maximum entropy (POME). This derivation sheds further light on the nature of the distribution and provides an alternative method for estimation of its parameters. The TCEV-POME estimation procedure is shown to be also suitable for regionalized use. Its performance is assessed by using the Monte Carlo technique and the results are compared with those of the MLE method.

2. THE PRINCIPLE OF MAXIMUM ENTROPY (POME)

Entropy is defined as a measure of uncertainty. The entropy function H(x) of a distribution, with pdf f(x), of a random variable X quantifies this uncertainty. When x is a continuous variable and assumes values in a domain c, H(x) is given as

$$H(x) = - \int_c f(x) \ln f(x)\, dx \tag{3}$$

This form was first used by Jaynes (1968) and represents an extension to the continuous case of the entropy function applied in communication theory by Shannon and Weaver (1949).

The principle of maximum entropy (POME) was formulated by Jaynes (1957, 1961) and states that "the minimally prejudiced assignment of probabilities is that which maximizes the entropy subject to the given information." Mathematically, "the given information" is quantified by some linearly independent constraints to be imposed while maximizing equation (3). POME has been applied in several hydrologic fields. Sonuga (1972, 1976) was probably the first who applied POME in hydrologic frequency analysis. Jowitt (1979), and Singh and Singh (1985) successfully exploited it to derive the extreme value type I and the Pearson type III distributions respectively. Furthermore, the function H(x) has been given for a number of probability distributions by Verdugo Lazo and Rathie (1978), and Singh, et al. (1985).

In all of these cases, POME has been shown to uniquely specify the amount of information in terms of constraints required to derive a given distribution. The parameter estimation methods based on POME almost always result in an easy, practical solution, and their performance is generally found to be comparable with that of MLE method.

Maximizing H(x) is conceptually easy when constraints are formulated as population means of specified functions of x. In fact, in such a case the required pdf f(x) has been shown (Singh, et al., 1986) to result in

$$f(x) = \exp[- a_0 - \sum_{j=1}^{r} a_j\, y_j(x)] \tag{4}$$

where $y_j(x)$ are some functions whose population means are the selected constraints, and a_j (j = 0, 1, ..., r) are Lagrange multipliers. Since, by definition,

$$\int_c f(x)\, dx = 1 \tag{5}$$

for equation (4), the following relationships hold:

$$\frac{\partial a_0}{\partial a_j} = - E[y_j(x)]$$

$$\frac{\partial^2 a_0}{\partial a_j^2} = \text{var}[y_j(x)] \qquad j = 1, 2, \ldots, r \quad i = 1, 2, \ldots, r \quad i \neq j \tag{6}$$

$$\frac{\partial^2 a_0}{\partial a_i \partial a_j} = \text{cov}[y_i(x), y_j(x)]$$

where E[•], var[•] and cov[•] indicate population mean, variance and covariance respectively.

Equations (4), (5) and (6) enable us to derive Lagrange multipliers as functions of apriori information provided by constraints. Finally, for equations (3) and (4), the entropy function can be written as

$$H(x) = a_0 + \sum_{j=1}^{r} a_j \, E[y_j(x)] \tag{7}$$

3. DERIVATION OF THE TCEV DISTRIBUTION

For simplicity, the assumption that equation (1a) also holds for non-positive values of x will be made in the following discussion. This approximation is reasonable in all practical applications of the TCEV distribution (Rossi, et al., 1984). If x has a TCEV distribution then the entropy function H(x) is given by

$$\begin{aligned} H(x) &= - \int_{-\infty}^{\infty} \ln f(x)\, f(x)\, dx = (\ln \theta_1 - \ln \Lambda_1) \int_{-\infty}^{\infty} f(x)\, dx + \\ &+ \frac{1}{\theta_1} \int_{-\infty}^{\infty} x\, f(x)\, dx + \Lambda_1 \int_{-\infty}^{\infty} \exp(- x/\theta_1)\, f(x)\, dx + \\ &+ \Lambda_2 \int_{-\infty}^{\infty} \exp(- x/\theta_2)\, f(x)\, dx - \\ &- \int_{-\infty}^{\infty} \ln\left[1 + \frac{(\Lambda_2/\theta_2) \exp(- x/\theta_2)}{(\Lambda_1/\theta_1) \exp(- x/\theta_1)}\right] f(x)\, dx \end{aligned} \tag{8}$$

On comparison with equation (7), the equations of constraints can then be written in the following form:

$$\int_{-\infty}^{\infty} f(x)\, dx = 1 \tag{9}$$

$$\int_{-\infty}^{\infty} x\ f(x)\ dx = E[x] \tag{10}$$

$$\int_{-\infty}^{\infty} \exp(-\ x/\theta_1)\ f(x)\ dx = E[\exp(-\ x/\theta_1)] \tag{11}$$

$$\int_{-\infty}^{\infty} \exp(-\ x/\theta_2)\ f(x)\ dx = E[\exp(-\ x/\theta_2)] \tag{12}$$

$$\int_{-\infty}^{\infty} \ln(1 + \frac{(\Lambda_2/\theta_2)\ \exp(-\ x/\theta_2)}{(\Lambda_1/\theta_1)\ \exp(-\ x/\theta_1)}\ f(x)\ dx$$

$$= E[\ln(1 + \frac{(\Lambda_2/\theta_2)\ \exp(-\ x/\theta_2)}{(\Lambda_1/\theta_1)\ \exp(-\ x/\theta_1)})] \tag{13}$$

The least biased pdf f(x) is determined by POME according to equations (4) and (9) - (13), and takes the form:

$$f(x) = \exp\{-\ a_0 - a_1\ x - a_2\ \exp(-x/\theta_1) - a_3\ \exp(-x/\theta_2) -$$

$$-\ a_4\ \ln[1 + \frac{\Lambda_2\ \theta_1}{\theta_2\ \Lambda_1}\ \exp[-\ x(\frac{1}{\theta_2} - \frac{1}{\theta_1})]]\} \tag{14}$$

The zeroth Lagrange multiplier a_0 is determined as follows. Inserting equation (14) in equation (9),

$$\exp(a_0) = \int_{-\infty}^{\infty} \exp[-\ a_1\ x - a_2 \exp(-\ x/\theta_1) -$$

$$-\ a_3\ \exp(-x/\theta_2)]\ [1 + \frac{\Lambda_2\ \theta_1}{\theta_2\ \Lambda_1}\ \cdot$$

$$\cdot\ \exp(-\ x(\frac{1}{\theta_2} - \frac{1}{\theta_1})^{-a_4})]dx \tag{15}$$

Let $z = \Lambda_1\ \exp(-x/\theta_1)$, $\theta = \theta_2/\theta_1$, $\Lambda = \Lambda_2/(\Lambda_1^{1/\theta})$. After simple manipulation,

$$a_0 = \ln\ \theta_1 - a_1\ \theta_1\ \ln\ \Lambda_1 + \ln \int_0^{\infty} z^{a_1\theta_1 - 1}\ \cdot$$

$$\cdot\ \exp(-\ a_2\ x/\Lambda_1)\ \exp[-\ a_3\ z^{(1/\theta)}\ \Lambda_1^{(-1/\theta)}]\ \cdot$$

$$\cdot\ [1 + \frac{\Lambda}{\theta}\ z^{((1/\theta)-1)}]^{-a_4}\ dz \tag{16}$$

Let us refer to the integral in equation (16) as I_0. Inserting equation (16) in equation (14) we get,

$$f(x) = \frac{1}{\theta_1} \Lambda_1^{a_1 \theta_1} \exp[- a_1 x - a_2 \exp(- x/\theta_1) -$$

$$- a_3 \exp(-x/\theta_2)] \left[1 + \frac{(\Lambda_2/\theta_2) \exp(- x/\theta_2)}{(\Lambda_1/\theta_1) \exp(- x/\theta_1)}\right]^{-a_4} / I_0 \quad (17)$$

When

$$a_1 = 1/\theta_1, \; a_2 = \Lambda_1, \; a_3 = \Lambda_2, \text{ and } a_4 = -1 \quad (18)$$

integral I_0 becomes unity and equation (17) becomes

$$f(x) = \left[\frac{\Lambda_1}{\theta_1} \exp(-x/\theta_1) + \frac{\Lambda_2}{\theta_2} \exp(- x/\theta_2)\right] \cdot$$

$$\cdot \exp[- \Lambda_1 \exp(- x/\theta_1) - \Lambda_2 \exp(-x/\theta_2)] \quad (19)$$

which, according to the approximation mentioned above, is the pdf of the TCEV distribution. Then this distribution can be thought to be least-biased, and consistent with the information given by equations (9) - (13).

3.1 Relation between Constraints and Parameters

The relationships between the parameters of TCEV distribution and the constraints are specified by exploiting equation (6) as follows. Partially differentiating a_0 w.r.t. a_1, a_2, a_3 and a_4 respectively, and taking into account equation (18),

$$\frac{\partial a_0}{\partial a_1} = - E[x] = - \theta_1 \ln \Lambda_1 + \theta_1 \int_0^\infty \ln y \exp(- y - \Lambda y^{(1/\theta)}) \cdot$$

$$\cdot (1 + \frac{\Lambda}{\theta} y^{((1/\theta)-1)}) \, dy \quad (20)$$

$$\frac{\partial a_0}{\partial a_2} = - E[\exp(- x/\theta_1)] = - \frac{1}{\Lambda_1} \int_0^\infty y \exp(- y - \Lambda y^{(1/\theta)}) \cdot$$

$$\cdot (1 + \frac{\Lambda}{\theta} y^{((1/\theta)-1)}) \, dy \quad (21)$$

$$\frac{\partial a_0}{\partial a_3} = - E[\exp(- x/\theta_2)] = - \frac{1}{\Lambda_1^{1/\theta}} \int_0^\infty y^{(1/\theta)} \cdot$$

$$\cdot \exp(- y - \Lambda y^{(1/\theta)}) \ (1 + \frac{\Lambda}{\theta} y^{((1/\theta)-1)}) \ dy \qquad (22)$$

$$\frac{\partial a_0}{\partial a_4} = - E[\ln(1 + \frac{(\Lambda_2/\theta_2) \exp(- x/\theta_2)}{(\Lambda_1/\theta_1) \exp(- x/\theta_1)}]$$

$$= - \int_0^\infty \ln(1 + \frac{\Lambda}{\theta} y^{((1/\theta)-1)}) \exp(- y - \Lambda y^{1/\theta}) \cdot$$

$$\cdot \ (1 + \frac{\Lambda}{\theta} y^{((1/\theta)-1)}) \ dy \qquad (23)$$

Solving integrals in equations (20) to (22) provides,

$$E[x] = \theta_1 \ln \Lambda_1 + \theta_1 \gamma - \theta_1 \sum_{j=1}^{\infty} \frac{(-1)^j \Lambda^j}{j!} \Gamma(j/\theta) \qquad (24)$$

$$E[\exp(- x/\theta_1)] = \frac{1}{\Lambda_1} [1 + \frac{1}{\theta} \sum_{j=1}^{\infty} \frac{(-1)^j \Lambda^j}{(j-1)!} \Gamma(j/\theta) \qquad (25)$$

$$E[\exp(- x/\theta_2)] = - \frac{1}{\theta \Lambda_2} \sum_{j=1}^{\infty} \frac{(-1)^j \Lambda^j}{(j-1)!} \Gamma(j/\theta) \qquad (26)$$

where $\gamma = 0.5772$ is the Euler constant and $\Gamma(\cdot)$ is the gamma function. θ and Λ are dimensionless parameters, already defined in terms of the four parameters of the TCEV distribution. The integral in equation (23) cannot be solved explicitly. But, for $\theta > 1.5$, it is closely approximated by the following function

$$\int_0^\infty \ln(1 + \frac{\Lambda}{\theta} y^{((1/\theta)-1)}) \exp(- y - \Lambda y^{1/\theta}) \ (1 + \frac{\Lambda}{\theta} h^{((1/\theta)-1)}) \ dy$$

$$= 0.1 \exp(-1)(3 + \theta)^{2.059} \Lambda^{\ln 3 - 2(5.5^{-\theta})} \qquad (27)$$

Therefore, the fourth constaints can be related to the parameters by

$$E[\ln(1 + \frac{(\Lambda_2/\theta_2) \exp(- x/\theta_2)}{(\Lambda_1/\theta_1) \exp(- x/\theta_1)})] = 0.1 \exp(-1)(3 + \theta)^{2.059} \cdot$$

$$\cdot \Lambda^{\ln 3 - 2(5.5)^{-\theta})} \qquad (28)$$

Equations (24) - (26) and (28) show that constraints are somehow related to the moments or moment-ratios of the distribution. In fact, besides the obvious case of the constraint $E[y_1(x)]$ representing the population mean of x, it is clear that $E[y_4(x)]$ depends on the dimensionless parameters θ and Λ only, while both $E[y_3(x)]$ and $E[y_2(x)]$ depend on θ, Λ and Λ_1 (note that Λ_2 is a function of Λ_1 via θ and Λ). A similar dependence is exhibited by the theoretical coefficients of skewness (and kurtosis) and variation respectively as can be easily shown using expressions of the moments given by Beran, et al. (1986). This implies that the estimation of the constraints will likely have a variability increasing with the rank.

Furthermore, equation (26) shows that $E[y_3]$ is also related to the probability, P_2, that the annual maximum value of x comes from the outlying component, P_2 having been derived by Beran, et al. (1986) as

$$P_2 = -\frac{1}{\theta}\sum_{j=1}^{\infty}\frac{(-1)^j\Lambda^j}{(j-1)!}\,\Gamma(j/\theta) \tag{29}$$

In fact, combining equations (26) and (29) gives $E[\exp(-x/\theta_2)] = P_2/\Lambda_2$. Analogously, it can be shown that $E[\exp(-x/\theta_1)] = P_1/\Lambda_1$, where P_1 represents the chance that the annual maximum value of x comes from the basic component.

4. ESTIMATION OF PARAMETERS

4.1 Point Estimation

Equations for estimation of parameters can be obtained by substituting sample values for the population means on the left-hand side of equations (24) to (26) and (28). The system of equations to be solved for giving estimates of the four parameters of the TCEV distribution thus takes the form:

$$\bar{x} = \theta_1 \ln \Lambda_1 + \theta_1 \gamma - \theta_1 \sum_{j=1}^{\infty}\frac{(-1)^j\Lambda^j}{j!}\,\Gamma(j/\theta) \tag{30}$$

$$\overline{\exp(-x/\theta_1)} = \frac{1}{\Lambda_1}\left[1 + \frac{1}{\theta}\sum_{j=1}^{\infty}\frac{(-1)^j\Lambda^j}{(j-1)!}\,\Gamma(j/\theta)\right. \tag{31}$$

$$\overline{\exp(-x/\theta_2)} = -\frac{1}{\theta\Lambda_2}\sum_{j=1}^{\infty}\frac{(-1)^j\Lambda^j}{(j-1)!}\,\Gamma(j/\theta) \tag{32}$$

$$\overline{\ln[1 + \frac{(\Lambda_2/\theta_2)\ \exp(-\ x/\theta_2)}{(\Lambda_1/\theta_1)\ \exp(-\ x/\theta_1)}]} = 0.1\ \exp(-1)(3 + \theta)^{2.059} \cdot$$

$$\cdot\ \Lambda^{\ln 3\ -\ 2(5.5^{-\theta})} \tag{33}$$

where the bar indicates that the sample mean of the underlying function is considered. For simplicity, the left-hand sides of equations (30) - (33) will be hereafter referred to as $\bar{Y}_1, \ldots, \bar{Y}_4$ respectively. Eliminating Λ_2 by way of θ, Λ and Λ_1, and rearranging,

$$\Lambda_1 = \frac{1}{\bar{Y}_1}\ [1 + \frac{1}{\theta} \sum_{j=1}^{\infty} \frac{(-1)^j \Lambda^j}{(j-1)!}\ \Gamma(j/\theta)] \tag{34}$$

$$\theta_1 = \bar{Y}_1/\{\ln[1 + \frac{1}{\theta} \sum_{j=1}^{\infty} \frac{(-1)^j \Lambda^j}{(j-1)!}\ \Gamma(j/\theta)]/\bar{Y}_2\} +$$

$$+ \gamma - \sum_{j=1}^{\infty} \frac{(-1)^j \Lambda^j}{j!}\ \Gamma(j/\theta)\} \tag{35}$$

$$\Lambda = \frac{1 - \Lambda_1 \bar{Y}_2}{\Lambda_1^{1/\theta}\ \bar{Y}_3} \tag{36}$$

$$0.1\ \exp(-1)(3 + \theta)^{2.059}\ \left(\frac{1 - \Lambda_1 \bar{Y}_2}{\Lambda_1^{1/\theta}\ \bar{Y}_3}\right)^{\ln 3 - 2(5.5^{-\theta})} = \bar{Y}_4 \tag{37}$$

Putting $\bar{Y}_3$ and $\bar{Y}_4$ respectively in the form

$$\bar{Y}_3 = \overline{\exp[-\ x/(\theta_1\theta)]} \tag{38}$$

$$\bar{Y}_4 = \overline{\ln\{1 + \frac{1}{\theta\Lambda_1} \frac{1 - \Lambda_1 \bar{Y}_2}{\bar{Y}_3}\ \exp[-\frac{x}{\theta_1}\ (\frac{1}{\theta} - 1)]\}} \tag{39}$$

One obtains that the only unknown in equation (37) is θ. On the other hand: (1) in equation (36), Λ does not appear on the right-hand side; (2) in equation (35), θ_1 is the only unknown once θ and Λ have been evaluated; and (3) Λ_1 does not appear on the right-hand side in equation (34). Therefore, a successive substitution iterative scheme can be developed for estimating the four parameters as follows. Assign tentative values to θ and Λ, then successively estimate θ_1 by equation (35), Λ_1 by equation (34), θ by equation (37), and Λ by equation (36). Substitute the last values of θ and Λ to those previously obtained and start again from estimation of θ_1. Stop when θ and Λ no longer change. Note that the procedure is fast because equation (34) and (36) admit solution in closed form and equations (35) and (37), though not explicit, can be easily solved numerically, for each exhibits one unknown only.

4.2 Regional Estimation

A regional flood frequency estimation algorithm can be developed using equations (34) - (37) (obviously together with equations (38) and (39)), which can also be used to validate the regionalization model proposed by Fiorentino, et al. (1985) and also described in Fiorentino, et al. (1986). In short, this model assumes that dimensionless parameters θ and Λ do not change over extensive regions, while parameter Λ_1 is constant in smaller areas. In this paper, a regionalization algorithm, based on POME, to estimate θ and Λ is presented.

Suppose there are k gauged sites in a selected region which is assumed to be homogeneous with respect to θ and Λ. Let each site have an annual flood series (AFS) with n years of record. At each site, one must estimate the basic component parameters θ_1 and Λ_1, which vary from site to site, plus the two regional values of θ and Λ. There are in practice 2k + 2 unknowns. An equal number of independent equations is then needed.

The first 2k equations of the algorithm herein proposed arise from writing equations (34) and (35) k times, once for each available AFS. The other two equations are derived by taking the average of left-hand sides of equations (32) and (33) over all k sites,

$$\frac{1}{kn} \sum_{r=1}^{k} \sum_{i=1}^{n} \Lambda_{2r} \exp(-\frac{x_{ir}}{\theta_{2r}}) = -\frac{1}{\theta} \sum_{j=1}^{\infty} \frac{(-1)^j \Lambda^j}{(j-1)!} \Gamma(j/\theta) \qquad (40)$$

$$\frac{1}{kn} \sum_{r=1}^{k} \sum_{i=1}^{n} \ln[1 + \frac{(\Lambda_2/\theta_2)_r \exp(-x_{ir}/\theta_{2r})}{(\Lambda_1/\theta_1)_r \exp(-x_{ir}/\theta_{1r})}]$$

$$= 0.1 \exp(-1)(3 + \theta)^{2.059} \Lambda^{\ln 3 - 2(5.5)^{-\theta}} \qquad (41)$$

Equations (40) and (41) can be written in a different manner taking into account the transformation

$$y = \frac{x}{\theta_1} - \ln \Lambda_1 \qquad (42)$$

which makes their left-hand sides dependent on θ and Λ only. Since the values of θ and Λ are assumed to be constant at every site, the following forms are thus obtained

$$\frac{1}{kn} \sum_{i=1}^{kn} \exp(-\, y_i/\theta) = -\frac{1}{\theta} \sum_{j=1}^{\infty} \frac{(-1)^j \Lambda^j}{(j-1)!} \Gamma(j/\theta) \tag{43}$$

$$\frac{1}{kn} \sum_{i=1}^{kn} \ln\{1 + \frac{\Lambda}{\theta} \exp[-\, (\frac{1}{\theta} - 1)\, y_i]\}$$

$$= 0.1 \exp(-1)(3 + \theta)^{2.059}\, \Lambda^{\ln 3 - 2(5.5)^{-\theta}} \tag{44}$$

The two procedures are mutually equivalent, at least when the available AFS's have the same length at every site. Anyway, whichever is used, estimates of θ_1 and Λ_1 at any site need to be obtained together with regional estimates of θ and Λ. In fact, both sets of equations depend on the basic component parameters, the former in an explicit manner and the latter through the transformation of equation (42).

The iterative scheme proposed by Fiorentino and Gabriele (1985) for the regionalized TCEV-MLE procedure can also successfully work using the POME-based estimation method. Details of this scheme can be readily found in Fiorentino, et al. (1986).

For a comparison between the proposed procedure and the MLE method, two features of the former look favorable: (1) Estimation of the basic component parameters, once θ and Λ have been evaluated, is relatively simple, for only equation (35) needs to be solved numerically. (2) The equations contain a smaller number of exponentials to be solved. However, only a large number of Monte Carlo experiments covering a wide range of situations, can confirm whether the TCEV-POME estimation procedure is competitive. In order to make an assessment, herein shown are the results derived by a limited number of computer simulation experiments.

5. EXPERIMENTAL DESIGN AND RESULTS

The proposed estimation procedure was assessed, if only approximately, using the Monte Carlo technique, and generating synthetic series from a TCEV distribution with parameters $\theta_1 = 10$, $\Lambda_1 = 10$, $\theta = 3.067$, and $\Lambda = 0.173$, which is equal to that used to appraise the TCEV-MLE procedure (Fiorentino and Gabriele, 1985; Arnell and Gabriele, 1986).

Two measures of performance were used, viz, the standardized bias:

$$\text{BIAS} = E(\hat{x} - x)/x$$

and the standardized root mean square error:

$$\text{RMSE} = E(\hat{x} - x)^2/x$$

where $\hat{x}$ and x denote the estimated and population values respectively of the statistic under examination.

Since the regionalization is the natural field for application of a distribution with a large number of parameters such as four, the attention was principally devoted to the assessment of the regionalized estimators.

One hundred repetitions of forty synthetic series, each with forty years of record, were generated, i.e., 100 homogeneous regions, each with 40 gauged sites, were simulated. Then the regionalization algorithm described above was applied. For each repetition, a regional estimate of θ and Λ together with forty on-site elements of θ_1 and Λ_1 were obtained. BIAS and RMSE of parameter and quantile regional estimators were then evaluated. Of course, due to the very short number of experiments, these results are not expected to reproduce the true values of BIAS and RMSE, but they do provide a first order approximation of the likely results.

In Table 1, statistical properties of the regionalized TCEV-POME estimators are compared with those obtained (when available) by

Table 1. BIAS and RMSE of quantile and regional parametric estimates.

	POME		MLE	
	BIAS	RMSE	BIAS	RMSE
Regional Parameters				
θ	-0.10	0.13	-0.01	0.15
Λ	0.04	0.19	0.32	0.72
Quantiles x_F				
Standardized Quantiles $x_F/\bar{x}$				
F = 0.5	0.01 0.01	0.09 0.02		
0.9	0.00 0.00	0.12 0.05	0.00 0.00	0.12 0.04
0.99	-0.05 -0.06	0.16 0.10	-0.02 -0.02	0.14 0.09
0.999	-0.07 -0.07	0.17 0.12	-0.03 -0.03	0.17 0.13

Fiorentino and Gabriele (1985), who applied the MLE procedure to the same case. The comparison was also extended to the standardized quantiles $x_F/\bar{x}$, which are important when an index flood regionalization scheme is considered. As usual $\bar{x}$ denotes the sample mean.

As regards quantiles (standardized or not), it can be seen that both methods exhibited practically the same RMSE, while the MLE procedure produced less BIAS than the POME method. On the other hand, the POME procedure produced favorable RMSE with respect to the regional parameter estimates, particularly as regards Λ, whose MLE estimator was highly variable. Regarding the BIAS of θ and Λ estimators, the two competing methods exhibited opposite behavior, POME being outperformed with respect to θ but showing superiority in estimating Λ.

The results shown in Table 1 gave rise to the suspicion that relative to the MLE method, POME provided poorer estimators of the TCEV basic component parameters and that only for this reason the superiority substantially exhibited in estimating θ and Λ decayed when quantile estimates were considered. To substantiate it, a large number of experiment, five thousand synthetic series each with forty years of record were generated, and parameters θ_1 and Λ_1 were estimated for each series using equations (34) and (35) while keeping θ and Λ constant and equal to the population values. BIAS and RMSE of the so obtained estimates are shown in Table 2 where they were also

Table 2. BIAS and RMSE of site-specific estimates obtained keeping θ and Λ constant.

	POME		MLE	
	BIAS	RMSE	BIAS	RMSE
Parameters				
θ_1	-0.01	0.18	-0.02	0.14
Λ_1	0.20	0.64	0.18	0.52
Quantiles				
F = 0.5	0.00	0.09		
0.9	0.00	0.12	-0.01	0.10
0.99	-0.01	0.15	-0.01	0.12
0.999	-0.01	0.16	-0.01	0.13

compared with the respective values given by the MLE procedure (S. Gabriele, personal communication).

As had been suspected, in this case POME performed worse than MLE with regard to both parameters and quantiles, though no significant difference was shown with respect to BIAS. This is important because it suggests that valuable estimates of TCEV quantiles could be attained by combining a different basic component estimator with the POME one, the latter used to reach good estimates of the regional parameters θ and Λ.

6. CONCLUSIONS

The use of the principle of maximum entropy (POME) for deriving the two-component extreme value distribution (TCEV) sheds further light on this distribution, which has recently been shown to offer a practical approach to the regional flood frequency analysis. POME specifies the constraints which quantify the information exploited by the TCEV distribution and with respect to which the TCEV distribution can be thought to be least-biased and consistent.

The estimation method based on the POME permits both site-specific and regional estimation. The equations to be solved for giving parameter estimates seem relatively simple when compared with those of the MLE method.

The regionalized TCEV-POME estimation procedure, proposed for application in a homogeneous region with respect to the shape parameters θ and Λ, performs comparably with the analogous regionalization algorithm which exploits the MLE method. In particular, regional estimates of θ and Λ obtained by using the proposed procedure are found to exhibit favorable root mean square error (RMSE) in the examined case. Nevertheless, quantile elements obtained using at-site estimates of the basic component parameters together with regional estimates of θ and Λ do not show, relative to the MLE method, any improvement in RMSE. Also, they are, even if only slightly, more biased.

As a consequence, one can argue that valuable regional estimates of TCEV quantiles can be attained by combining the POME estimators of θ and Λ (or at least of Λ when only good performance in BIAS is sought) with another estimation procedure, e.g., the MLE method, which provides good estimates of the basic component parameters.

7. REFERENCES

Arnell, N. W. and Gabriele, S., 1986. 'Regional flood frequency analysis using the two-component extreme value distribution: an assessment using computer simulation experiments.' Proceedings of the Seminar, Combined Efficiency of Direct and Indirect Estimates for Point and Regional Flood Prediction, held at Perugia, Italy, December 16-20, 1985 (in press).

Beran, M. A., Hosking, J. R. M. and Arnell, N. W., 1986. Comment on 'Two-component extreme value distribution for flood frequency analysis,' by Rossi et al., 1984. Water Resources Research (in press).

Fiorentino, M. and Gabriele, S., 1985. 'Distribuzione TCEV: metodi di stima dei parametri e proprieta statistiche degli stimatori.' Consiglio Nazaionale delle Ricerche-1RP1, Geodata n. 25, Cosenza, Italy (in Italian).

Fiorentino, M., Rossi, F. and Versace, P., 1985. 'Regonal flood frequency estimation using the two-component extreme value distribution.' Hydrological Sciences Journal, **30**, 51-64.

Fiorentino, M., Gabriele, S., Rossi, F. and Versace, P., 1986. 'Hierarchical approach for regional flood frequency analysis.' Paper presented at the International Symposium on Flood Frequency and Risk Analyses, held May 14-17, Louisiana State University, Baton Rouge, Louisiana.

Jaynes, E. T., 1957. 'Information theory and statistical mechanics, I.' Physical Review, Vol. 106, 620-630.

Jaynes, E. T., 1961. Probability Theory in Science and Engineering. McGraw-Hill Book Co., New York.

Jaynes, E. T., 1968. 'Prior probabilities.' IEEE Trans. Syst., Man Cybern., **3**(SSC-4), 227-241.

Jowitt, P. W., 1979. 'The extreme-value type-I distribution and the principle of maximum entropy.' Journal of Hydrology, **42**, 23-38.

Rossi, F., Fiorentino, M. and Versace, P., 1984. 'Two-component extreme value distribution for flood frequency analysis.' Water Resources Research, **20**(7), 847-856.

Rossi, F., Fiorentino, M. and Versace, P., 1986. Reply to the Comment on 'Two-component extreme value distribution for flood frequency analysis.' Water Resources Research (in press).

Shannon, C. C. and Weaver, W., 1949. The Mathematical Theory of Communication. University of Illinois Press, Urbana, IL, 117 pp.

Singh, V. P. and Singh, K., 1985. 'Derivation of the Pearson type (PT) III distribution by using the principle of maximum entropy (POME).' Journal of Hydrology, **80**, 197-214.

Singh, V. P., Singh, K. and Rajagopal, K., 1985. 'Application of the principle of maximum entropy (POME) to hydrologic frequency analysis.' Report of the Office of Water Policy, U.S. Department of the Interior, Washington, DC.

Sonuga, J. O., 1972. 'Principal of maximum entropy in hydrologic frequency and analysis.' Journal of Hydrology, 17, 117-191.

Sonuga, J. O., 1976. 'Entropy principle applied to rainfall-runoff process.' Journal of Hydrology, 30, 81-94.

Verdugo Lazo, A. C. Z. and Rathie, P. N., 1978. 'On the entropy of continuous probability distributions.' IEEE Transactions on Information Theory, Vol. IT-24, 1, 120-122.

Versace, P., Fiorentino, M. and Rossi, F., 1982. 'Analysis of flood series by stochastic models.' in Time Series Methods in Hydrosciences, edited by A. H. El-Shaarawi and S. R. Esterby, 315-324, Elsevier, New York.

HISTORICAL FLOOD FREQUENCY DATA: ITS VALUE AND USE

Jery R. Stedinger and Timothy A. Cohn
Department of Environmental Engineering
Cornell University
Ithaca, New York 14853
U.S.A.

ABSTRACT. Proper interpretation of cultural, biological, and geological records of historical floods can result in improved estimates of flood quantiles. Maximum likelihood, quasi-maximum likelihood, adjusted moments, and curve fitting methods for the estimation of 1, 2, and 3 parameters are considered. For a historical-flood perception threshold near the 10-year flood (i.e., on average one flood in ten is above the threshold and hence is recorded), a 100-year historical period of observation provides the equivalent of roughly 40, 65, and 80 years of systematic record in terms of the variability of 100-year flood estimates when fitting 1, 2, and 3 parameter distributions, respectively. The comparable numbers for a perception threshold at the 100-year flood are 15, 20, and 45 equivalent years of systematic data when estimating 1, 2, and 3 parameters, respectively.

1. INTRODUCTION

The demand for accurate flood estimation procedures has continued to increase as a consequence of conflicting social objectives: we expect to be protected from natural disasters, while simultaneously trying to minimize the cost, upheaval, and environmental consequences that accompany flood-mitigation programs. Unfortunately, the science of physical hydrology cannot generally provide an adequate description of the distribution of flood flows based on the physical characteristics of a catchment. Thus the commonly employed flood-quantile estimation methods rely on general statistical estimation techniques which make use of available gage records. The precision of such flood-quantile estimators is limited by the length of the available gage records.

Records of historical floods -- floods that occurred before the beginning of systematic streamflow gaging -- can be used to circumvent this limitation. Such historical information is often available; large floods carve shoulders in alluvial soils, move boulders, deposit datable sediments, damage trees, and catch the attention of people who mark the crest stage of such large floods. Recent developments in paleohydrologic techniques have substantially improved our ability to interpret physical and botanical flood evidence [Baker, 1985; Stedinger and Baker, 1987].

This paper examines the statistical properties of flood-quantile estimation methods that utilize historical flood information. The term "historical information" is used loosely to refer to any knowledge we may obtain about floods that occurred before the beginning of systematic gaging. In some cases, historical flood information may consist of the size and date of all large floods that occurred during a particular period. In other cases, it might be

V. P. Singh (ed.), Regional Flood Frequency Analysis, 273–286.

knowledge of the number of times that a threshold was exceeded (e.g. how many times a street was flooded) in a given period.

2. LITERATURE REVIEW

Paleohydrology has been defined by Costa [1984] as "the study of the movements of water and sediment in channels before the time of continuous hydrological records or direct measurements." The literature focuses on two issues: techniques for acquiring historical flood information (e.g. the date and size of large floods), and statistical methodologies for incorporating such information in subsequent analyses. Proper utilization and interpretation of historical flood information requires a clear understanding of both issues [Gerard and Karpuk, 1979; Stedinger et al., 1986; Hirsch and Stedinger, 1987].

2.1. Statistical Censoring

In the late 1940's statisticians developed a general framework for analyzing "censored data" [Hald, 1949; Cohen, 1950]. By censored we mean that in a potential sample of size N, a known number of the smallest or largest observations are missing [David, 1970]. Statisticians distinguish between two types of censored samples, depending on the process that created the sample: Type I censoring occurs whenever the censoring takes place above or below a known and fixed threshold. It corresponds to the "threshold exceedance" model for historical flood information recommended by Leese [1973]. The number of items in a type I censored sample is a random variable, equal to the number of events that exceeded the threshold. With type II censoring, a fixed number of the smallest or largest observations are removed, regardless of their magnitude; with type II censoring there is no fixed threshold. The distinction between type I and type II censoring is important; statistical procedures and assumptions appropriate for one type of censoring are not always appropriate for the other. The data sets and flood information provided by the two types of experiments also differ.

Several researchers have investigated procedures for utilizing type II historical flood information [Tasker and Thomas, 1978; Zhang, 1982, 1985; Cong et al., 1979, 1985; Hosking and Wallis, 1986a,b]. Others have considered or recommended the fixed-threshold, or type I censoring model [Leese, 1973; Gerard and Karpuk, 1979; Condie and Lee, 1983; Cohn, 1984; Hirsch, 1985; Condie, 1986; Stedinger and Cohn, 1986; Hirsch and Stedinger, 1987]. In order to determine which type of censoring is appropriate to describe a given situation, and what information a historical record provides, it is necessary to understand the processes that may lead to the creation of historical flood records. In the following section several sources of historical flood information are examined.

2.2. Sources of Historical Information: Human Records

Perhaps the most familiar forms of historical flood information are the flood-markers that can be seen on buildings near rivers. In New England river valleys one finds shingles or painted lines on the sides of buildings indicating the peak stage of large floods. Similar flood marks can be seen in many parts of the world; flood-marking is a common human response to large and unusual floods. Chen et al. [1974] report that:

> it has been found at many places along the upper reaches of the Yangtze River [that there are] plenty of stone carvings and monuments with inscriptions of the

> successive floods occurring in the years of 1153, 1227, 1560, 1788, 1796, 1860, 1870...13 places with stone carvings and monuments recording the 1788 flood.

Similarly, Thomson et al. [1964] document hundreds of floods that occurred in New England since 1620. The data have been preserved in historical society documents, newspaper photographs and articles, and even in people's memories, diaries and correspondence.

One cannot be sure why a given flood crest was marked. It seems reasonable to assume that citizens take note of floods that are important events in their lives. One frequently finds markers associated with the inundation of structures. This suggests that a "threshold exceedance," or type I model is appropriate for describing the sampling process. One can consider the flow level, Q_0, that just touches the building to be a threshold; anything above that level is assumed to result in a permanent record, while floods below Q_0 are forgotten. One can also define a "historical record length" corresponding to the number of years, h, during which an exceedance of Q_0 would have been observed if it had occurred. For example, h might be the number of years since a building was constructed. However, before utilizing such information one should verify that the building has not been moved, and that the stage-discharge relationship in the river has not changed substantially during the historical period.

Human records could also be the product of a type II censoring process. One can imagine that people remember the largest flood that occurred, but have less interest in smaller floods. In such a situation, there would always be a record of the largest flood during the historical period, regardless of whether or not it exceeded a particular threshold.

2.3. Sources of Historical Information: Botanical and Geological Records

Botanical and geological evidence can also provide information documenting the occurrence of large floods. Sigafoos [1964], Helley and LaMarche [1973], and Yanosky [1983] discuss techniques for exploiting abnormal tree ring growth to identify both the stage and date of floods in a watershed. Similarly, large floods may tip tree trunks, leaving a visible indication of flood level [Yanosky, 1983]. Costa [1978] points out that "the age of mature trees destroyed by recent floods may give a minimum time since an event of similar magnitude occurred, provided the tree was broken or toppled by floodwaters, not just undermined." This method was employed by Hupp [1982].

There are also indirect methods for identifying and establishing the date of floods. Lichen, which induces the weathering of boulders, grows on the shaded side of a boulder [Hupp 1983]. Since floods are responsible for many large-scale movements of boulders, one can measure the time since the last flood by measuring lichen-induced weathering. Hupp [1983] considers the ecological consequences of floods in terms of species variability caused by flooding in the vegetation beside a river. He also notes that the age of most shrubs or trees in each zone corresponds to the number of years since the last inundation of that zone.

Costa [1974, 1978], Patton and Baker [1977], and Kochel et al. [1982] describe techniques for analyzing eroded and deposited sediments that accompany extremely large floods. In favorable circumstances, these records can provide evidence of floods that occurred during the past 10,000 years. Deposited sediments, called slack-water deposits, typically contain organic materials (seeds, leaves, or other biological material) which can be dated using either conventional radiocarbon analysis or the tandem accelerator mass spectrometer for direct measurement of ^{14}C [Baker, 1985]. The latter technique can provide accurate dates with very small samples -- a single seed or bleb of charcoal [Baker, 1985]. However, the dates measured in this manner may not be the date of flooding: seeds may have come from previous years.

Botanical and geological evidence may present a different statistical sampling problem than do records based on human observations. In general, the threshold-exceedance, or type I concept is often appropriate, since the stage of each tree-trunk or boulder corresponds to a fixed threshold, Q_0. A historical flood record is created whenever the water exceeds that level. However, plants and boulders are not like people: they remain in the floodplain when the water rises. Thus a complication is introduced since a large recent flood can obliterate information about older, smaller floods.

Baker et al. [1986] and Stedinger et al. [1986] describe type II censoring situations involving flood scars on canyon walls. In these circumstances, one may only be able to identify the stage of the single largest flood. In such cases, the historical record length is taken to be the number of years that recognizable soil differences defining a flood scar would survive following a large flood, which may be on the order of 2000 years in some cases.

3. FLOOD-QUANTILE ESTIMATION PROCEDURES

3.1. Graphical Procedures

Hydrologists have considered several methods for incorporating historical information into flood frequency studies. Much of the early work in this area, as in flood frequency analysis in general, concerned proper graphical presentation of flood data, or "plotting positions." Benson [1950] discusses the use of historical information to improve flood estimates on the Susquehanna River at Harrisburg. In addition to a 72-year systematic gage record, Benson had measurements of the 7 historical flood flows whose stage exceeded 18 feet that had occurred during the 88 years prior to the installation of a streamflow gage in 1873.

Benson recognized that he had estimates of all floods that exceeded the 18-foot stage during the entire 162-year period. He then assumed that the "missing" below-18-foot floods that occurred before 1874 came from the same population as the below-18-foot floods between 1874 and 1947. Thus the latter floods were weighted to represent the missing floods from the earlier period. Graphical methods which are essentially equivalent to Benson's have been employed both to display historical flood data and to estimate the flood-frequency relationship at a site [Beard, 1975; Natural Environmental Research Council, 1975; Chen et al., 1974; Gerard and Karpuk, 1979; Cong et al., 1979; Zhang, 1982, 1985; Wall et al., 1983; Wang and Adams, 1984; Cong, 1985]. Hirsch and Stedinger [1987] examined the theoretical underpinnings and performance of such plotting positions. They emphasized, as do Gerard and Karpuk [1978], that historical flood records may be the product of a threshold-exceedance, or type I censoring experiment; most published plotting positions, and related curve-fitting and adjusted-moments methods, assume type II censored samples. Hirsch and Stedinger developed a modified plotting position formula for historical flood data from a fixed-threshold type I censoring process (see also Hirsch, 1985).

3.2. Curve-Fitting Procedures

Cong et al. [1979] developed and tested several curve-fitting procedures for estimating flood quantiles with historical information. In particular, they investigated a least-absolute-value fitting criterion, which selects the values of a gamma distribution's parameters so as to minimize the sum of the absolute distances between the observed floods and their anticipated or expected values, given the assigned plotting positions, and the specified values of the postulated distribution's parameters.

3.3. Method-of-Moments Procedures

A method-of-moments analog to Benson's graphical procedure was adopted by the U. S. Water Resources Council [1982] in its Bulletin 17, which suggests uniform procedures for federal flood studies in the United States. As described by Kirby [1981], the procedure is also based on the concept of a threshold Q_0, though its value need not be specified. The Bulletin 17B procedure weights the moments of the historical and systematic data to obtain adjusted-moments estimates of the Pearson type III distribution it employs. These adjusted moments are then used in a standard method-of-moments procedure. Its performance with historical data was explored by Tasker and Thomas [1978], Condie and Lee [1982], Hirsch [1985] and Stedinger and Cohn [1986].

3.4. Maximum Likelihood Estimation

Maximum likelihood estimation techniques have long been favored by statisticians for incorporating censored data into statistical analyses [David, 1970]. In this case, the likelihood function, L, has three factors: One reflecting the observed values in the gage record, one reflecting the fact that k floods during the h-year historical period exceeded the threshold, Q_0, and one reflecting the magnitudes of the observed floods. Thus one obtains [Leese, 1973]:

$$L \;\alpha\; [\,\Pi\, f(Q_i)\,]\;[\,F(Q_0)^{(h-k)}\,(1-F(Q_0))^k\,]\;[\,\Pi\, f(Q_j|Q_j>Q_0)\,] \tag{1}$$

where α stand for "proportional to," and f and F are, respectively, the pdf and CDF of the postulated flood distribution. The MLE estimators are generally found by a numerical search over the parameter space.

Censored-data maximum likelihood estimators were used by Leese [1973] for estimating the two parameters of the Gumbel distribution with type I censored data and a gage record. Condie and Lee [1982] and Condie [1986] consider using historical information when fitting the three-parameter lognormal distribution. Condie and Pilon [1983] present MLE estimators for the log-Pearson type III distribution. Hosking and Wallis [1986a,b] consider MLEs for the Generalized Extreme Value (GEV) distribution with type II censored data. Stedinger and Cohn [1986] provide estimators for the two-parameter lognormal distribution when the values of the historical peaks are observed, and when only exceedances of a threshold were recorded.

3.5. Hybrid Maximum Likelihood Estimator

True maximum likelihood estimators do not exist for the three-parameter lognormal distribution. Several researchers have employed local maximum likelihood estimators

[Condie and Lee, 1982; Hoshi et al., 1984]. An alternative to local MLEs can be obtained by using Stedinger [1980]'s quantile-lower-bound estimator for the lower bound parameter using the systematic data alone. MLEs for the remaining two parameters can be obtained using the algorithm described by Cohn [1984] and Stedinger and Cohn [1986].

4. THE VALUE OF HISTORICAL INFORMATION

Monte Carlo experiments have begun to provide a general understanding of the behavior of flood-quantile estimation procedures that utilize historical information [Cong et al., 1979; Condie and Lee, 1982; Cohn, 1984; Hirsch, 1985; Hosking and Wallis, 1986a,b; Tasker and Stedinger, 1985; Stedinger and Cohn, 1986]. Thus one can consider the benefits that might be obtained, at least in theory, from proper utilization of historical information.

We will focus on parameter estimation with the three-parameter lognormal distribution (LN3) with a coefficient of variation (CV) of 1.0 (skew = 4.0), and a type I censoring model. The LN3 distribution has been frequently employed in flood frequency studies, and the results obtained with this distribution appear to be comparable to those obtained with others [Cohn, 1986].

4.1 Three-Parameter Estimation

Figure 1 reports the mean square error (MSE) of three flood-quantile estimators -- hybrid maximum likelihood (MLE/QLB), method of moments, and curve-fitting -- for the 100-year flood given a 50-year gage record and a 200-year period of historical observation. The methods are compared for 5 censoring thresholds equal to the 10-, 20-, 50-, and 100-year flood, and in the case where no historical information is available.

The first bar in each group corresponds to the hybrid MLE (MLE/QLB) technique described above. In general, this procedure was found to be equivalent to local MLEs with respect to MSE in situations with historical data, and somewhat better in the absence of historical information. The second bar corresponds to an adjusted-moments procedure, where the three sample moments are calculated in real space. This procedure differs from the method for estimating the parameters of the log-Pearson type III distribution in Bulletin 17B [USWRC, 1982], which employs the moments in log space. The third bar is for a least-absolute-value curve-fitting method [Cong et al., 1979] using the threshold-exceedance plotting positions given by Hirsch and Stedinger [1987]. Hirsch [1985] found that use of these plotting positions yields smaller biases and MSEs than the plotting positions recommended by Cong et al. [1979]. The results in Figure 1 are taken from Cohn [1986].

All the methods show large improvements from the use of historical information. Still, the hybrid MLE (MLE/QLB) provides the lowest MSE at all thresholds. At the 10-year flood threshold, the MLE/QLB has a MSE equal to 1.0; the curve-fitting procedure is nearly the same (1.1), and the method-of-moments procedure is least efficient with a MSE of 1.3. As the threshold rises, the hybrid MLE becomes relatively more efficient than either the method of moments or the curve fitting techniques. In the absence of historical information, the MLE/QLB and method-of-moments techniques have about the same MSE, while the curve-fitting routine is substantially less efficient.

Following Stedinger and Cohn [1986], these results can be expressed in terms of effective record length and average gain. The effective record length is proportional to the reciprocal of the mean square error statistic:

$$\mathrm{ERL}[s, h] = s \cdot \mathrm{MSE}[\, X^*_{0.99} \mid s, h{=}0] \,/\, \mathrm{MSE}[\, X_{0.99} \mid s, h] \qquad (2)$$

where $\mathrm{MSE}[X^*_{0.99} \mid s, h{=}0]$ is the mean square error of the most efficient estimator in the absence of historical information, in this case MLE/QLB. For estimators whose variance is approximately proportional to the inverse of the record length, the ERL statistic corresponds to the number of years of gage data that an efficient estimator would require to yield a 100-year-flood estimate with the same precision as the historical-data estimator obtains with a combination of systematic and historical information. This allows one to discuss the value of historical information in terms of the effective number of years of gage data that it adds to the systematic flood record. Similarly, the average gain (AG) is defined as the average increase in the effective record length per year in the historical period:

$$\mathrm{AG} = (\mathrm{ERL}[s, h] - s) \,/\, h \qquad (3)$$

The average gain measures the marginal improvement in each estimator's precision from the addition of historical information.

Figure 2 reports the AG (in percent) from historical information for the cases in Figure 1. The average gain for the MLE/QLB routine varies from 46% for a threshold at the 100-year flood, to 78% for a threshold at the 10-year flood. In the latter case, a 100-year period of historical observation would be equivalent to 78 years of additional gage record! The other estimators, though less efficient than the MLE, none the less show substantial improvements. The relative differences also depend on the parameters of the underlying distribution [Cohn, 1986].

4.2. One- and Two-Parameter Estimation

Cohn [1986] considers maximum likelihood estimation with historical information of the single scale parameter of a lognormal distribution. Stedinger and Cohn [1986] address maximum likelihood estimation with historical information of the two parameters of a lognormal distribution; they also report results for an adjusted-moments estimator that is conceptually similar to the estimator employed in Bulletin 17B [USWRC, 1982]. However, the latter was found to be substantially less efficient than their MLE routines.

Figure 3 reports the average gains attributed to historical information when estimating one, two, and three parameters of the lognormal distribution. The efficiency of the estimators rises substantially as the number of parameters increases. For a threshold at the 10-year flood line, the average gains are 43%, 64%, and 78%, respectively, when estimating one, two, and three parameters. Similar, but more modest results are obtained for higher censoring thresholds. In general, the value of historical information increases with the number of parameters that need to be estimated.

4.3. Flood-Flow Estimation Errors and Binomial-Censoring Results

Several hydrologists have expressed concern about the accuracy of historical flood magnitude estimates. Historical flood estimates, especially those from alluvial channels subject to erosion and geological shifting, should be considered suspect. Cong [1985] and Hosking and Wallis [1986a,b] investigated the impact that measurement errors might have on flood-quantile estimators. Hosking and Wallis [1986a,b] found that such errors, if not accounted for in the analysis, could greatly reduce the value of historical flood data.

Statistical estimators are available which are relatively insensitive to measurement errors and can be used with more qualitative information. Russell and Bonser [1980], Russell

[1982] and Stedinger and Cohn [1986] employ estimation procedures that utilize only the number of times that a threshold was exceeded (possibly zero) during the historical period. Thus one need not estimate the magnitudes of the historical floods: one only needs to know how often a threshold was exceeded. Such data has been called "binomial threshold exceedance" information.

Binomial data do not readily lend themselves to curve-fitting and moments methods. However, maximum likelihood methods perform well with such data. The likelihood function, L, is proportional to the first two terms in equation 1, so that:

$$L \propto [\Pi f(Q_i)]\ [F(Q_0)^{(h-k)}(1-F(Q_0))^k] \qquad (4)$$

The MLEs must be found by numerical methods.

Figure 4 reports the MSE of the hybrid MLE with binomial data (MLE/BIN) and the MLE/QLB for the cases that appear in Figure 1, when three parameters were estimated. MLE/BIN's performance is nearly the same as MLE/QLB for high censoring thresholds. When the threshold is at the 10-year flood, MLE/BIN's MSE is approximately twice that of the MLE/QLB procedure, and is more than it was for a censoring threshold at the 50-year flood. Still, with binomial data a substantial improvement in MSE results and the effective record length is more than double the 50 years provided by the gage record.

Figure 5 reports the average gains associated with binomial-threshold-exceedance historical data when estimating one, two, and three parameters. These results must be interpreted with care, since the mean square errors of the binomial-data estimators do not generally fall inversely with record length. Still, a comparison of Figure 3 and Figure 5 shows that the binomial-threshold-exceedance information provides almost the same average gain as the complete historical data set when estimating one or two parameters, and is competitive in some cases when estimating three parameters.

These results suggest that errors in estimated flood magnitudes, when properly considered in the analysis, may not seriously reduce the value of historical information. A more sophisticated procedure could estimate the precision of each historical flood by an interval, or perhaps an estimate with a standard error. Such a presentation of the data would allow measurement uncertainties to be incorporated directly into the statistical analysis, as was done by Stedinger et al. [1986].

4.4. Choice of Distribution

A hydrologist can never be sure, and would be foolish to believe, that a fitted distribution is the same as nature uses to generate flood flows. Stedinger and Cohn [1986] tested the performance of 2-parameter MLE estimators for the lognormal distribution when data were drawn from log-Pearson type III distributions. They found that inclusion of historical information substantially reduced the bias and MSE of flood-quantile estimates. The calculated average gains when fitting the wrong distribution were roughly the same, and for high perception thresholds were generally larger, than they were when fitting the correct distribution. Cohn [1986] reports that MLE estimators that utilize historical flood information are also robust to use of the wrong model when estimating one parameter. These results should not be a surprise; little else is as relevant in flood-frequency investigations that the frequency of large floods, whether or not the magnitudes of the smaller and unimportant floods are observed.

Figure 1. MSE of 100-Year Flood Estimators with a 50-Year Gage Record and a 200-Year Historical Period

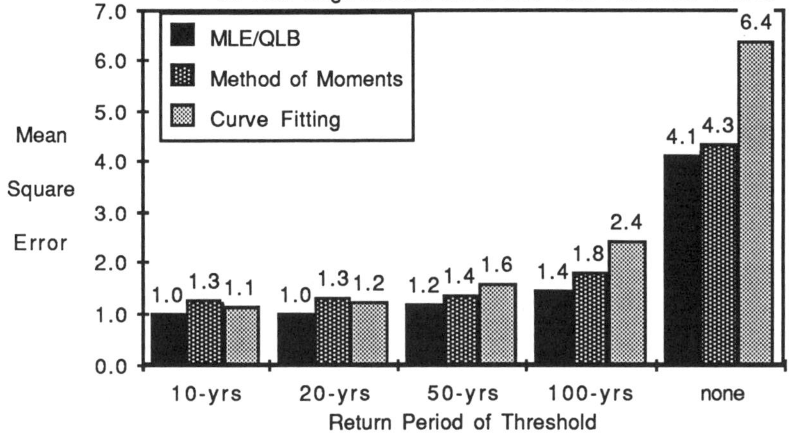

Figure 2. Average Gain (in percent) of 100-Year Flood Estimators when Estimating 3 Parameters

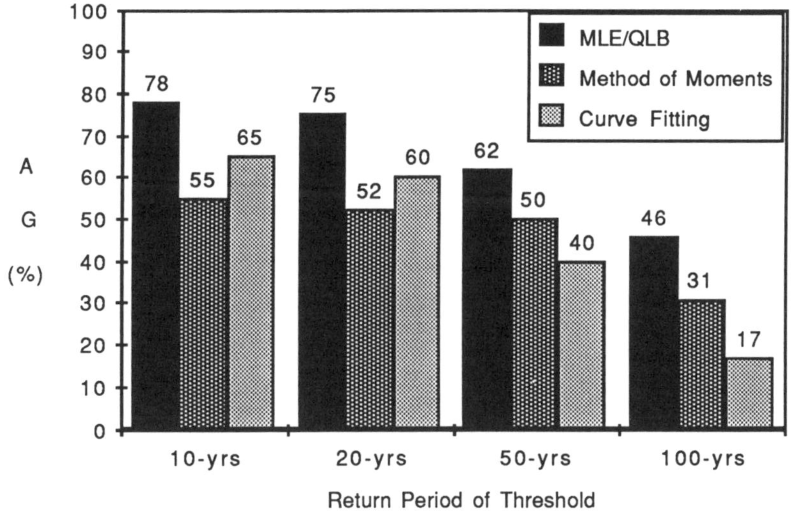

Figure 3. Average Gain (in percent) of 100-Year Flood Estimators when Estimating 1, 2, and 3 Parameters

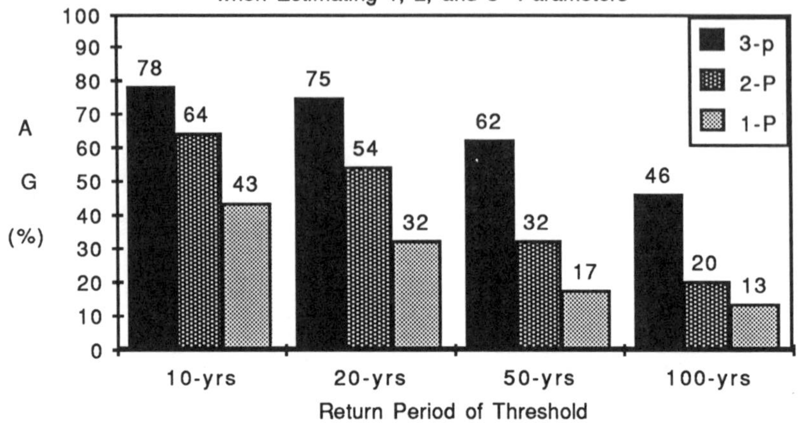

Figure 4. MSE of 100-Year Flood Estimators with a 50-Year Gage Record and 200-Year Historical Period of Observation

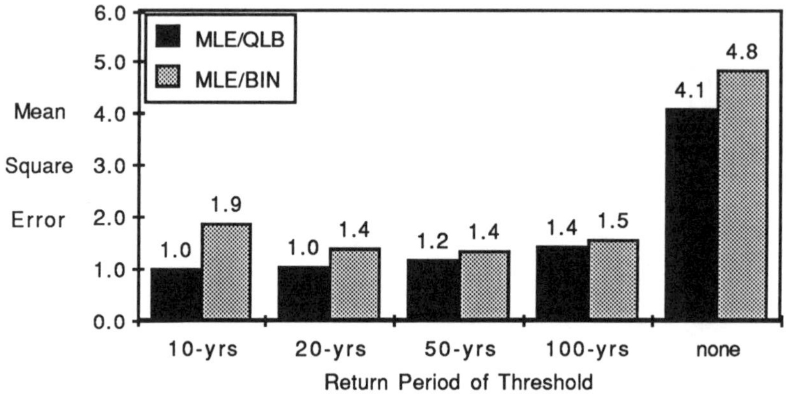

Figure 5. Average Gain (in percent) of 100-Year Binomial Flood Estimators when Estimating 1, 2, and 3 Parameters

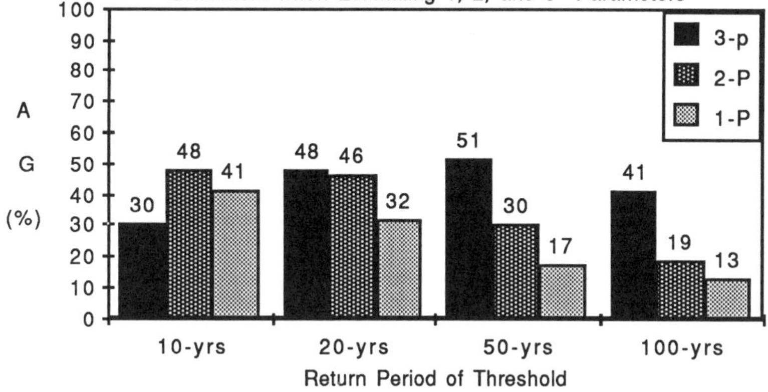

4.5. Regional Analysis

Several studies have considered the value of historical flood data with respect to the development of regional flood quantile models. Tasker and Stedinger [1985] show that both censored data and binomial-threshold-exceedance data can result in better estimation of a regional model's parameters. However, Hosking and Wallis's [1986a,b] analysis employing the index flood method and type II censoring found that improvements in the regional flood model did not translate into better at-site quantile estimators. Of course, as shown in Figure 3, at-site historical data would still be of use with such methods to improve estimates of the parameter employed at each site to scale the regional growth curve.

5. CONCLUSIONS

This paper considers the origins, value, and proper interpretation of several kinds of historical flood information that can be employed in flood frequency studies. The performance of several flood-quantile estimators that utilize such information is also considered. In the cases reviewed here, maximum likelihood procedures are found to make the best use of historical flood information, although other methods also benefit substantially from inclusion of such data.

Use of an imprecise estimate of the single largest historical flood can be detrimental if the imprecision of the estimated flow is neglected in the analysis [Hosking and Wallis, 1986a,b]. However, flood quantile estimators are available which protect against such errors, at a relatively small loss of efficiency, by utilizing only binomial-threshold-exceedance information. Likewise, when the fitted distribution differs from the actual flood distribution, historical flood data's value may increase because it should describe the actual frequency of the larger floods of interest. In conclusion, historical information, when properly interpreted and utilized, should be of substantial value in the estimation of flood quantiles and flood risk.

6. ACKNOWLEDGEMENTS

The support provided by the U.S. Geological Survey, and also the National Science Foundation through Presidential Young Investigator award CEE-8351819, were greatly appreciated. The authors also wish to thank V. Baker, J. Costa, S. Gerould, R. Hirsch, C. Hupp, C. Logan, and R. Therivel, for their valuable comments, assistance, and encouragement.

7. REFERENCES

Baker, V. 1985. Paleoflood Hydrology of Extraordinary Flood Events. U.S.-China Bilateral Symposium on the Analysis of Extraordinary Flood Events, Oct. 21-23, Nanjing, China.

Baker, V., L. L Ely, J. E. O'Connor, and J. B. Partridge. 1986. Paleoflood Hydrology and Design Applications. Presented at the International Symposium on Flood Frequency and Risk Analyses, May 17-21, 1986, Louisiana State University, Baton Rouge, Louisiana.

Beard, L. R. 1975. Hydrologic Frequency Analysis. In: Hydrologic Engineering Methods for Water Resources Development, U.S. Army Corps of Engineers.

Benson, M. A. 1950. Use of Historical Data in Flood Frequency Analysis. *Trans. American Geophysical Union,* **31**(3), pp. 419-24.

Chen, C., Y. Yeh, and W. Tan. 1974. The Important Role of Historical Flood Data in the Estimation of Spillway Design Floods. Peking, China, 15 pp.

Cohen, A. C. 1950. Estimating the Mean and Variance of Normal Populations for Singly Truncated and Doubly Truncated Samples. *Annals of Math. Statistics,* **21**(4), pp. 557-69.

Cohn, T. A. 1984. The Incorporation of Historical Information in Flood Frequency Analysis. M.S. thesis, Cornell University, Ithaca, N.Y.

Cohn, T. A. 1986. Flood Frequency Analysis with Historical Flood Information. Ph.D. thesis, Cornell University, Ithaca, N.Y.

Condie, R. 1986. Flood Samples from a Three-parameter Lognormal Population with Historic Information: The Asymptotic Standard Error of Estimate of the T-year Flood, *J. of Hydrology,* **58**(1/2), pp. 47-61.

Condie, R. and K. Lee. 1982. Flood Frequency Analysis with Historic Information. *J. of Hydrology,* **58**(1/2), pp. 47-61.

Condie, R., and P. J. Pilon. 1983. Fitting the Log-Pearson Type III Distribution to Censored Samples: An Application to Flood Frequency Analysis with Historic Information. *Proc. of the 6th. Canadian Hydrotechnical Conference,* Ottawa, Canada, pp. 11-22, June 2-3.

Cong, S. 1985. Measurement Error of Flood Flow and Its Effects on the Results of Flood Flow Frequency Analysis. U.S.-China Bilateral Symposium on the Analysis of Extraordinary Flood Events, Oct. 21-23, Nanjing, China.

Cong, S., W. Tan, S. Huang, Y. Xu, W. Zhang, Y. Chen. 1979. Statistical Testing Research on the Methods of Parameter Estimation in Hydrologic Computation. East China College of Hydraulic Engineering, Nanjing, China, 26 pp.

Costa, J. E.. 1974. Stratigraphic, Morphologic, and Pedologic Evidence of Large Floods in Humid Environments. *Geology,* **2**(6), pp. 301-303.

Costa, J. E., 1978. Holocene Stratigraphy in Flood Frequency Analysis. *Water Resources Research,* **14**(4), pp. 626-32.

Costa, J. E. 1984. Fluvial Paleoflood Hydrology. *EOS, Trans. American Geophysical Union,* **65**(45), p. 890.

David, H. A. 1970. *Order Statistics.* John Wiley and Sons, New York.

Gerard, R. and E. W. Karpuk. 1979. Probability Analysis of Historical Flood Data. *J. of the Hydraulics Division, ASCE,* **105** (HY9), pp. 1153-65.

Hald, A. 1949. Maximum Likelihood Estimation of the Parameters of a Normal Distribution which is Truncated at a Known Point. *Skandinavisk Aktuarietidskrift,* **32**(1/2), pp. 119-34.

Helley, E. J., and V. C. LaMarche. 1973. Historic Flood Information for Northern California Streams from Geologic and Botanical Evidence. *U.S. Geological Survey Professional Paper 485-E,* 16 pp.

Hirsch, R. M. 1985. Probability Plotting Position Formulas for Flood Records with Historical Information. U.S.-China Bilateral Symposium on the Analysis of Extraordinary Flood Events, Oct. 21-23, Nanjing, China.

Hirsch, R. M., and J. R. Stedinger. 1987. Plotting Positions for Historical Floods and their Precision. To appear, *Water Resources Research.*

Hoshi, K., J. R. Stedinger, and S. J. Burges. 1984. Estimation of Log-Normal Quantiles: Monte Carlo Results and First-Order Approximations. *J. of Hydrology,* **71**(1984), pp. 1-30.

Hosking, J. R. M., and J. R. Wallis. 1986a. Paleoflood Hydrology and Flood Frequency Analysis. *Water Resources Research,* **22**(4), pp. 543-550.

Hosking, J. R. M., and J. R. Wallis. 1986b. The Value of Historic Data in Flood Frequency Analysis. *Water Resources Research,* **22**(11), pp. 1606-12.

Hupp, C. R. 1982. Extension of Flood Record for Passage Creek, Virginia, Utilizing Tree-Ring Analysis. *Bull. of Ecological Soc. of America,* **63**(2), p. 78.

Hupp, C. R. 1983. Geo-Botanical Evidence of Late Quaternary Mass Wasting in Block Field Areas of Virginia. *Earth Surface Processes and Landforms,* **8**, pp. 439-50.

Kirby, W. 1981. *Instructions for Peak Flow File.* United States Geological Survey, Vol. 4, Chapter 1, Reston, Virginia.

Kochel R. C., and V. R. Baker, and P. C. Baker. 1982. Paleohydrology of Southwestern Texas. *Water Resources Research,* **18**(8), pp. 1165-83.

Leese, M. N. 1973. Use of Censored Data in the Estimation of Gumbel Distribution Parameters for Annual Maximum Flood Series. *Water Resources Research,* **9**(6), pp. 1534-42.

Natural Environmental Research Council. 1975. British Flood Studies Report, Vol. I, Hydrological Studies. 550 p., London, England.

Patton, P. C. and V. C. Baker. 1977. Geomorphic Response of Central Texas Stream Channels to Catastrophic Rainfall. In: *Geomorphology of Arid and Semi-Arid Regions,* pp. 189-217, D. O. Doehring (editor), State University of New York, Binghamton.

Russell, S. D. 1982. Flood Probability Estimation. *J. of Hydraulics Division, ASCE,* **108**(HY1), pp. 63-73.

Russell, S. D. and J. D. Bonser. 1980. *Peak Flood Estimation Program.* Civil Engineering Program Library, Dept. of Civil Engineering, University of British Columbia.

Sigafoos, R.S. 1964. Botanic Evidence of Floods and Flood Plain Development. *U.S. Geological Survey Professional Paper 485-A,* 35 pp.

Stedinger, J. R. 1980. Fitting Log-Normal Distributions to Hydrologic Data. *Water Resources Research,* **16**(3), pp. 481-90.

Stedinger, J. R., and T. A. Cohn. 1986. The Value of Historical and Paleoflood Information in Flood Frequency Analysis. *Water Resources Research,* **22**(5), pp. 785-93.

Stedinger, J., R. Therival, J. Grygier, and R. Wu. 1986. *Analysis of the Frequency of Large Floods on the Salt and Verde Rivers using Historical and Paleoflood Information.* Report to the Salt River Project, Phoenix, Arizona ; Dep. of Environ. Engr., Cornell Univ., Ithaca, N.Y., October.

Stedinger, J. R., and V.R. Baker. 1987. Surface Water Hydrology: Historical and Paleoflood Information. IUGG Quadrennial review for 1983-1986, to appear, *Reviews of Geophysics and Space Physics.*

Tasker, G. D., and J. R. Stedinger. 1985. Regional Regression of Flood Characteristics Employing Historical Information. U.S.-China Bilateral Symposium on the Analysis of Extraordinary Flood Events, Oct. 21-23, Nanjing, China.

Tasker, G. D., and W. O. Thomas. 1978. Flood Frequency Analysis with Pre-Record Information. *J. of the Hydraulics Division, ASCE,* **104**(2), pp. 249-59.

Thomson, M. T., W. B. Gannon, M. P. Thomas, G. S. Hayes. 1964. Historical Floods in New England. *United States Geological Survey Water Supply Paper 1779-M,* 105 pp.

United States Water Resources Council Hydrology Committee. 1982. *Guidelines for Determining Flood Flow Frequency,* Bulletin 17B (Revised). U.S. Govt. Printing Office, Washington, D.C.

Wall, D. J., J. C. Herrin, D. W. Newton, and D. F. Kibler. 1983. Use of Limited At-Site Flood Data in Estimating Flood Peaks. Presented at the Fall Meeting of the American Geophysical Union, San Francisco, December, 1983.

Wang, S-X and B. J. Adams. 1984. *Parameter Estimation in Flood Frequency Analysis.* Publication 84-02, Department of Civil Engineering, University of Toronto, April, 1984.

Yanosky, T. M. 1983. Evidence of Floods on the Potomac River from Anatomical Abnormalities in the Wood of Flood-Plain Trees. *U.S. Geological Survey Professional Paper 1296,* 42 p.

Zhang, Y. 1982. Plotting Positions of Annual Flood Extremes Considering Extraordinary Values. *Water Resources Research,* **18**(4), pp. 859-64.

Zhang, Y. 1985. On the Role and Treatment of Outliers in Probability Estimation Method of Flood Frequency Analysis. U.S.-China Bilateral Symposium on the Analysis of Extraordinary Flood Events, Oct. 21-23, Nanjing, China.

PALEOHYDROLOGIC DATA AND FLOOD FREQUENCY ESTIMATION

William L. Lane
Hydraulic Engineer
Engineering and Research Center
Bureau of Reclamation
PO Box 25007
Denver, CO 80225

ABSTRACT

The Pecos River is used to illustrate the use of paleohydrologic data in estimating flood frequency distributions. In this study, systematic, historical, and paleohydrology records are examined. The value of each of these types of records in the estimation of rare flood frequencies is demonstrated. The Pecos River is characterized by very long (1900-present) systematic records which, except for two very significant floods, indicate a very well behaved frequency distribution. The peak flood flows in both 1954 and 1974 were outstandingly large and would not have been expected given only the other systematic records. Those two floods create both problems and rather unique opportunities for alternative frequency analyses. The methodology of the Interagency Advisory Committee on Water Data Bulletin 17B is used to estimate the frequency distribution under various assumptions of data availability. A serious limitation was discovered in the Bulletin 17B procedures for handling historic flood information. The limitations found in this study only affect the historic flood adjustments as outlined in appendix 6 of Bulletin 17B. They do not in any way affect the basic flood frequency procedures set forth in Bulletin 17B and do not reduce the general usefulness of those procedures. An alternative approach is proposed and applied. The existence of paleohydrology information regarding peak flows provides a basis for additional input reflecting some of the highest flows of the last 4,500 years. These data, based upon sediment depositions, are used as a form of historic records to augment the systematic records.

INTRODUCTION

This study represents an attempt to estimate the frequencies associated with extremely rare floods. By their very nature, there are very little data normally available to make these estimates. At the same time, although data to support these estimates are scarce, the estimates are of great importance. For water resource projects, the sizing of major structural features, such as spillways, is dictated by the larger rare floods. Several sources of information can be utilized in making flood frequency estimates. This case study is unusual in that several types of information are available and several alternative frequency analyses can be performed.

V. P. Singh (ed.), Regional Flood Frequency Analysis, 287–298.

Systematic records are available for the Pecos River dating from 1900. This is among the longest systematic records found in the United States. Most gaged records are much shorter. In addition, the gaged records contain two very large floods (1954 and 1974) which can be treated either as part of the gaged record, as unrepresentative data and thrown out, or as historic data. Paleohydrology data, based on sediment depositions, provide a source of even more information. The paleohydrology data, by their nature, require that several interpretations and assumptions be made before they can be utilized. Once these requirements are met, the paleohydrology data may also be used as a form of historic data.

For clarity, the term "systematic" refers to data which have been collected in a systematic or consistent manner such as gaged data. Data of this type would be continuous over periods of time and reflect both large and small annual floods. By contrast, historic records are data which only include rather extreme events and as a result data are not continuous. Historic data usually take the form of "the largest flood during the last 150 years," or "the largest three floods during at least the past 300 years." Historic data usually consist of a set of magnitudes of large events associated with a fairly long time period, the historic period. Paleohydrology data can be considered to be a form of historic data, although often the paleohydrology data do not include the flood magnitudes. In that case, the data may be in the form of a count of the number of floods exceeding a certain flood magnitude during a fairly long time period.

Because of the extensive data and the variety of sources of data, the Pecos River offers perhaps one of the best sites in the United States for demonstrating the frequency estimation of very rare floods. This case study offers some insight into the potential and drawbacks associated with statistical methods. Increasingly, efforts are being made to extend frequency curves far beyond the period of systematic records. Case studies, such as this, shed some light on the question of how reliable are these estimates.

The basic methodology used for comparison is that of Bulletin 17B, which provides primary guidance to all United States Federal agencies in carrying out flood frequency calculations. A serious drawback in the Bulletin 17B approach to historic data was discovered during this study. An alternative approach is proposed for use with historic data. This drawback does not appear in the normal application of Bulletin 17B. It only manifests itself in the treatment of historic information.

Three different frequency curves are selected for comparison:

Case 1. Analysis of the entire systematic records through 1982 except for the exclusion of the floods of 1954 and 1974. This analysis provides results that would be obtained if the two largest events of record were thrown out as being unrepresentative.

Case 2. Analysis of the entire systematic records with additional historic data based upon paleohydrology evidence. In this case, 1954 and 1974 are also treated as historic data points. The historic time period used in this case is 1000 years. The procedure used for historic data analysis is that of appendix 6 of Bulletin 17B.

Case 3. The same analysis as for case 1 except that an alternative approach is used for the analysis of the historic data.

SETTING

The Pecos River Basin is located in the South-Central United States and is part of the Rio Grande River drainage. The Pecos River flows in a generally south by southeast direction starting in northern New Mexico near Sante Fe. It flows south through New Mexico and then southeast through Texas until joining the Rio Grande River at the Mexico border between the Texas towns of Comstock and Langtry. The basin is characterized by a moderate relief with maximum elevations of nearly 1,200 meters and an elevation at the mouth of slightly more than 300 meters. Although a large basin of over 90,000 square kilometers, the runoff is less than 25 millimeters per year. The climate is generally hot and dry and the vegetation is typical of that found in arid and desert areas. The river itself, like most streams in arid areas, is subject to low flows most of the year with flooding typically of short duration with high peak flow values. This flashy nature of the flooding coupled with rather infrequent flooding is typical of basins of this type.

RECORDS

The systematic or gaged records for this river are among the longest in the United States, dating from 1900 to present. For this study, the location of concern is very near the mouth of the Pecos River. Although the records used in this study come from several different gages, they all represent essentially the same drainage area and the same basic location on the river. More detailed information regarding the flood values and gage locations is contained in Lane (1985).

Included in the systematic records are two very large floods which are considered to be historic records. That is, they are considered to be the largest floods to have occurred over some time span larger than that of the systematic records. The magnitude of the 1954 flood (26,800 m^3/s) is over 8 times the next largest flood of record at that time, the 1932 flood (3,300 m^3/s). The 1974 flood (16,300 m^3/s) is the only other flood to exceed the 1932 flood. Normal historic treatment of these two floods would be limited to considering them the two largest floods since 1746, when the first permanent settlements were established in the area. However, paleohydrology methods indicate that these were the largest floods in a much longer time interval and this information is of much more value. (Patton and Baker, 1977; Baker et al., 1979).

Paleohydrology methods have been applied to the Pecos River in several locations and have been reported in several publications (Patton and Baker, 1977; Baker et al., 1979; Kochel, et al., 1982; Kochel and Baker, 1982; Baker, 1982; Baker et al., 1983). This work done at the University of Texas, and later other universities, seems to hold great promise in providing additional information which can be used to extend frequency curves far beyond that possible using the systematic records. The methods of gathering and interpreting the paleohydrology data are well documented by these references and will not be covered in this paper.

The paleohydrology data from several reports (principally Kochel, et al., 1982; Baker et al., 1983) have been examined and compiled as shown in Table I (Lane, 1985). These data come from a site on the lower Pecos River known as the Arenosa Shelter, just upstream of the mouth of the Pecos River.

PALEOHYDROLOGY DATA

As shown, in Table I, the raw paleohydrology data are in the form of counts of floods exceeding various levels over different time periods for each set of counts. To be useful for frequency analysis, it is desirable to convert this raw data into an equivalent number of historic data peaks associated with an equivalent historic record length. The following procedure was used in this study. First, the return periods for the truncation levels are estimated as shown in Table I. Note that the floods of 1954 and 1974 are included in the analysis as historic data. The return periods are estimated using the standard plotting position formula given in Bulletin 17B with the two parameters, a and b, both set equal to four-tenths. Thus, there are now five points to plot on a log probability plot, Figure 1. This results in a paleohydrology implied frequency curve. In this study, the fitted curve is constructed by eye, although a mathematical fit could easily be developed. Also, in this study, only a simple log-normal distribution is used for the paleohydrology frequency curve. It is fairly reasonable to assume that this simple approach does give reasonable results which would be very close to those obtained with more sophisticated methods. Up to this point, this procedure follows closely that used by Patton and Baker (1977). Since each of five points used is represented by a different number of years or different historic periods, a weighted average of the five historic periods is used to define the overall historic period to be used with the paleohydrology data. If all of the points had been based on a common length period, say 700 years, then the curve's historic period would have been that length, 700 years. However, since each point is based on a different length and since each point is based on a different number of counts, a weighted average is used, weighting the total period for each point by the number of counts involved. The weighted average is close to 1,000 years and that equivalent historic period is adopted.

With the historic period defined at 1,000 years, it is necessary to define the equivalent number of "historic" flows to be furnished from paleohydrology records. For this case study, the number of flows to be defined by the paleohydrology data is taken as the number of flows above the lowest truncation level that would be expected in a 1000-year period based upon the paleohydrology frequency curve.

Table I. Paleohydrology Data

Number of Floods	Time Period (yrs before 1982)	Estimated Exceedance Discharge (m^3/s)	Plotting PositionB/ (yr)
4	4,450 - 4,150	5,100	83
4	4,150 - 3,350	6,100	222
2	3,600 - 3,220	6,800	238
5	3,220 - 2,230	8,500	215
5A/	2,440 - 0	10,800	530

A/ Three Paleohydrology floods plus 1954 and 1974 floods.

B/ Calculated by the formula: (N + 0.2)/(M - 0.4), where N = length of time period in years, M = number of floods during the time period. This plotting position formula is that contained in Bulletin 17B, (N-a-b+1)/(M-a), where the parameters, a and b, are both set to four tenths.

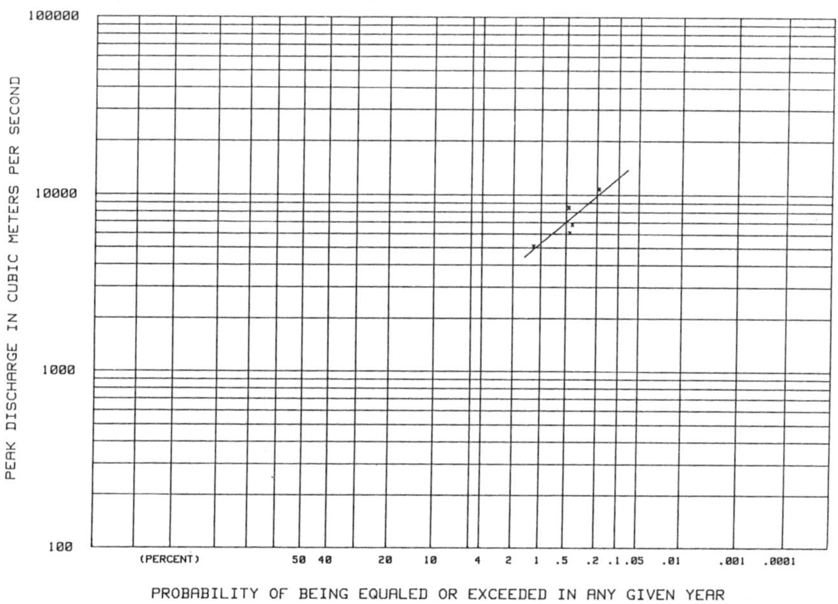

Figure 1. Frequency Distribution Estimated from Paleohydrology Based Historic Data

Based on this, the 11 highest flows for the historic period are defined by the paleohydrology data frequenc curve.

HISTORIC DATA ANALYSIS

The first historic data treatment approach chosen for this case study is to analyze the available data using the standard procedures given in Bulletin 17B. The paleohydrology data were interpreted and modified as just described before being used in parameter estimation for the log-Pearson distribution. It was apparent that the Bulletin 17B treatment of historic data contained a very significant flaw. That procedure is outlined in appendix 6 of Bulletin 17B. It is to be emphasized that this flaw only affects the treatment of historic data and not the typical frequency analysis. The equations fail to produce the desired results. This flaw was found to cause erroneous results when using the long extensions associated with paleohydrology data. In the case of the Pecos River, it was found that the inclusion of the paleohydrology data actually caused the estimated frequency curve to be worse than if the data had not been used at all. It was found that the use of historic data according to appendix 6 of Bulletin 17B will always increase the mean, standard deviation, and skew of the log flows. This happens in all cases, even those where it is quite obvious that the historic data indicated the opposite should take place.

The basic philosophy of Bulletin 17B with regard to historic data needs to be discussed in order to clarify the flaws that were discovered. The historic data adjustment is claimed to be a method for improving frequency estimates (Interagency Advisory committee on Water Data, 1982, page 5). On page 19, it is stated that "The underlying assumption to this adjustment is that the data from the systematic records are representative of the intervening period between the systematic and historic record lengths." Further information on the philosophy used is given on page 6-1 where it is stated, "Flood information outside that in the systematic record can often be used to extend the record of the largest events to a historic period much longer than that of the systematic record."

In application of this philosophy, two problems were discovered. The first involves the assumption that the systematic records are representative of the entire historic period less the historic data years.

An example is helpful in illustrating the problem. Assume that a set of data exists which includes five historic peaks which are the largest floods out of a historic period of 2,000 years. Also assume that 20 years of systematic records are available. In this case, the systematic records would define a distribution which 1,995 years out of the 2,000-year historic period would be assumed to follow. Bulletin 17B treats the systematic records as representative of the 1,995 years and in effect fits a distribution to the systematic records which then is used to provide 1,995 data points. Had this distribution been used to furnish only the lowest 1,995 points out of

2,000, the 17B procedure would be reasonable. In effect, the largest five points out of 1,995 are provided and they are conceptually equivalent to the five historic data points, if the systematic distribution is indeed correct and the 17B approach double counts the largest values. In all cases, by not eliminating the double counting, the flow estimates for rare floods are artifically enlarged. It must be pointed out that the 17B procedure does not in fact calculate the 1,995th data point, but accomplishes the same effect by weighting the systematic records in a procedure that has exactly the same effect.

The second problem with the 17B approach is that very little weight is given to the historic data. In the example, the systematic records are given nearly 400 times as much consideration as is given the historic data, in spite of the fact that one-fifth of the data is historic.

An interesting, although disturbing, observation can be made that the relative weights of the historic data set and of the systematic data set remain unchanged regardless of the number of years of systematic records. Common sense tells one that the weight given historic data should decrease as the number of years of systematic records increases and should increase if fewer years of systematic records are available.

As an alternative to the historic data approach of Bulletin 17B, an approach is proposed in this paper. Basically, the approach is to construct an artificial set of data equivalent in length to the historic period. A standard 17B frequency analysis is then performed on this set of data. The set of data is developed by (1) using the systematic records to define the lower values up to a certain return period, (2) using the historic values to define the highest values, and (3) using a weighting procedure to define the input points between the upper and lower values. If there are H historic values, those values are simply used as the largest values of the artificial data set. The frequency distribution fit to the systematic records is used to provide values up to a point which is felt to be approximately of the same reliability as the lowest value of the historic data. In the example, the reliability of the 5th largest out of a 2,000-year period is used to select an equally reliable point on the distribution based upon 20 years of systematic record. Because of the possibility of lost historic data and the higher potential for inaccuracies in the historic flow estimates, it is not clear how the relative reliabilities should be addressed. As it turns out, this is not a very critical decision. As a form of compromise between several potential approaches, the position on the systematic distribution corresponding to the (H+1)/2 value among the systematic records is used as the upper limit for artificial data from that distribution. For points between there and the lower point of the historic flow distribution, points are estimated using a straight line drawn on log-Normal probability paper between these two end points.

PROBLEM APPROACH

As discussed already, the approach is to perform a frequency analysis for various combinations of data. The first of these is to assume the systematic records were available, but no other data.

The analysis is to use all of the systematic records with the exception of the two largest events, 1954 and 1974. This is an analysis showing what a hydrologist might calculate today. The two extremely large events could well be assumed to be unrepresentative of the flood flow distribution. For these quite lengthy systematic records, both of these events appear to be much more rare than what one would usually expect. Figure 2 illustrates frequency curves based on the systematic records, both excluding and including the 1954 and 1974 values.

The second and third cases make full use of all of the paleohydrologic data and also treat 1954 and 1974 as historic records. These might represent the "best" estimated frequency curves possible using flow records of all types.

The second case uses these data with the 17B approach while the third case is the modified approach. Figure 3 illustrates the combining of the systematic and historic distributions according to the modified approach.

RESULTS OF ANALYSIS

The three cases were analyzed and a frequency distribution was fit to each. The log-Pearson type III distribution parameters are shown in Table II and the distributions are plotted in Figure 4. The log-Pearson type III distribution is defined by three parameters, the mean, standard deviation, and coefficient of skewness, all calculated in term of the logarithms of the flows. Bulletin 17B describes the estimation of these parameters and the basic frequency analysis in detail.

The mean, standard deviation, and skew all increase with the historic data approach as outlined in Bulletin 17B. What is meant by an increase in skew is that the skew becomes more positive. On the other hand, the historic data treatment proposed, decreases those statistics. This is as expected, since the paleohydrology data indicate that estimated frequencies (exceedance) for large floods are too large and that the frequency curves are too high.

It is noted, that the extremely high flows observed in 1954 and 1974 appear to be extemely rare. A frequency distribution based only upon the systematic records including 1954 and 1974 is much higher than any of the other estimated frequency curves.

CONCLUSIONS

In this study, the paleohydrology data reduced the magnitude of rare flood estimates. The paleohydrology data appear to have improved the

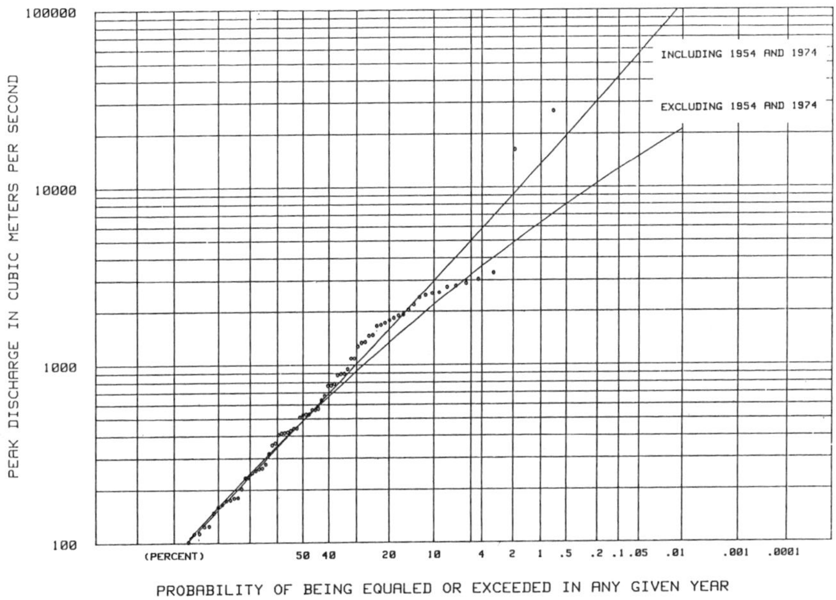

Figure 2. Frequency Distribution Estimated from Systematic Data

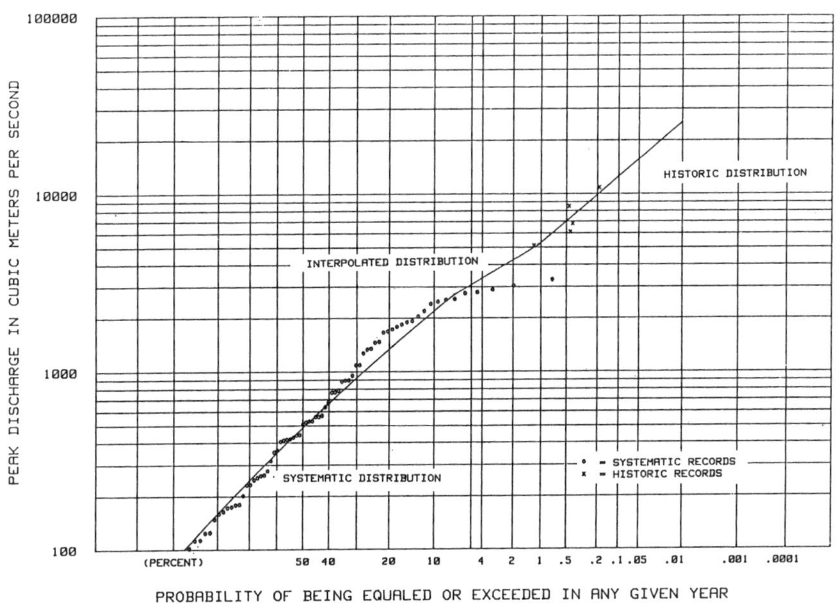

Figure 3. Distribution of Synthetic Data Representing the Combination of Historic and Systematic Data

Table II. Parameters of Estimated Frequency Distributions

Description	Mean	Standard Deviation	Skew
Case 1. 1900–1982 Without 1954, 1974	2.6557	0.5520	-0.3400
Case 2. 1900–1982 With Paleohydrology, Bulletin 17B Approach	2.6691	0.5604	-0.2670
Case 3. 1900–1982 With Paleohydrology, Proposed Approach	2.6534	0.5468	-0.3732

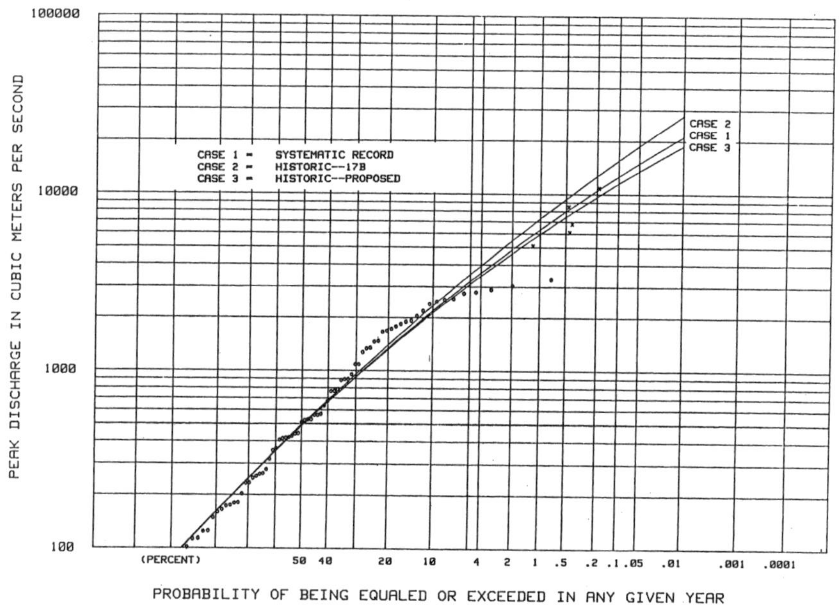

Figure 4. Final Estimated Frequency Distributions

frequency estimates. This conclusion must be tempered with two facts. One is that paleohydrology methods may underestimate flood magnitudes and frequencies if any past flooding deposits have been destroyed. Secondly, the confidence intervals are very wide and what appears to be improvement may well be simply a misinterpretation. Additional experience in using paleohydrology data is needed.

The Bulletin 17B treatment of historic data is not applicable to this type of historic data. As shown in this study, the Bulletin 17B approach in general gives too little weight to paleoflood data, and because of the mathematics involved, produces biased results which in all cases skews the distribution towards higher flow values regardless of what the historic information implies. The methods used in this case study, while adequate for this study, indicate one possible approach but cannot yet be considered as a replacement for the historic data approach of Bulletin 17B. It is emphasized that the basic flood frequency procedures outlined in Bulletin 17B are not in question, only the treatment of historic information.

The general treatment of paleohydrology data requires substantial analysis before the data are in a form amenable to frequency analysis. Much more work is required to develop general guidelines in this area.

REFERENCES

Baker, V. R. 1982. Geology, 'Determinism, and Risk Assessment.' Chapter 10 in _Scientific Basis of Water-Resource Management_, National Academy Press, Washington, D.C.

Baker, V. R., R. C. Kochel, P. C. Patton. 1979. 'Long-Term Flood Frequency Analysis Using Geological Data.' IAHS-AISH Publ. No. 128, _Proceedings of the Canberra Symposium_.

Baker, V. R., R. C. Kochel, P. C. Patton, and G. Pickup. 1983. 'Paleohydrologic Analysis of Holocene Flood Slack-Water Sediments.' Special Publications of the International Association of Sedimentologists.

Interagency Advisory Committee on Water Data. 1982. 'Guidelines for Determining Flood Flow Frequency,' Bulletin 17B.

Kochel, R. C., and V. R. Baker. 1982. 'Paleoflood Hydrology.' _Science_, V. 215, pp. 353-361.

Kochel, R. C., V. R. Baker, and P. C. Patton. 1982. 'Paleohydrology of Southwestern Texas.' _Water Resources Research_, V. 18, pp. 1165-1183.

Lane, W. L. 1985. 'Rare Flood Frequency Estimation, A Case Study of the Pecos River.' _American Geophysical Union Annual Spring Meeting_, Baltimore.

Patton, P. C. and V. R. Baker. 1977. 'Geomorphic Response of Central Texas Channels to Catastrophic Rainfall and Runoff.' In, D. O. Doehring (editor), *Geomorphology in Arid Regions*, pp. 189-217, State University of New York, Binghamton.

THE 100-YEAR FLOOD

Emmett M. Laursen
Professor Emeritus
Department of Civil Engineering and Engineering Mechanics
THE UNIVERSITY OF ARIZONA
Tucson, Arizona 85721 U.S.A.

ABSTRACT. The standards for hydraulic design have changed over the years so that now the 100-year flood is a standard for many, if not most, flood and drainage projects. It being the standard, other possible concepts tend to be ignored. Rather than a design to pass the 100-year flood, a design which curtails or suspends operations but sustains only minor damage might, in the long run, be better. A related factor is that the difference between the catastrophe which can only happen once and the recurring loss is often neglected.

The need to know the magnitude of the 100-year flood leads to great efforts to predict that flood from inadequate records, usually by assuming a probability density function with an infinite tail. Because there is in actuality an upper limit (variously defined), such procedures result in undue reliance on the characteristics of the unrealistic infinite tail and can lead to misleading conclusions. An extremely interesting line of research by geomorphologists is the investigation of paleofloods which can establish the maximum flood to be truly expected.

Especially in arid and semi-arid regions, the flooding problem can be bank erosion, avulsion, aggradation, or degradation as well as inundation. There is a need to be able to assess the damage potential due to sedimentation effects on a probability basis similar to standard flood analysis. Sedimentation effects can be cumulative in many small floods as well as due to a single large flood.

UNANSWERED QUESTIONS AND UNQUESTIONED ANSWERS

A 5-year flood will make the local newspaper and TV; a 10-year or 20-year flood will probably make the National news. If there are some good visuals -- like people being rescued from rooftops by hovering helicopters, or bridges or buildings falling into a roaring torrent -- if the estimated losses are many million of dollars, if the rescue centers are crowded with the dispossessed, and if there are a few deaths, the media will herald the news for a certainty. Bad things happen in floods which are not really unusual. Within a memory span,

V. P. Singh (ed.), Regional Flood Frequency Analysis, 299–308.

everyone will have experienced floods of moderate magnitude and frequency either directly as a participant, indirectly as an onlooker, or vicariously on TV. Although few will understand adequately the many problems inherent in a flood, most will be impressed with the uncontrolled power of the flood and the devastation resulting from the flood. No wonder, then, that the public feels safe when they are told that flood-control works, drainage systems, or bridges are designed for the 100-year flood. All lesser floods should also be harmless--and they do not expect to live a hundred years.

A recent cartoon in the American Scientist (Harris, 1985) pictured the entrance to a research institute with directional signs for unanswered questions to the right and unquestioned answers to the left. One would be torn to know which direction to go if seeking advice on the 100-year flood. There are many fascinating unanswered or inadequately answered questions about 100-year floods, but it is the thesis of this paper that it is more important to look at the unquestioned answers. During an inquiry, new unanswered questions would probably surface, but their answers might not be fruitful unless and until the rationale for design is established on a substantive basis. Some of the things some people believe and do without question need to be questioned. Creative, innovative site-specific solutions should be encouraged, but not to the exclusion of good, old, standard solutions.

BRIDGES

A bridge is a good example to illustrate some of the ramifications of the 100-year rule. A bridge is expensive enough to be a real concern, but simple enough to grasp the essentials of the problem. Prior to the Interstate Highway System, most bridges were designed for the 25-year flood, or the flood of record--unless they were somehow special. As higher standards were being set for the Interstate System, Carl Izzard, Mr. Hydraulics of the Bureau of Public Roads, (who simply knew that the 25-year rule was not good enough) succeeded in getting the standard increased to the 50-year flood by the force of his personality. When the Federal Flood Insurance Program was enacted, it spoke of the 100-year flood; shortly thereafter a Presidential Order instructed all agencies to consider the 100-year flood in the design of structures in flood plains. Most bridges, thenceforth, were designed for the 100-year flood.

Putting aside for the moment the question of what is meant by the words "designed for", a simple analysis will show that this is still not good enough (Laursen, 1970). The reason is that if a bridge is assigned a life of 25 years, there is a 0.25 chance it will fail in a flood greater than the 100-year flood during its lifetime, and that it will not cost 25 percent more to build the bridge to withstand the maximum flood truly expected. Of course, one should admit there may be some bridge that should be designed for the 100-year flood, but then there may also be some bridge that should be designed for the 25-year flood, or the 500-year flood. The Federal Highway Administration seems flood, or the 500-year flood. The Federal Highway Administration seems to have adopted the viewpoint that the bridge should not fail (Corry, et.al. 1980) although it is not clear just how explicit they have been in their instructions to the States.

One must also admit that in designing a bridge in respect to hydraulics, nothing can be done precisely or with complete confidence. Several steps are involved and the questions involved in each step are not answerable as well as one would like. First, there is the flood magnitude-frequency relationship for the particular bridge site. Then there is the hydraulics and stream behavior of the particular site (this is the hardest part). After these first two steps, it is comparatively easy to decide on the scour and the hydrodynamic forces (if the geometry of the bridge has been tested by someone). Finally, there should be an economic analysis of the sum of construction cost and probable loss to find the most economical solution in the long run. The process can be repeated many times if many variations of geometry, foundation type, and superstructure type are considered. Such repetition may make the analysis and design process long and tedious, but it is essentially this simple. Of course, one must put up with the same inadequately answered questions over and over, and choose the "best" solution with some repetitions. It seems, however, that bridges are expensive enough so it would be an unusual situation in which a bridge should be designed to fail. If the value of the bridge itself in some instance would not be sufficient to arrive at this conclusion, there are always the losses of traffic delay, traffic accidents and traffic deaths which can also be considered and which will push the decision in the direction of safety.

In the abstract there is always the possibility of not being able to justify the cost of a bridge which will not fail. Consider the "Tale of the John Day Bridge" which served as an example of Bayesian risk analysis (Benjamin & Cornell, 1970). The John Day Bridge was constructed on a relocated stretch of highway because the John Day Dam, under construction, was going to drown out the old highway; it consisted of a main span and several approach spans on each side. One of the piers of the main span turned out to be located on an unexpected, difficult-to-excavate formation. The contractor was allowed to pour the footing higher than called for in the design. A flood occurred; the pier and two spans fell down. In an early report on the failure (ENR 1965), it was indicated that $150,000 was "saved"; the Government Accounting Office (GAO) later said that because of less excavation and less concrete there was a reduction in cost of $9,200. The textbook example used the $150,000 figure and "fictitious" probabilities in concluding (for illustrative purposes) that the decision was correct.

The possibilities of the decision were that either wise or unwise decision makers could be either lucky or unlucky. If lucky, no one would ever know the extent of wisdom. If unlucky, the unwise decision maker would -- at the very least -- be severely chastised. If unlucky, the wise decision maker would, nevertheless, have to "prove" that he was wise. The textbook example concluded that Bayes would say the decision maker was wise even though the bridge fell down. But the textbook writers did not seem to know that there was a railroad bridge being built parallel to the highway bridge and that it encountered that same foundation problem. However, the foundation engineer for the railroad bridge asked the contractor the right question: "How much would you charge to move an approach span from one side to the other?" As I heard tell, "Nothing" was the answer. This was obviously a better solution since the bridge did not fall down when the flood came.

What is to be learned from this tale? That risk analysis is likely to be misleading if the ingredients being analyzed are not adequately understood and if imaginative engineering is not practiced.

OLD BRIDGES

In the case of new bridges it is a good bet that proper consideration of probable loss and the cost of insuring against loss will result in designing for the maximum expected flood. However, there are many more old bridges across alluvial rivers than there are new bridges; what of the case of old bridges? The numbers are all different: the nominal life of the old bridge is less, its value is less, the cost of making it less vulnerable is more. What to do, if anything, is not intuitively apparent. Each old bridge is a case unto itself, but there is a huge investment in old bridges out there that (almost) nobody is looking at*. The bridges in a transportation system can vary from one of limited utility to traffic which would not be replaced if destroyed to one of full serviceability and, therefore, having a value equal to the cost of its replacement. For the first, the choice is between tearing it down or keeping a sharp eye on it during floods. For the second, the choice might be as to which of several remedial measures is best. All aspects involved in assessing the vulnerability of bridges to floods need to be re-examined (Laursen, 1984): (a) the flood magnitude-frequency relationship, (b) the existence, definition, and magnitude of the maximum flood, (c) the probability of occurrence of recurring events, (d) the incorporation of an inflation factor in economic analysis, (e) the hydraulics of stream flow, stream behavior, scour, aggradation and degradation, (f) the cost of traffic delay, (g) the probability and cost of traffic accidents and deaths, (h) the remaining life and value of existing structures, and (i) possible remedial measures and their costs.

If the flood to be "designed for" is not the maximum flood, the sum of costs and probable losses has a flat minimum (not a cusp), - which means that the numbers must be correct (relatively rather than absolutely) to be able to choose the best solution. If damage or destruction is an admitted possibility in a best solution, the question of what is meant by the expression "designed for" needs to be questioned. Is it necessary that the bridge sustain no damage and the

---

* The Arizona Department of Transportation is trying to identify existing bridges which are vulnerable to floods and to take corrective actions.

traffic be uninterrupted, or is it only necessary that the bridge be standing and repairable and that traffic can be restored soon after the flood? In some situations, the first definition is most suitable; in others, the second.

What should be clear from this brief consideration of old bridges is that the 100-year flood is not a flood of particular interest--basically for the same reason as it is not for new bridges. The flood any bridge should be designed for is the flood which results in the greatest benefit for the least cost; either it should not fail, or there are a number of designs which are of almost equal merit. Finding the best solution may depend on the imagination of the designer in dreaming up a variation that is both safer and cheaper while it still does the job for which it is needed.

OTHER STRUCTURES AND SYSTEMS

The design of other flood related hydraulic structures and systems should undergo a similar analysis. Whether they should be designed for a frequent, a moderate, a rare, or the maximum flood is a matter of cost and benefit (or possible loss). It is even possible that there can be two "best" designs: one for a frequent flood and one for a rare flood. The designs of the structure itself and of such improvements as might depend on the structure would probably be very different, and the choice might be a matter of financing.

In any event, <u>none</u> should be designed for the 100-year flood, or any other standard flood, <u>unless</u> it happens that this turns out to be the best design. To have confidence in the end result of an economic risk analysis, one must have confidence in the information entered into the analysis.

A culvert seems like a little thing, but the relative values of initial cost and probable loss of culvert and road should not be much different than for a bridge. A culvert can be replaced quicker than a bridge, but the relative cost of each day's delay to traffic may be greater than for a bridge. If the effect on traffic must be considered, it must be done carefully. An investigation of design flood frequency of culverts (Young, et al., 1974) gave a benefit rather than a loss for driving at 55 mph on the detour instead of 75 mph on the Interstate. The conclusion should have been to keep the detour and abandon the Interstate because that benefit would grow with every day. Another flaw in the reasoning was the assumption that the nearest road could serve as a detour even though it very likely crossed the same stream over a culvert of less hydraulic capacity. Since lesser roads could only justify smaller culverts, any detour is more likely to be long than short.

The correct capital recovery factor is also a problem in times of sizable inflation. If the going interest rate is used, the probable losses should be increased by the inflation rate. This will come close to a wash with the same inflation factor on both sides of the equation.

However, there is another feature of culvert design which is very important but seldom addressed: debris. If addressed, it is usually with a suggestion to place debris barriers upstream of the culvert -- but who has seen a debris barrier except an ordinary fence, which would not be strong enough to function as a debris barrier in a major flood? Although debris is obviously random in many ways, it is also obviously highly correlated with floods. Maybe debris is one of those things one must put up with, but that is not to say it should be ignored. Perhaps if only oversized, expensive culverts can be counted on for major, rare floods, it would be better to design the road to be overtopped but passable in anything greater than a moderately frequent flood -- something like a 5-year flood.

Similar questions can be asked about storm sewers and drainage channels. Where does the water go when the design flood is exceeded? The buildings of a shopping mall or the houses of a real-estate development may have a limited life of 25 or 50 years, but they will be replaced and the city has a life of hundreds or thousands of years -- even forever. Could deposition reduce the capacity of the sewer or channel? Will upstream development increase flood magnitudes?

The same questions asked of flood control works and of floodplain delineation and management would reveal that a lot of people someday are going to be very surprised and say, "I thought it was supposed to be safe!" Some people will be lucky during their lifetime, but eventually someone will be the unlucky one. A hundred years is not forever, and the probability of occurrence of the 1,000-year flood in a lifetime is not insignificant.

THE FLOOD MAGNITUDE-FREQUENCY RELATIONSHIP

In order to perform an analysis comparing the costs and benefits of alternative designs, it is necessary to know the flood magnitude--frequency relationship. A few years of flood data, supplemented by rainfall data and flood data from neighboring streams, will give an adequate notion of frequent floods. Larger, rarer floods require more data than is available, if they are to be defined with confidence. Most mathematical functions used to describe flood probabilities have an infinite tail; e.g., Log Pearson III, denying that there is a finite upper limit to the magnitude of the biggest flood that could occur. It is manifestly impossible to design for an infinite flood, therefore, there has to be some rare flood that must serve as the basis for design. But the data are not sufficient to specify the probability of occurrence of the large floods. Izzard wanted a higher standard and settled for the 50-year flood; the Federal Flood Insurance Program felt the need for a larger, rarer flood and settled for the 100-year flood; the Alaska pipeline had to satisfy the environmentalists and politicians and seized on the "5,000-year flood". A procedure for obtaining numbers for the magnitude of these floods was needed that everybody would agree on. The numbers would not need to be the real magnitudes of those rare floods; they would need only to be numbers interpreted and extrapolated from a given set of inadequate data in an agreed-upon procedure, thereby discouraging any questioning since everyone would get the same numbers.

The problem is illustrated in Fig. 1 where the heavy curve is obtained by the standard procedure and the stippled area is where the true curve might be; the problem is only a problem if someone asks questions.

How the picture changes when there is a finite limit to the magnitude of the maximum expected flood is shown in Fig. 2. With such a limit, the cost curve for rarer and rarer floods will flatten off to some value and the optimum design will move towards the rarer floods or the maximum flood. The difference is not merely a trading of the uncertainty of the value of the maximum flood for the uncertainty of the values of all the rare floods, it is rather a change in the range of uncertainty from infinite to finite. Moreover, the uncertainty in the value of the maximum flood truly to be expected can be reduced through research. A number of geomorphologists interested in paleofloods (Kochel, et al., 1974; Webb, 1985) have found evidence of major floods which in effect extend our records back thousands of years. A number of papers or paleofloods are in these Proceedings. If one can be confident that a design is for the largest flood in something like 2,000 to 10,000 years, one should be able to sleep soundly. Of course, one might be a little more conservative if the project is a very large reservoir immediately upstream of a city of millions built on a low floodplain than if the project is a storm drain two feet in diameter.

Does the economic decision change if the magnitude-frequency relation is Curve A or Curve B in Fig. 2? This amounts to asking how well the magnitude of the 100-year flood needs to be known. A considerable number of real projects would have to be investigated in order to answer this question. However, it is reasonable to expect that Curve A would always result in a design based on the maximum flood, but that Curve B might sometimes result in a design based on the 100-year flood. There will always be uncertainty, but then would be the time to ask whether there is some other way to accomplish the desired end at less cost with -- at most -- a minor lessening of benefits. The beauty, elegance, and rightness of a solution that can come from a comprehensive view of the problem plus an imaginative concept was dramatized for several projects by Arthur Morgan (1971) years ago. His cases demonstrated that imagination is the true genius of engineering.

SEDIMENT

Rivers are the geological agents which transport the weathered rock of the mountains to the sea. To ignore the sediment eroded, transported and deposited by the river is folly. The sediment is there and more likely than not when something is done to a river, what happens with the sediment is bad rather than good. Reservoirs fill with sediment and even if it takes a thousand years; there should be a plan of what to do next. In 500 years, the storage capacity is half what it was (maybe less if the deposition is in the shallow headwaters of the pool)

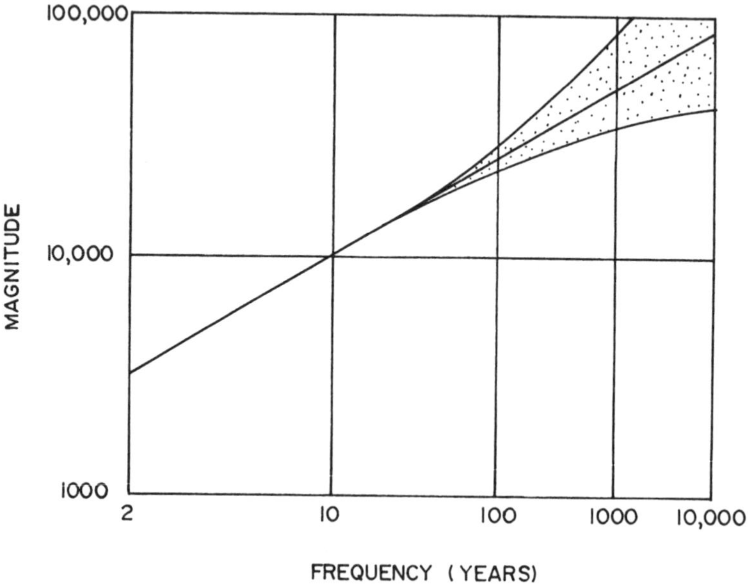

Figure 1. The unlimited magnitude-frequency relationship for floods.

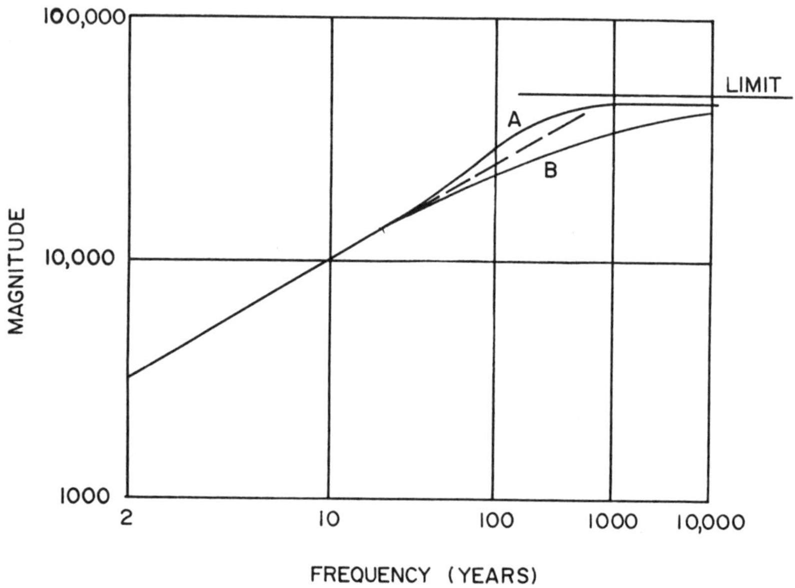

Figure 2. The limited magnitude-frequency relationship for floods.

and most water resource projects will be in trouble whether the goal is hydroelectric power, flood control, irrigation, municipal and industrial water supply, or navigation. In 100 years, there can be a significant detrimental effect on the project, especially if there are (by chance) more than expected very large, very rare flood events.

Upstream of the reservoir there will be aggradation -- and trouble. Downstream of the reservoir there will be degradation -- and trouble. First the aggradation and degradation -- and troubles --will be on the mainstream, then the tributaries will experience the same reactions -- and troubles. Other structures on the mainstream and the tributaries can be affected -- and have troubles.

The other most common river "improvement" is to contain it within levees. Since the sediment being deposited in the sea continually lengthens the river, the river bed and the alluvial valley in which it flows must perforce continually rise. When the land beyond the levees can no longer be a part of the depositional plain, the river and the land inside the levees gradually rise to be above the surrounding land elevation -- a condition surely courting disaster one fine, rainy day.

Textbooks on hydrology and hydraulic engineering pay much more attention to water than they do to sediment. As a consequence, many hydrologists and hydraulic engineers pay less attention to sediment and sedimentation processes than they should. (The greatest mistakes are usually made by people who are oblivious to the fact that they are making mistakes.)

SUGGESTIONS

Presently, many jurisdictions, agencies and firms adopt the 100-year flood rule for many, or most, projects without due consideration because nobody is likely to raise questions. In lieu of this attitude it is suggested that:

1. The 100-year flood should be declared persona non grata; if by chance it is the correct flood to design for, it should be designated the 99-year or the 101-year flood.

2. For all projects, the question should be asked and answered of what happens if the biggest flood truly to be expected occurs.

3. For all projects, a prediction should be made of what will (or might) happen to the sediment load being carried by the flow and to the bed and banks of the river channel.

4. Only if a project is almost like another project can the analysis of costs and benefits be almost perfunctory.

5. For all projects, there should be an examination of what costs what, what results in benefits, and why losses could occur, and then a review of what could be done differently to minimize costs, maximize benefits, and reduce possible losses.

6. The monies now spent trying to find the magnitude of the 100-year flood should be spent on finding the magnitude of the maximum flood truly to be expected.

7. The difference between a recurring loss and a one-time-only catastrophe should be recognized.

8. The effect of inflation should be explicitly recognized in regard to both costs and benefits.

9. The difference between economics and finance should be appreciated.

Too many organizations and agencies (or the people in them) find it too easy to go by some set of rules -- like "design" for the 100-year flood -- rather than to think about answers to unanswered or inadequately answered questions, or to (Heaven forbid) challenge "officially" answered questions.

REFERENCES

Benjamin, J. R. and C. A. Cornell, Probability, Statistics, and Decision for Civil Engineering, McGraw-Hill, New York, 1970.

Corry, M. L., J. S. Jones and P. L. Thompson, 'The Design of Encroachments on Flood Plains Using Risk Analysis,' Hydraulic Engineering Circular No. 17, Federal Highway Administration, USDOT, October 1980.

Engineering News-Record, January 7, 1965, pp. 20-21, March 11, 1965, p. 25.

Harris, S. "Cartoon" in American Scientist, **Vol. 73**, No. 2, March-April 1985, p. 150.

Laursen, E. M., 'Bridge Design Considering Scour and Risk,' ASCE Transportation Engineering Journal, **Vol. 96**, No. TE2, May 1970.

Laursen, E. M., 'Assessing the Vulnerability of Bridges to Floods,' Second Bridge Engineering Conference, Transportation Research Record 950, TRB, NRC, Washington, D.C., 1984.

Morgan, A., Dams and Other Disasters, P. Sargent & Company, Boston, Massachusetts, 1971.

Young, G. R., M. R. Childrey and R. E. Trent, 'Optimal Design for Highway Drainage Culverts,' ASCE Journal of the Hydraulics Division, **Vol. 100**, No. HY7, July 1974.

CHANGES IN HYDROLOGIC CONDITIONS RELATED TO LARGE FLOODS ON THE ESCALANTE RIVER, SOUTH-CENTRAL UTAH

ROBERT H. WEBB
U.S. Geological Survey
300 W. Congress Street
Tucson, AZ 85701

VICTOR R. BAKER
Department of Geosciences
University of Arizona
Tucson, AZ 85721

ABSTRACT. The history of flooding on the Escalante River in south-central Utah illustrates many problems typical of flood-frequency analysis for other rivers in the southwestern United States. Floods between 1909 and 1932 eroded an arroyo as much as 300 m in width and 15 to 17 m in depth from a previously unincised flood plain. A shift to increased amounts, intensity, and variability of summer precipitation corresponds with this period of channel change. Large floods have not occurred since 1940, and the channel within the arroyo walls has narrowed in response to fluctuating hydrologic conditions and the spread of introduced riparian vegetation. Paleohydrologic analysis revealed that four large floods occurred during the last 400 years, with all possibly occurring between 1909 and 1932. The evidence could be equivocally interpreted to represent a stationary or nonstationary distribution of large floods, and the 100-year flood could range from 190 to 630 m^3/s, because of uncertainty in the appropriate length of the historic record. Future research on flood frequency for southwestern rivers would benefit from a robust statistical test for nonstationarity, particularly with emphasis on the effect of use of historic data in flood-frequency analyses.

INTRODUCTION

The estimation of flood frequency for ephemeral or intermittent rivers is difficult in arid regions. Most rivers in the southwestern United States have short gaging records that limit the usefulness of statistical flood-frequency analyses. Recurrent large floods of unexpected magnitude in normally dry channels (e.g. Saarinen et al, 1984) have caused property damages and loss of structures that might have been avoided if the frequency of large floods were better understood. The occurrence of inordinately large flood events in recent years has led some observers to conclude that the annual flood series may be nonstationary for some rivers (Saarinen et al, 1984; Reich and Davis, this volume).

Rivers in the Southwest have a long history of channel change related to flooding. All major tributaries of the Colorado River in

V. P. Singh (ed.), Regional Flood Frequency Analysis, 309–323.

TABLE I. Dates of historic floods that initiated arroyos or channel changes in southern Utah (from Webb, 1985)

River system	Date of flood	Significance
Virgin River	January 8, 1862	Destroyed four towns, initiated an arroyo
Santa Clara River	January 18, 1862	Initiated a 150-m-wide, 4-m-deep arroyo
Kanab Creek	July 29, 1883	Initiated 30-m-wide, 17-m-deep arroyo
Paria River	March 13, 1884	Damaged townsites and farm lands; initiated an arroyo
Escalante River	August 31, 1909	Initiated an arroyo
Fremont River	September 22, 1896	Damaged three towns; initiated an arroyo
San Rafael	August 31, 1909	Changed roads into deep gullies
Pack and Mill Creeks	September 22, 1896	Initiated arroyos; damaged town of Moab
San Juan River	June 18, 1884	Destroyed farmland at Bluff; largest flood in record

southern Utah changed their channels in response to floods between 1862 and 1909 (Table I). The Santa Cruz River at Tucson, Arizona, has changed its channel repeatedly in response to flood events (Betancourt and Turner, in press). The Salt River near Phoenix, Arizona, has migrated laterally as much as 1.6 km in response to large flood events over a 112-year period (Graf, 1983a). The Gila River in east-central Arizona has changed its channel in response to "high flow" and "low flow" regimes (Burkham, 1976b). Southwestern rivers adjust their channel geometry to the time series of flood events and therefore exhibit a memory for flooding.

The Escalante River (Figure 1) provides an example from the Southwest to illustrate the magnitude of watershed and hydro-climatological changes and their influence on the temporal stationarity of large floods. Upper Valley Creek, the headwaters tributary of the Escalante River, is an arroyo that was initiated during a large flood in 1909. Downstream from this arroyo, the bedrock canyon of the Escalante River provides an ideal setting for the reconstruction of peak discharges using "slackwater sediments"

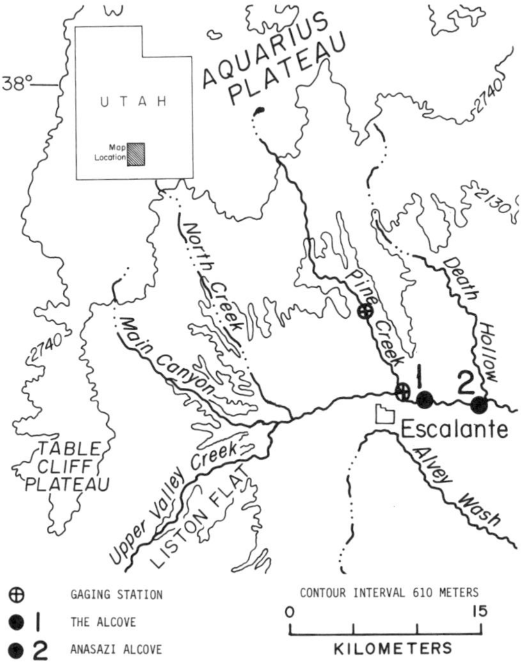

FIGURE 1. The Escalante River and tributaries, south-central Utah.

(Baker, 1985). Gaging stations that have been maintained on Pine Creek and the Escalante River near Escalante since 1943 (Figure 1) provide a systematic record for flood-frequency analysis. Climatic fluctuations, land-use practices, and the spread of introduced riparian species provide a physical basis for questioning the stationarity of the historic flood frequency of this river.

HISTORY OF CHANNEL CHANGE

Pre-Arroyo Channel Conditions

Upper Valley was first described in 1872 during a scientific expedition led by Almon Harris Thompson, Chief Topographer of the Powell Survey (Gregory, 1939). According to Thompson's diary, the headwaters were characterized by grassy meadows and excellent range conditions. Upper Valley Creek was not continuously incised but instead was a discontinuous ephemeral stream above its confluence with Main Canyon. A prominent headcut had become established near the confluence of Upper Valley Creek and Main Canyon, although this headcut apparently did not extend into Upper Valley.

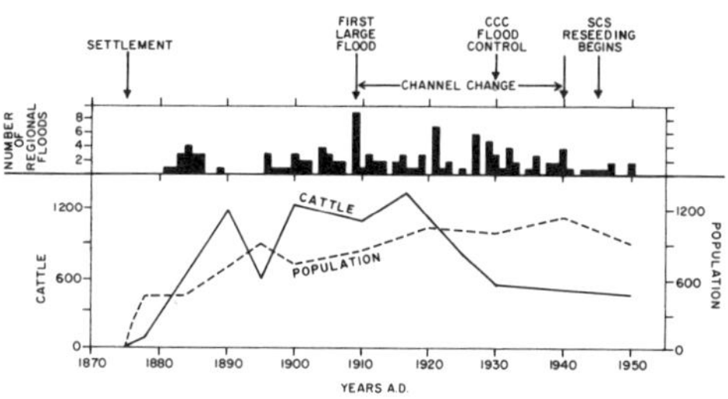

FIGURE 2. Temporal correlation of population changes and livestock numbers with channel change on the Escalante River.

Mormon pioneers settled along the Escalante River and founded the town of Escalante in 1875 (Woolsey, 1964). The flood plain was cleared of vegetation for agriculture immediately after settlement. Canals were dug in the flood plain to allow irrigation. Livestock grazing increased as the population grew and large herds of cattle were introduced by 1880 and sheep by 1886 (Figure 2). Residents observed a large increase in average streamflow after 1882 (Gregory and Moore, 1931), which was thought to be in response to land-use practices (Woolsey, 1964).

William Lewman, surveyor for the General Land Office, described Upper Valley and measured the stream width in 1893-94 (Lewman, 1893-94). The grassy meadows that existed along Upper Valley Creek in 1872 were replaced by sagebrush and greasewood by 1893. The average channel width for Upper Valley Creek and the Escalante River was 8.4 m in 1893, compared with an average of 82 m in 1985.

Floods and Channel Change in Upper Valley

The cutting of the continuous arroyo began during a large flood on August 31, 1909, with a large headcut forming at Escalante (Edson Alvey, local resident, written communication, 1985). This flood destroyed the first staff gaging station (U.S. Geological Survey, 1910) and was the first large flood since settlement in 1875. The headcut was extended during floods in 1910, 1911, 1914, and 1916. Gregory (1918) reported the rapid erosion, and his photographs (Webb, 1985) show that the arroyo below the confluence of Upper Valley Creek and Main Canyon had formed by 1918.

Flooding and associated channel change continued in the 1920's and 1930's. During several floods in the summer of 1921, Upper Valley Creek eroded its flood plain locally to the canyon walls (Deseret News, August 31, 1921). Other floods that damaged structures occurred in August 1927, August 1931, and July 1932 (Garfield County News, September 13, 1927; Garfield County News, August 7, 1931; Garfield County News, July 15 and August 26, 1932). A large flood on August 27, 1932, destroyed a power plant, houses, and bridges and caused extensive erosion of agricultural land near Escalante (Garfield County News, September 2, 1932).

Reports of flooding for the Escalante River are rare after 1932. The largest flood recorded at the gaging station near Escalante was 98 m^3/s in August 1953. Floods were widespread in southern Utah in mid-August 1957 and a flood of 158 m^3/s occurred in Upper Valley on August 2, 1959 (Butler and Marsell, 1972). Grazing intensities declined after 1918, channel stabilization structures were built after 1930, the U.S. Forest Service initiated bank-protection and channel-stabilization projects in conjunction with the Civilian Conservation Corps in the early 1930's, and the Soil Conservation Service began range-improvement projects in the 1940's (Figure 2; Woolsey, 1964). These changes in land-use practice may have had an influence on the decline in incidence and size of floods.

Comparison of aerial photography of Upper Valley Creek taken in 1940 and 1974 revealed that the morphology of the channel changed little over 34 years (Webb, 1985). Local channel widening, the formation of low terraces within the arroyo walls at Escalante, and the spread of introduced riparian species were the only observable changes. Therefore, the arroyo of Upper Valley Creek was initiated and developed fully between 1909 and 1940. This rapid change in channel morphology--a change experienced throughout the region (Graf, 1983b)--illustrates the instability of southwestern river channels. The channel changes were synchronous with land-use practices that could have promoted erosion (Figure 2). Also, the deposition of the low terraces after 1940 is related to a lack of large floods, as was reported also for the Gila River by Burkham (1976b).

HISTORIC FLOODS

Dating Historic Floods

Flood deposits were studied at two sites in the bedrock canyon downstream from Escalante (Figure 1) to quantify the discharges of historic floods (Webb, 1985). At The Alcove, 3.2 km downstream from the gaging station, high stages result in deposition of flood sediments under low-velocity conditions in a large cave (alcove) upstream from a contraction. At Anasazi Alcove, 7.8 km downstream from the gaging station, flood sediments were deposited under a 100-m-long section of overhanging rock upstream of another contraction. Silt lines have been preserved on the overhanging canyon walls that define the outside of river bends at both sites. Details of the flood evidence and hydraulic calculations are given in Webb (1985) and Webb et al (in press).

The flood deposits at The Alcove and Anasazi Alcove are correlative and provide evidence for four recent floods (Webb et al, in press). Deposits at the two sites were correlated on the basis of stratigraphic position, radiocarbon (^{14}C) dates, lithology, grain size, and color (Table II). Five radiocarbon dates indicate that these floods are "modern" (after A.D. 1650). At least two floods are historic because cattle dung is entrained in layer B at Anasazi Alcove (Table II). The highest flood layer (A) at The Alcove represents the flood of 1932 and has graffiti dated "1936" carved into its top.

The flood of 1909 was the first large flood since settlement in 1875 and is possibly the only large flood to occur in a much longer time period. Laing and Stockton (1976) studied growth suppressions in ponderosa pine growing adjacent to Pine Creek (Figure 1), which constitutes less than 20 percent of the Escalante River drainage area. Growth suppressions are a decrease in annual ring width (as viewed in core samples) related to exposure of roots by stream erosion during floods (Laing and Stockton, 1976). The essentially permanent nature of the reduction suggests that the first suppression may be the only reliable flood indicator. Growth suppressions indicate the frequency of channel-changing floods but do not necessarily indicate the relative magnitudes of events. Laing and Stockton (1976) present data for years between A.D. 1700 and 1970, but the number of available cores for a given year increased from 6 in 1705 to 106 in 1955. We consider these data to be reliable only after 1800 because 12 or more cores are available for the years after this date.

Flood-induced channel changes were infrequent on Pine Creek between 1800 and 1909 (Laing and Stockton, 1976). The flood of 1909 caused the most pervasive growth suppressions in the 170-year record, with 36 percent of the available cores recording suppressions. Suppressions continued in a relatively high percentage of the cores until the 1950's; Laing and Stockton (1976) speculated that

Table II. Characteristics of historic flood evidence at The Alcove and Anasazi Alcove, Escalante River, Utah

Flood layer	Year	Sediment sorting & size	Munsell color	Stratigraphic position	Dating evidence[1]
A	1932	medium sand, well-sorted	10 YR 7-6/4	Highest flood layer	"1936" graffiti; 0±40 yrs B.P.
B	1927(?)	fine sand, well-sorted	10 YR 6/4-3	Underlies layer A	cattle dung; 350±110 yrs B.P.
C	1916(?)	fine sand, well-sorted	10 YR 6/4-3	Underlies layer B	none
D	1909	fine-medium sand, well-sorted	10 YR 6/4-3	Lowest flood layer	historical and tree-ring evidence; 150±80, 210±90, and 260±110 yrs B.P.

1. ^{14}C dates are given in yrs B.P. (years before A.D. 1950). ^{14}C dates between 0 and 300 yrs B.P. cannot be assigned absolute ages and are considered "modern."

overgrazing may have been the reason because livestock numbers were reduced after 1955. Neither the 1932 flood nor two other known floods were as definite in the series as the 1909 flood. Either these floods were not recorded in the tree cores or the floods originated from the other 80 percent of the watershed.

Additional data on flood occurrence and entrenchment of alluvial flood plains have some bearing on the dating of the four floods. Before the four recent floods, the next oldest flood deposit is radiocarbon dated at 430±100 years before present (yrs B.P.) (Webb et al, in press). Assuming that large floods would leave depositional evidence, only four large floods have occurred after 400 yrs B.P. The alluvial channel of the Escalante River was shallow and narrow from 400 yrs B.P. (A.D. 1550) until settlement. The absence of channels in the headwaters suggests that no large floods have occurred in the last 400 yrs B.P. because the occurrence of large floods and channel changes are temporally related on the Escalante River (Webb, 1985).

In the absence of other evidence of large floods and major channel changes between 0 and 400 yrs B.P. and given a probable infrequent occurrence of channel-changing events between A.D. 1800 and 1909 (Laing and Stockton, 1976), the four floods are reasonably assumed to have occurred in historic time. The lowermost deposit (layer D, Table II) represents the flood of 1909. Possible dates for layers B and C, based on climatic records, are 1916, 1927, and July 1932 (Webb, 1985). Alternatively, these layers could represent multiple peaks of a single flood. The flood-frequency implications of these assumptions will be discussed in a later section.

Historic Flood Discharges

The discharge for the flood of 1932 was calculated using step-backwater analysis (Webb et al, in press). A discharge of 600 m^3/s at The Alcove best fit the high-water marks (Webb et al, in press). A stage-rating curve was developed to estimate discharges required to deposit the other historic flood layers, assuming a similar channel morphology and roughness. The discharges for the flood of 1909 and flood layers B and C were estimated to be 570, 550, and 550 m^3/s, respectively. The discharge for the flood of 1909 could have been as low as 500 m^3/s on the basis of photogrammetric evidence of channel change (Webb et al, in press). Discharges could not be estimated for flood layers at Anasazi Alcove because of a short reach length (Webb et al, in press). The discharges calculated for deposits at The Alcove would be expected to have a poor accuracy rating.

CHANGES IN HYDROCLIMATOLOGICAL CONDITIONS

Precipitation Changes

The climatic record for Escalante (1902-82) was examined for possible fluctuations or changes that could have induced summer floods on the Escalante River. Although annual precipitation (195 mm) has not changed, the amount of precipitation during July and August decreased after 1932 (Webb, 1985). The mean summer precipitation for 1909 to 1932 was 36 percent greater than for the rest of the record (Figure 3). Precipitation shifted to the fall months after 1932, probably falling as snow at higher elevations (Webb, 1985). Precipitation from June to September is highly variable between 1909 and 1932 and is less variable during the rest of the record. Coefficients of variation of 0.54 for 1909 to 1932 and 0.39 for the rest of the record denote the difference in variability between the two periods.

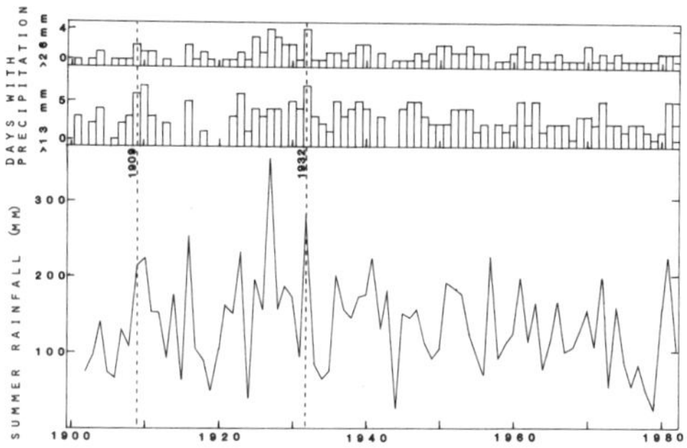

FIGURE 3. Summer (July-September) precipitation and daily rainfall intensities at Escalante, Utah (1902-1982).

The period of 1909 to 1932 had more frequent large storms than the rest of the record (Figure 3). Four daily precipitation totals exceeded 26 mm during the summers of 1927 and 1932;

the largest historic flood on the Escalante River occurred in 1932. The years with the highest number of storms exceeding 13 mm are 1909, 1910, 1923, and 1932 (Figure 3). High daily precipitation totals do not necessarily indicate the occurrence of a flood; for example, only 41 mm were recorded on the day of the disastrous flood in 1909. Point precipitation records do not necessarily reflect basinwide precipitation because of the strong orographic components of summer precipitation. The combination of wet summers and frequent high-intensity precipitation, however, suggests a significant though subtle climatological shift that may have heralded an increase in large floods between 1909 and 1932.

Similar increases in high-intensity summer storms have been proposed as the cause for arroyos in New Mexico and southern Arizona; however, the timing of the shifts is not in phase throughout the Southwest. The period of 1855 to 1880 is anomalous, whereas the period of 1909 to 1932 does not appear unusual in the New Mexico record (Leopold et al, 1966). Significantly higher intensities of precipitation, with no change in annual precipitation, occurred from 1866 to 1916 compared with lower intensities of precipitation from 1917 to 1966 for southern Arizona (Cooke, 1974). The subtle shift in summer precipitation at Escalante may have begun before 1909 but persisted slightly longer (until the mid-1930's) than shifts for other southwestern areas (until the late 1910's).

Watershed Changes

Several other environmental changes that may have influenced flooding have occurred. Grazing and clearing of vegetation from the flood plain are responsible for a conversion of grasslands to shrublands. This conversion would be expected to cause some unknown but possibly significant increase in storm runoff. Tamarisk and Russian olive trees, introduced from Eurasia in the late 1800's (Robinson, 1965), have become established in the arroyo (Figure 4) and replaced the pre-arroyo stands of native willows and cottonwood trees over the past 53 years. Low terraces have formed within the arroyo walls and vary in number along the length of the channel; two terraces have formed near the town of Escalante since 1932.

All rivers in the Southwest have experienced a similar spread of introduced riparian vegetation and have altered their channel dimensions since 1940. Tamarisk has spread upstream along the Colorado River from northern Arizona through Utah at a rate of 20 km/year (Graf, 1978). Dense tamarisk stands on flood plains decrease flow velocities, increase flow depth and roughness characteristics (Burkham, 1976a), increase channel storage, change the width-depth characteristics to a narrow, deep configuration (Hadley, 1961), and significantly attenuate any flood peak larger than can be contained in the channel. The Green River in Utah and Oraibi Wash in northern Arizona have narrowed their channels apparently in response to the tamarisk invasion (Graf, 1978; Hadley, 1961). The formation of terraces and narrowing of the channel on the Little Colorado River in Arizona has been attributed to climatic fluctuations (Herefore, 1984).

FIGURE 4. The Escalante River downstream of the crossing at Escalante Utah. Upper) in July 1932 (R.W. Bailey photograph #270658, courtesy of the Utah Historical Society). Light flash on original negative appears at upper left. Aerial photography taken in 1940 also shows the denuded channel. Lower) in October 1985 (R.H. Webb, R. M. Turner collection, stake number 1241). Photograph is an approximate match because dense vegetation growth obscures part of the view. Vegetation is mainly introduced tamarisk and Russian olive trees with scattered native willows and cottonwoods.

The spread of tamarisk (Graf, 1978) or adjustments in sediment transport related to the formation of the arroyo (Womack and Schumm, 1977), however, also could have caused these terraces. Regardless of cause, formation of the inner-arroyo terraces attenuates flood peaks (Burkham, 1976b).

These changes may be in opposite directions with no net effect on flood frequency. Conversion of grasslands to shrublands would be expected to increase runoff and erosion. Confinement of flow inside arroyo walls increases velocities and peak discharges. Narrowing of channels and increases in roughness owing to introduced vegetation should significantly attenuate flood peaks. Whether or not an actual net effect occurs must be determined through hydraulic modeling and is not addressed in this paper. Knowledge of the changing hydrologic environment, however, is necessary for the evaluation of the stationarity of the annual flood series on the Escalante River.

FLOOD FREQUENCY ON THE ESCALANTE RIVER

The foregoing hydrologic information can be incorporated in a flood-frequency analysis for the Escalante River. No sizable tributaries enter the river between the gaging station and The Alcove (Figure 1), and the difference in drainage areas is 10 km^2 (1.2 percent). Therefore, any historic flood information derived from The Alcove can be used to augment the gaging record. The variability in flood information will be emphasized to illustrate the sensitivity of flood-frequency analysis for the Escalante River.

The gaging record for the Escalante River near Escalante is short and discontinuous (Table III). A log-Pearson type III analysis (Water Resources Council, 1982) was used to estimate a 100-year flood of 190 m^3/s for this station (Figure 5A). The discharges calculated for historic flood deposits at The Alcove were included in the analysis as historic floods (Table III), and a generalized skew coefficient of -0.2 was determined from a regional flood-frequency analysis (Webb, 1985). Discharges of 500 to 570 m^3/s were used for the flood of 1909 to assess the effect of potential channel changes during floods on the frequency curves.

Addition of the four floods as historic events to the flood-frequency analysis (Water Resources Council, 1982), with the first flood dated at 1909, increased the estimated 100-year flood to 630 m^3/s (Figure 5A). Decrease of the 1909 discharge from 570 to 500 m^3/s decreased the estimated 100-year flood by an insignificant 10 m^3/s. With the addition of the historic floods, the computed skew coefficient changed more than the other two moments of the distribution. Using the 26-year systematic record only, 630 m^3/s is larger than the estimated 5,000-year flood; however, 630 m^3/s is within the 95-percent confidence interval for the estimated 500-year flood (Figure 5A).

The Water Resources Council (1982) guidelines use the date of the first historic flood as the start of the historic record. Alternative lengths of historic record were used in the analysis because the date of the flood that deposited layer D is uncertain and a longer time span with lack of large floods can be reasonably assumed. The alternative record lengths used are (1) 76 years (four floods postdating 1909), (2) 110 years (four floods postdating settlement in 1875), (3) 185 years (four floods postdating the first reliable tree-ring records of 1800), and (4) 435 years (four floods postdating the next oldest flood evidence about A.D. 1550). The flood-frequency curves converge with increasing historical record length on the curve estimated using only the systematic record (Figure 5B). Without considering the implications of the changes in the watershed, one is tempted to conclude that the historic flood information does not significantly improve the estimates of flood frequency obtained from the systematic record alone.

The evidence for flooding suggests that four large floods greater than or equal to 500 m^3/s occurred in the 23 years between 1909 and 1932. Using record lengths of 76, 110, and 435 years, the binomial probability of four floods exceeding 500 m^3/s in 23 years decreases from 0.031 for a 76-year record to 0.000055 for a 435-year record. The former scenario indicates a 1 in 30 chance that the four floods occurred randomly between 1909 and 1932. The latter scenario, albeit extreme, indicates that 1909-32 was a discrete nonstationary period of flooding that could recur.

Three conclusions are possible concerning flood frequency on the Escalante River. Depending on the interpretation of the dating evidence, a historic record of 76 years could be used and the 100-year flood estimate could be increased by as much as 220 percent more than that estimated from the systematic record alone. Acceptance of this conclusion would allow use of a 100-year flood of 630 m^3/s. In contrast, the four large floods could be assumed to have occurred randomly within a 435-year historic record, and their inclusion with the systematic record does not significantly change the estimated 100-year flood. Acceptance of this conclusion would result in a 100-year flood of 190 m^3/s. Alternatively, four large floods occurred during a 23-year period and the frequency of large floods was nonstationary between 1909 and 1932. Acceptance of this conclusion would also yield a 100-year flood of 190 m^3/s under the assumptions that the period of nonstationarity ended by 1940 and watershed changes since 1940 had no net effect on flood frequency. No valid statistical means exists for resolving the ambiguity caused by the imprecise dating of the historic record and the potential nonstationarity of the annual flood series.

Several recent studies have identified problems with the Water Resources Council (1982) method for treatment of historic floods in flood-frequency analysis. Lane (this volume) found that the recommended procedures overly weight the contribution of the systematic record to the determination of the moments of the distribution. The Water Resources Council (1982) method extracts little statistically useful information from historic data (Stedinger and Cohn, 1986). Perception (or censoring) thresholds that define time periods over which large, infrequent floods of different magnitudes would have been observed had they occurred is a promising approach to use of historic data (Gerard and Karpuk, 1979). Censored data are particularly

TABLE III. Discharges used in flood-frequency analysis of the Escalante River near Escalante gaging record (09337500; 810 km^2), with historic discharges estimated from flood deposits in The Alcove and Anasazi Alcove.

ANNUAL PEAK DISCHARGES							
Year	Discharge (m^3/s)	Year	Discharge (m^3/s)	Year	Discharge (m^3/s)	Year	Discharge (m^3/s)
1909	500-570	1947	31	1955	84	1978	1
1917?	550	1948	9	----	--	1979	7
1927?	550	1949	79	1972	22	1980	40
1932	600	1950	5	1973	72	1981	78
1943	50	1951	44	1974	5	1982	6
1944	56	1952	35	1975	6	1983	5
1945	24	1953	98	1976	31	1984	57
1946	15	1954	15	1977	22		
Mean Logarithm		1.29		Generalized Skew			-0.23
Standard Deviation		0.51		Adopted Record Skew[1]			-0.63
Computed Record Skew[1]		-0.63		Adopted Skew[2]			0.40

1. Based on the systematic record only.
2. Modified using historical floods and 77-year record.

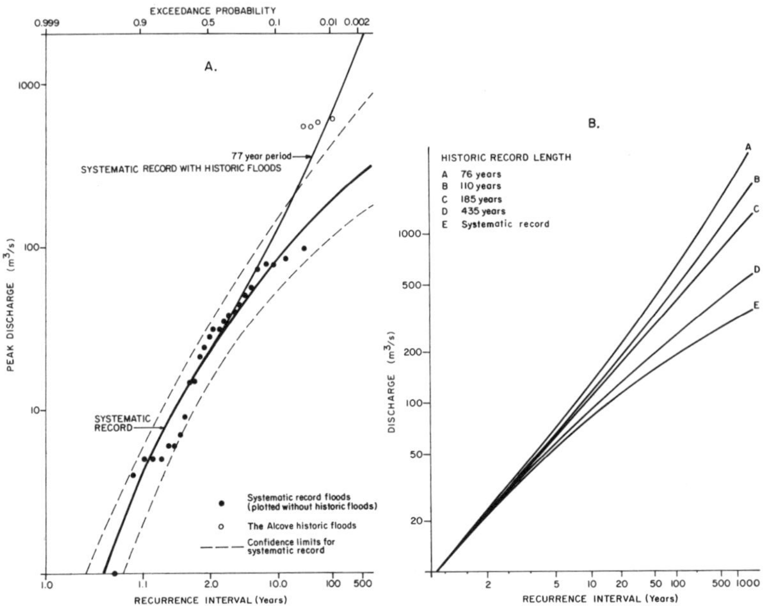

FIGURE 5. A flood-frequency analysis for the Escalante River. **A.** Analysis of the gaging record with historic floods. **B.** Variations in curve shape with changes in the length of the historic record.

amenable to analysis using maximum-likelihood methods (Condie, 1986). Stedinger and Cohn (1986) demonstrated with Monte Carlo simulation that maximum-likelihood methods effectively use historic floods in the estimation of flood frequency. Maximum-likelihood methods, however, cannot provide conclusive evidence as to whether the annual flood series for the Escalante River has been stationary or nonstationary.

IMPLICATIONS FOR FLOOD FREQUENCY IN THE SOUTHWESTERN UNITED STATES

The history of floods and changing hydrologic conditions on the Escalante River is a microcosm of hydrology in the Southwest. Large floods initiated arroyos on every major tributary of the Colorado River in southern Utah between 1862 and 1909 (Table 1). On some rivers, such as the San Juan, the discharges for these floods have been estimated as the largest in history. Channel changes have occurred throughout the Southwest in response to floods (Graf, 1983b), and problems similar to those described in this paper are encountered for many other rivers in the Southwest.

The interpretation of flood frequency on the Escalante River is equivocal. The estimated 100-year flood could range from 190 to 630 m^3/s, depending upon the interpretation of the historic record length. The annual flood series can be assumed to be stationary or nonstationary, depending on the possibility of another shift to increased summer precipitation. No flood-frequency relation for this river is completely unambiguous.

A robust test for stationarity is needed to help resolve this problem of potential nonstationary in flood frequency. Historic data, particularly paleohydrologic information (Baker, 1985), generally contains uncertainties in the magnitudes and dating of flood events. A method needs to be developed for the incorporation of historic information with an appropriate historic record length into systematic records and the testing of the resulting distribution for stationarity. Otherwise, a strictly qualitative nonstationarity assessment is likely after every future large flood (e.g. Saarinen et al, 1984). Statistical analysis of flood frequency requires the assumption of stationarity in the annual flood series, and this assumption is not necessarily valid in the southwestern United States.

The effect of climatic shifts, changes in channel dimensions, and increases in introduced vegetation on flood frequency needs to be quantitatively assessed for rivers in the Southwest. Climatic shifts may recur relatively frequently or may recur only every millennium. Drastic reductions in channel widths cannot be ignored in assessments of flood frequency. Hydrologic routing models are needed to assess the effect of the channel changes, because the needed historic information on channel configurations is usually available. Historic changes in hydrologic conditions need to be evaluated for rivers in the Southwest where nonstationarity in the annual flood series is suspected.

ACKNOWLEDGMENTS

Julio Betancourt and Jim Eychaner critically reviewed the manuscript. This work was supported partially under NSF Grants EAR-81-19981 and EAR-83-00183 and a States Water Resources Research Program Grant to V.R. Baker.

REFERENCES

Baker, V.R. 1985. 'Paleoflood hydrology of extraordinary flood events.' in: U.S. - China Bilateral Symposium on the Analysis of Extraordinary Flood Events, 48 pp. Nanjing, China.

Betancourt, J.L., and Turner, R.M. In press. 'Historic arroyo-cutting and subsequent channel changes at the Congress Street Crossing, Santa Cruz River, Tucson, Arizona.' in: Arid Lands, Today and Tomorrow. Office of Arid Lands, University of Arizona, Tucson.

Burkham, D.E. 1976a. Hydraulic effects of changes in bottom-land vegetation on three major floods, Gila River in southeastern Arizona. U.S. Geological Survey Professional Paper **655-J**, 14 pp.

Burkham, D.E. 1976b. Effects of changes in an alluvial channel on the timing, magnitude, and transformation of flood waves, southeastern Arizona. U.S. Geological Survey Professional Paper **655-K**, 25 pp.

Butler, E., and Marsell, R.E. 1972. Cloudburst floods in Utah, 1939-69. U.S. Geological Survey--Utah Division of Natural Resources Cooperative-Investigations Report No. **11**, 103 pp.

Condie, R. 1986. 'Flood samples from a three-parameter lognormal population with historic information: The asymptotic standard error of estimate of the T-year flood.' Journal of Hydrology, V. **85**, pp. 139-150.

Cooke, R.U. 1974. 'The rainfall context of arroyo initiation in southern Arizona.' Zeitschrift fur Geomorphologie Supplementband **21**, pp. 63-75.

Gerard, R., and Karpuk, E.W. 1979. 'Probability analysis of historical flood data.' Journal of the Hydraulics Division, American Society of Civil Engineers, V. **105**, pp. 1153-1165.

Graf, W.L. 1978. 'Fluvial adjustments to the spread of tamarisk in the Colorado Plateau region.' Geological Society of America Bulletin, V. **89**, pp. 1491-1501.

Graf, W.L. 1983a. 'Flood-related channel change in an arid-region river.' Earth Surface Processes and Landforms, V. **8**, pp. 125-139.

Graf, W.L. 1983b. 'The arroyo problem: paleohydrology and paleohydraulics in the short term.' in: Gregory, K.J. (editor), Paleohydrology. pp. 279-302. John Wiley and Sons, New York.

Gregory, H.E. 1918. Unpublished field notes. Denver, U.S. Geological Survey, Field Records Library, Gregory Field Notebook No. **296**, pp. 43 and 79.

Gregory, H.E. 1939. 'Diary of Almon Harris Thompson.' Utah Historical Quarterly, V. **7**, pp. 38-83.

Gregory, H.E., and Moore, R.C. 1931. The Kaiparowits Region, a geographic and geologic reconnaissance of parts of Utah and Arizona. U.S. Geological Survey Professional Paper **164**, 161 pp.

Hadley, R.F. 1961. 'Influence of riparian vegetation on channel shape, northeastern Arizona.' in: Geological Survey Research, 1961. U.S. Geological Survey Professional Paper **424C**, pp. 30-31.

Hereford, R. 1984. 'Climate and ephemeral stream processes: Twentieth century geomorphology and alluvial stratigraphy of the Little Colorado River, Arizona.' Geological Society of America Bulletin, V. **95**, pp. 654-668.

Laing, D., and Stockton, C.W. 1976. Riparian dendrochronology, a method for determining flood histories of ungaged watersheds. 18 pp., Final Report to the Office of Water Resources Technology, Project A-058-ARIZ, Washington, D.C.

Lane, W.L. This volume. 'Paleohydrologic data and flood frequency estimation.'

Leopold, L.B., Emmett, W.W., and Myrick, R.M. 1966. Channel and hillslope processes in a semiarid area, New Mexico. U.S. Geological Survey Professional Paper **352-G**, 253 pp.

Lewman, W. 1893-94. Unpublished field notes. Salt Lake City, Bureau of Land Management, General Land Office Survey, Book **A-238**.

Reich, B.M., and Davis, D.R. This volume. 'Estimating the regulatory flood on a degrading river.'

Robinson, T.W. 1965. Introduction, spread, and areal extent of saltcedar (Tamarix) in the western states. U.S. Geological Survey Professional Paper **491-A**, 12 pp.

Saarinen, T.F., Baker, V.R., Durrenberger, R., and Maddock, T., Jr. 1984. The Tucson, Arizona, flood of October 1983. 112 pp., National Academy Press, Washington, D.C.

Stedinger, J.R., and Cohn, T.A. 1986. 'The value of historical and paleoflood information in flood frequency analysis.' Water Resources Research, V. **22**, pp. 785-793.

U.S. Geological Survey. 1910. Surface water supply, part IX, Colorado River basin. Water Supply Paper **269**, pp. 183-184.

Water Resources Council. 1982. Guidelines for determining flood flow frequency. 28 pp., U.S. Water Resources Council, Bulletin **17B**, Washington, D.C.

Webb, R.H. 1985. Late Holocene flooding on the Escalante River, south-central Utah. 204 pp., unpublished Ph.D dissertation, University of Arizona, Tucson.

Webb, R.H., O'Connor, J.E., and Baker, V.R. In press. 'Paleohydrologic reconstruction of flood frequency on the Escalante River, south-central Utah.' in: Baker, V.R., Kochel, R.C., and Patton, P.C. (editors). Flood geomorphology.

Womack, W.R., and Schumm, S.A. 1977. 'Terraces of Douglas Creek, northwestern Colorado.' Geology, V. **5**, pp. 72-76.

Woolsey, N.G. 1964. The Escalante story, a history of the town of Escalante. Springville, Utah, Art City Publishing, 463 pp.

PALEOFLOOD HYDROLOGIC ANALYSIS AT UNGAGED SITES, CENTRAL AND NORTHERN AUSTRALIA

Victor R. Baker
Department of Geosciences
University of Arizona,
Tucson, Arizona 85721

Geoff Pickup
CSIRO Central Australian Laboratory
P.O. Box 2111,
Alice Springs, NT 5750 Australia

Robert H. Webb
Department of Geosciences
University of Arizona,
Tucson, Arizona 85721

ABSTRACT. Paleoflood hydrology is the study of past or ancient flood events that occurred prior to direct measurement by modern hydrologic procedures. Because of recent advances in flow modeling and geochronologic techniques, it is possible to perform relatively accurate evaluations of paleoflood magnitudes and ages in channels with resistant flow boundaries. Slack-water sedimentary deposits and other high-water indicators are correlated along study reaches to define maximum paleoflood discharges. We tested the procedure in two regions of the Northern Territory, Australia, an area of minimal gaging information. At the Katherine Gorge site, with a drainage area of 6390 km^2, the maximum paleoflood was 6300 m^3/s. At the upper Finke Gorge site, with a drainage area of 4400 km^2, the maximum paleoflood was 5200 m^3/s. The dating of multiple paleofloods at these sites was facilitiated by tandem accelerator mass spectrometry for determining ^{14}C activity in very small organic samples. For floods in the time periods 1950 to present it is possible to achieve dating to the calendar year by reference to known curves of anomalously high atmospheric ^{14}C activity. The treatment of such data in a flood-frequency analysis requires further research, especially on the role of censoring thresholds in stacked flood slack-water depositional sequences.

INTRODUCTION

Paleoflood hydrology is the study of past or ancient flow events which occurred prior to direct measurement by modern hydrologic procedures.

V. P. Singh (ed.), Regional Flood Frequency Analysis, 325–338.

New research techniques allow paleoflood hydrology to be accomplished at certain ungaged sites, extending estimates of flood magnitudes and frequency over a data base of centuries to thousands of years (Baker, 1983). This paper will describe a specialized type of paleoflood hydrology that has produced especially complete reconstructions of multiple floods. The method requires special combinations of geological circumstances that must be carefully evaluated in each application.

To test the evolving methodology of paleoflood hydrology, we investigated several sites in the Northern Territory, Australia (Figure 1), during 1980-1983. The Northern Territory has very few systematic flood records. The widely dispersed network of gages provides mostly short records, and the occasional very large flood occurrences are inadequately recorded. Gages may be destroyed during especially large floods, and access is often difficult during the storm conditions that promote flooding. Consequently, existing gages are not adequately rated for extremely large flood flows.

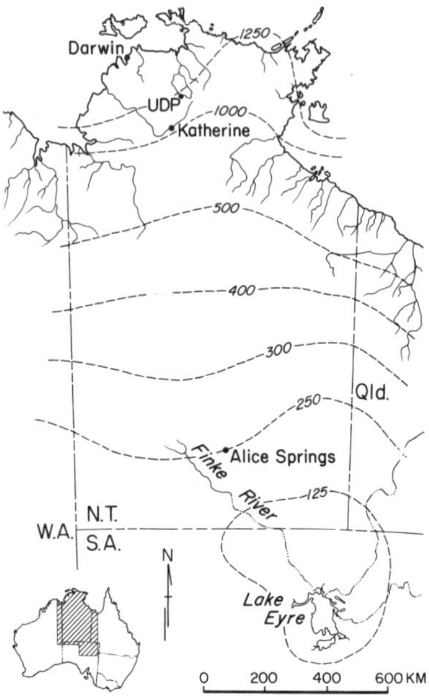

Figure 1. Location map showing study areas in the Northern Territory, Australia.

We have reported elsewhere on the geological aspects of slack-water sedimentation by large, rare floods at the Northern Territory study areas (Baker et al., 1983a, 1983b, 1985). The intention here is to emphasize the hydrologic aspects of the data and to discuss research needs in the statistical treatment of paleoflood information. First, however, we will review briefly the methodology.

PALEODISCHARGE ESTIMATION

The best paleoflood sites are located in channel reaches with flow boundaries constrained by bedrock, immobile sediment, or other resistant boundary materials. Such channels produce relatively large stage changes for changes in flood discharge (Baker, 1984). Moreover, they do not change their cross sections appreciably during major floods, as commonly occurs in alluvial channels. The relatively high stages achieved by floods in resistant-boundary channels produce several kinds of high-water indicators. Most common are deposits of sand, silt, and (rarely) gravel that accumulate relatively rapidly from suspension in highly energetic floods. These accumulate in local areas of reduced flow velocity and are called "slack-water deposits." Factors characterizing the accumulation, preservation, and analysis of slack-water deposits are discussed in detail by Baker et al. (1983a).

Computerized procedures for hydraulic flow modeling are used to tie the elevations of various paleoflood high-water marks to discharges routed through surveyed river cross sections (Ely and Baker, 1985). These procedures can yield very accurate estimates when appropriate assumptions in the method are adequately considered. In this study we employed the HEC-2 model for computing water-surface profiles by step-backwater analysis (Feldman, 1981).

PALEOFLOOD AGES

Stratigraphy is the branch of geology that deals with the arrangement of sedimentary strata, especially in relation to chronologic order of sequence. This aspect of a paleoflood study requires analysis by a specialist trained in stratigraphic geology. Individual flood sedimentation units are recognized, discriminated, and correlated between multiple sites along a paleoflood study reach (Kochel et al., 1982). Most important is the use of geochronologic procedures to provide absolute dates of the individual flood events. Numerous recent advances have been made in geochronology.

Perhaps the most exciting new tool for the dating of paleoflood deposits is tandem accelerator mass spectrometery (Hedges and Gowlett, 1986), which provides radiocarbon ages for organic samples as small as a few milligrams. A radiocarbon date is derived from a geochemical laboratory determination of the remaining present-day activity of the radiogenic isotope ^{14}C in an appropriate organic material (Faure, 1977). Tandem accelerator mass spectrometry can be accomplished on small quantities of charcoal, seeds, and other organics that are commonly intercalated with ancient flood deposits. Conventional radiocarbon dating can be accomplished by gas-proportion counting procedures or liquid scintillation on much larger samples (generally 1 to 2 grams or more).

Floods in the period 10,000 to 350 years ago are most amenable to radiocarbon dating. This is generally achieved with an error of approximately 90 radiocarbon years. It is important to emphasize that radiocarbon analytical laboratories report sample ages in radiocarbon years before the present, not in calendar years. For purposes of radiocarbon analyses, the "present" is considered calendar year 1950. Moreover, secular changes in $^{12}C/^{14}C$ reservoir characteristics have occurred, resulting in significant deviations of radiocarbon years from calendar years. Calibration curves exist for the last 7500 years (Damon et al., 1974), but these must be used with caution since they retain the uncertainty characteristics of conventional readiocarbon dates.

In general, paleoflood studies concern ages that do not pose major problems in relating calendar years to radiocarbon years. The critical date is the one defining the total length of the flood record. Since most paleoflood records are at least 1000 years in length, radiocarbon counting errors are small in comparison.

Because of loading of the atmosphere by CO_2 during the industrial revolution (Stuiver, 1982), ages in the period 1650 to 1950 A.D. are extremely difficult to resolve by radiocarbon analyses. However, numerous other dating procedures may be used for such relatively young floods, including historic documentation, distinctive archaeological materials, and dendrochronology.

Because of anomalously high ^{14}C activity in the atmosphere from nuclear testing it is possible to date floods in the period 1950 A.D. to present essentially to the calendar year (Baker et al., 1985). Measurements of bomb-produced ^{14}C in the atmosphere provide a method for calibrating the radiocarbon activities in this time period. The exciting potential of this tool is that very accurately dated flood sequences may be documented for 35+ year records at appropriate ungaged sites.

CASE STUDY: KATHERINE RIVER

The Katherine Gorge is a 20-km long canyon developed in jointed, resistant sandstone. It is located in the north-central part of Australia, 32 km northeast of the town of Katherine, Northern Territory. The gorge is a slot-like chasm averaging 50 to 150 m wide and 40 to 60 m deep. Linear side canyons join the main gorge; these tributary mouths are ideal sites for slack-water sediment accumulation during rare, great flow events.

During our field surveys in 1980 and 1983 we identified and described several slack-water sedimentation sites, dated the flood deposits, and surveyed the hydraulic geometry (Baker et al., 1985). The hydraulics of the lower gorge were evaluated with the HEC-2 computer routine to determine energy-balanced water surface profiles for surveyed cross sections. Results of the flow modeling were expressed as stage-discharge curves at the various cross sections and slack-water sites. Thus, elevations reached by flood sediments could be directly related to the responsible flood discharges by use of these rating curves.

Two general varieties of slack-water deposit were investigated in the Katherine Gorge: (1) low mounded deposits, and (2) relatively thin accumulations of slack-water sediment mantling colluvial slopes. Figure 2 shows an exposure in a low mounded deposit. A series of ultramodern radiocarbon dates (Baker et al., 1985) indicate that this deposit records floods in the period 1956-1980. Such low-level deposits record relatively frequent floods, indicating the frequencies at which relatively low flood stages are achieved.

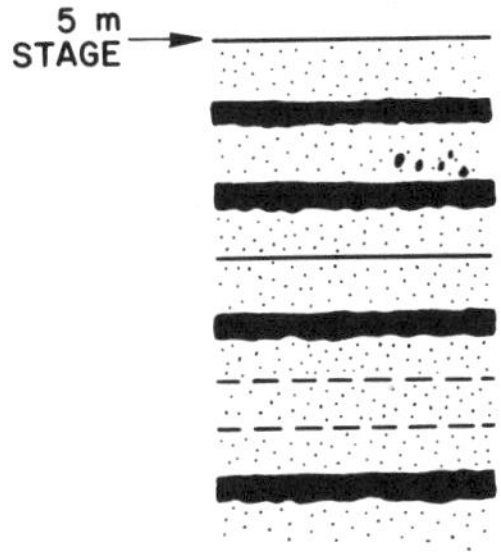

MATERIAL	$\Delta^{14}C$ ‰	DATE	LAB NUMBER
LEAF MAT ORGANICS	264.7 ± 35.4	1980	ANU-2440
CHARCOAL	92.4 ± 7.0	1958	ANU-2439
LEAF MAT ORGANICS	278.9 ± 8.9	1980	ANU-2441
LEAF MAT ORGANICS	109.8 ± 11.9	1958	ANU-2442
LEAF MAT ORGANICS	3.3 ± 7.4	1956	ANU-2443

Figure 2. Slack-water sedimentary layers and intercalated organic detritus in the Katherine Gorge, Northern Territory, Australia.

At another site we documented a thin mantle of slack-water sediment extending up to the maximum stage of a very large, rare flood. The stratigraphic relationships show that the responsible flood can be traced to the stage reached by 5000 m^3/s, according to the flow-modeled rating curve at this site. Radiocarbon dating indicates that the responsible flood was the March, 1957, event, which is known from historical observations at the town of Katherine, 25 km southwest (downstream) of the study area (Watson et al., 1970). A higher deposit, from an older, undated flood, can be traced to a maximum discharge level of 6300 m^3/s.

CASE STUDY: FINKE RIVER

The Finke River (Figure 1) heads in a semiarid region of crystalline rock about 75 km west of Alice Springs, Northern Territory. The study area is located in the James Ranges (Figure 3), where the Finke is confined to a narrow gorge. Just north of the study area at Hermansburg, 115 km west-southwest of Alice Springs, the Finke drains approximately 4400 km^2. A gage at Hermansburg has recorded low-flow events since 1966. However, only stage information is available for the 3 great floods in this period (Figure 4). Historical accounts indicate that these were the largest floods since at least 1895. However, gage failure during major floods and the lack of measured flows at the gage greater than 10 m^3/s result in unknown discharges for the events.

Flood slack-water depositional sites in the Finke River Gorge occur at the mouths of small tributaries or perched in bedrock meander scars. A typical cross section in the Gorge reveals a channel averaging

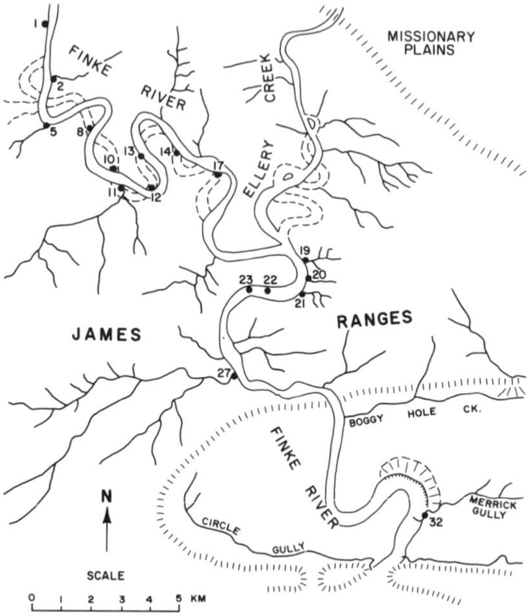

Figure 3. Location map showing Finke Gorge study area and slack-water sites (numbered). Dashed lines show bedrock paleomeanders.

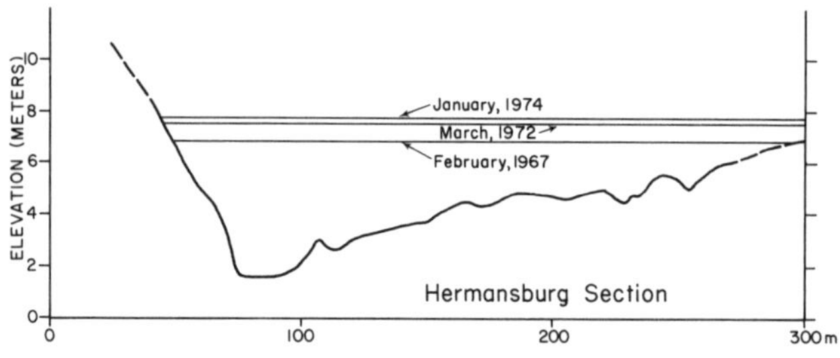

Figure 4. Cross section and stage data for the Hermansburg gage on the Finke River.

about 250 m wide confined by canyon walls of sandstone. The sandy slack-water deposits were exposed at pits dug into the large, mounded accumulations of paleoflood sediments. Organic materials incorporated in these deposits, mainly charcoal from aborginal hearths or bushfires, were used for radiocarbon dating.

The stratigraphic analysis of 12 Finke Gorge sites revealed evidence of 7 discrete flood layers. Using ultramodern radiocarbon dates the floods of 1967, 1972, and 1974 were clearly identified (Baker et al., 1983a). Four prehistoric floods were distinguished by radiocarbon dating and stratigraphic correlation. These will be referred to in subsequent discussion as flood layers F, A, S, and N. Layer F has radiocarbon ages in the range of 250 to 400 yrs B.P., a period when exact calendar age specification is not possible (Stuiver, 1982). Layer A was deposited approximately 500 years ago based on a date of 470±180 yrs B.P. (Laboratory Number AA-426). Layer S is well dated at 700 yrs B.P. because of bracketing radiocarbon dates of 670±100 yrs B.P. (A-3578) in a hearth post-dating the flood and 710±120 yrs B.P. (A-3795) on charcoal transported by the flood. Layer N is marked by pedogenic soil carbonate, indicating a much greater age than the other layers. Flood transported charcoal indicates a maximum age of 850±250 yrs B.P. for this layer.

The magnitudes of the various paleofloods were evaluated by flow modeling of water-surface profiles, as described above. Two reaches of the Finke were analyzed, one upstream of the junction with Ellery Creek and one downstream of this major tributary (Figure 3). The resulting profiles and maximum slack-water deposit heights (Figures 5 and 6) indicate maximum paleoflood discharges of 5200 m^3/s upstream of Ellery Creek (layer P of deposit 8 in Figure 7) and 8000 m^3/s downstream of Ellery Creek (layer F).

There are several important sources of error in this particular study. Because of the reconnaissance nature of this investigation, we only used 20 cross sections in the flow model, 14 for the upstream reach (Figure 5) and 6 for the downstream reach (Figure 6). The survey lengths of 8 and 6 km were approximately 27 and 20 times the respective average channel widths. More cross sections and greater attention to roughness variations would have improved the modeling accuracy. Although bedrock confines channel walls and locally crops out on the

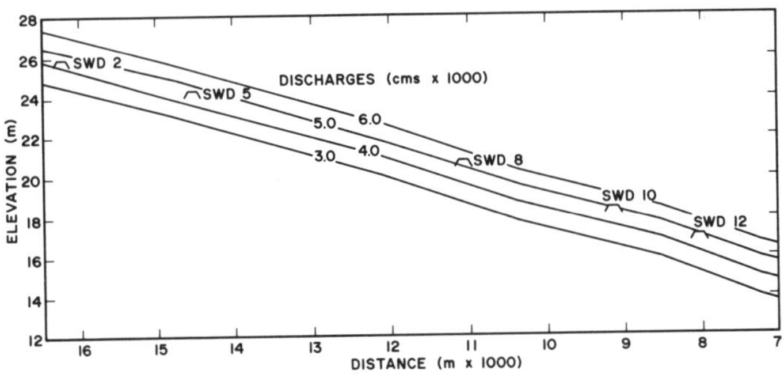

Figure 5. Water-surface profiles for indicated discharges along a reach upstream of Ellery Creek (sites 2 through 12 on Figure 3). Slack-water deposit sites (SWD) are numbered.

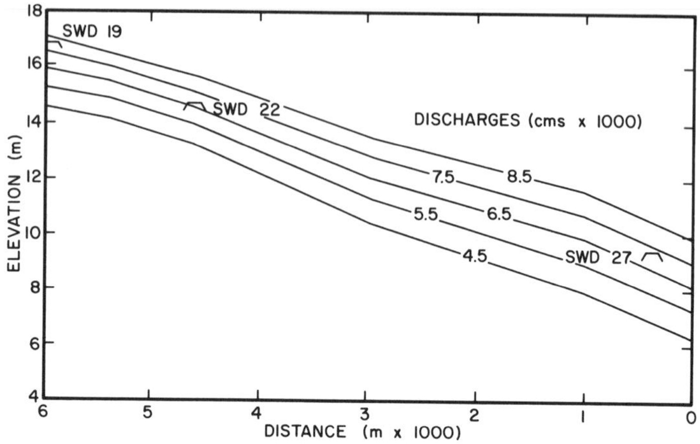

Figure 6. Water-surface profiles for indicated discharges along a reach downstream of Ellery Creek (sites 19 through 27 on Figure 3).

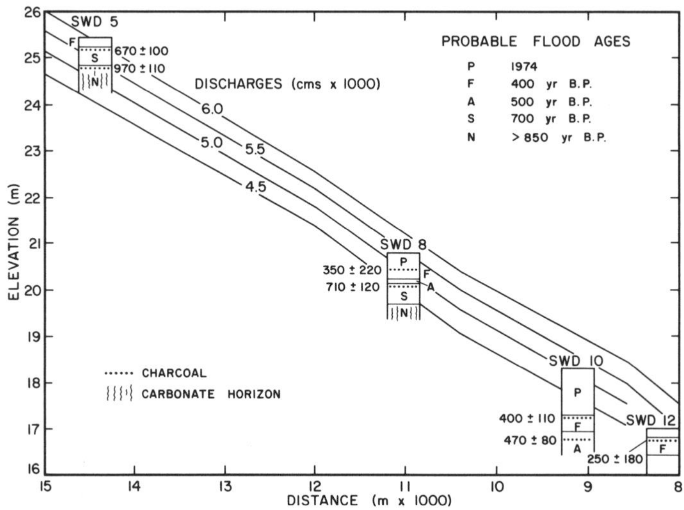

Figure 7. Generalized diagram showing the approximate relationship of slack-water deposit stratigraphy at various numbered sites to water-surface profiles upstream of Ellery Creek. Numbers with ± sign indicate radiocarbon dates for flood layers (upper case letters) as explained in the text.

bed, some bed scour is possible in major floods. Observation of scour effects from the March, 1983 flood and the presence of a gravel armor suggest that scour would not have been appreciable.

Despite the error in this specific application, the paleoflood data represent the only large-event discharge calculations for this portion of the Finke Fiver system. Ungaged sites may be analyzed with much more precision than achieved in this study when data needs are more restrictive and local conditions permit.

FLOOD-FREQUENCY DETERMINATION

Both the Finke and Katherine studies illustrate the use of paleoflood hydrology to specify the magnitude of extremely large rare floods at ungaged sites. In addition, the Finke study illustrates a sequence of multiple, large paleofloods that would be appropriate for flood-frequency analysis. However, several questions arise in the appropriate treatment of paleoflood data.

Early work in paleoflood hydrology (Kochel and Baker, 1982; Baker, 1983) concentrated on magnitude estimation and on dating individual flood events. Transformation of flood magnitudes and ages to frequency estimates was accomplished by simple procedures, as described by U.S. Water Resources Council (1982), treating the paleofloods as historic flow data. However, Lane (1985) finds that this approach introduces both mathematical and philosophical errors in its treatment of paleoflood data, such that it yields erroneous results in long record extension (W.L. Lane, personal communication, 1984).

Some idealized examples will illustrate the nature of paleoflood data. These are given to stimulate necessary research on extracting flood-frequency information from paleoflood sites. Consider first the "worst case" example in Figure 8. This is a single, vertically-stacked

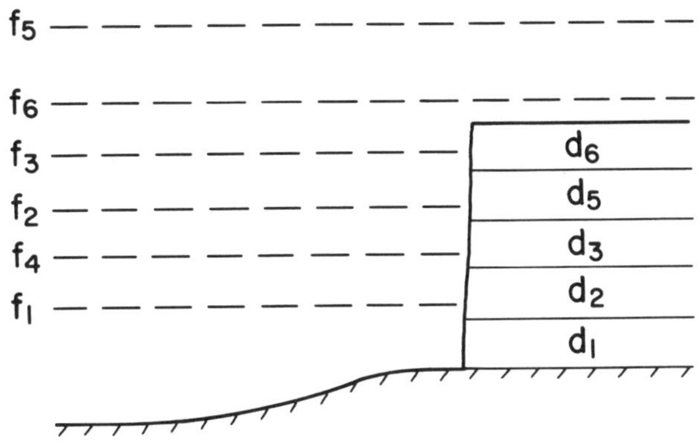

Figure 8. Single-site stacked sequence of slack-water deposits, d_1, d_2, etc., showing relationships to associated paleoflood stages, f_1, f_2, etc.

sequence of slack-water deposits. Since a new flood deposit cannot be emplaced unless the responsible flood stage exceeds the elevation of a previous deposit, the recorded flood series is censored. The censoring threshold is the elevation of the top of the deposit emplaced by the succeeding large flood. Note that this threshold increases when exceedences occur. However, the exact magnitude of exceedence is unknown, since various depths of flood water above the threshold are capable of emplacing a deposit. Note also that floods smaller than this upward moving threshold are not recorded in this example. For example, in Figure 8 floods f_1, f_2, and f_3 are recorded successively in the stack as deposits d_1, d_2, and d_3, but flood f_4 is not recorded, since it fails to exceed the rising threshold. Also note that the largest flood, f_5, is recorded but a subsequent smaller flood emplaces a deposit over it.

Analysis of this sequence begins with the problem of plotting position. As noted by Hirsch (1985) standard plotting position formulae are strictly applicable only to systematic records. For example, the Weibull formula requires estimation of the number of years of record. In Figure 8 the record length applies only to floods exceeding the stack height threshold. The record length for the whole stack would be the age of the surface on which deposit d_1 is emplaced. However, if only the age of deposit d_1 is known, there would be a prior time period, of unknown length, in which a flood was not recorded in the stack. The result would be that all return periods would be minimum values.

A possible approach to a frequency analysis of this example is to evaluate exceedences. Let t_0 be the age of establishing the stack base, ie. the first threshold. Let t_1 be the age of flood f_1. Then t_0-t_1 would be the period to first exceedence of that threshold by flood f_1 of magnitude Q_1. Similarly t_1-t_2 would be the period of first exceedence of the new threshold established by d_1, the deposit of f_1. Thus, information is recorded on the first time a given magnitude is achieved that exceeds various thresholds. It should be possible, therefore, given assumptions of stationarity and likely distributions, to extract frequency information from the stack. For example, using the results of Yen (1970) and equating the above-defined exceedence periods to record lengths for each threshold, one can calculate probabilities that the floods will represent an event of average return period T_R. For a record length of 400 years, Yen's (1970) results show that there is approximately 67% probability that the maximum recorded event (magnitude 1 for that period) represents an event that statistically should occur at least once in 400 years. For relatively long record lengths the probability becomes high that the largest event actually represents an event with a return period equal to the record length.

The problems of the single-site, vertical-stacked sequence are minimized by correlation and tracing through stratigraphic analysis (Kochel et al., 1982). In Figure 9 the slack-water site permits tracing of flood deposits laterally to their highest levels as they thin upstream along a tributary. An additional check on the maximum flood level is provided by a scour line that can be traced to a dated deposit in a cave. Moreover, an inset deposit has accumulated below the stack. Such relationships allow the identification of smaller floods in the inset, because a new, lower threshold is established by the river

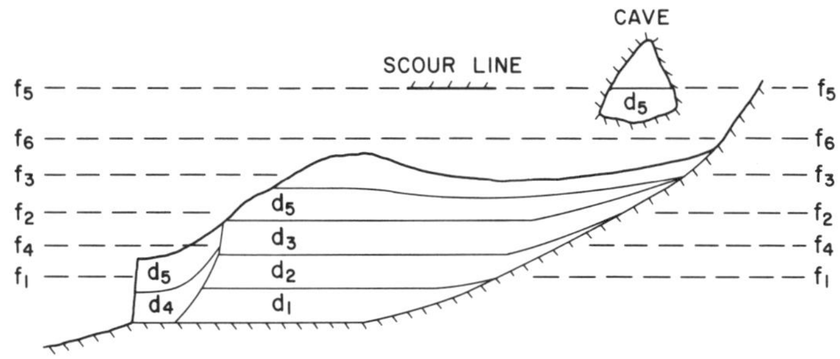

Figure 9. Complex slack-water depositional site showing evidence of maximum stages and inset deposits, as explained in text.

itself. The very high threshold established by the cave is especially useful in recording the magnitude of the largest flood. In this example the transformed magnitude data can be treated as precise discharge values, rather than minima. Moreover, the inset deposit preserves information on smaller, late series floods that fail to exceed thresholds in a single, vertically accumulating stack.

Figure 10 shows the most general case. Numerous vertical stacks, which began accumulating at different threshold positions, are correlated along a study reach. Scour lines or other direct high-water indicators are incorporated. Note that such studies rely on the highest deposit or mark of a given flood to define the magnitude of that flood. Thus, greater accuracy is achieved when the slack-water paleoflood study attempts to document as many sites as possible in an appropriate study reach. Floods that fail to leave deposits at one site, because they fail to exceed the local threshold, will be preserved at

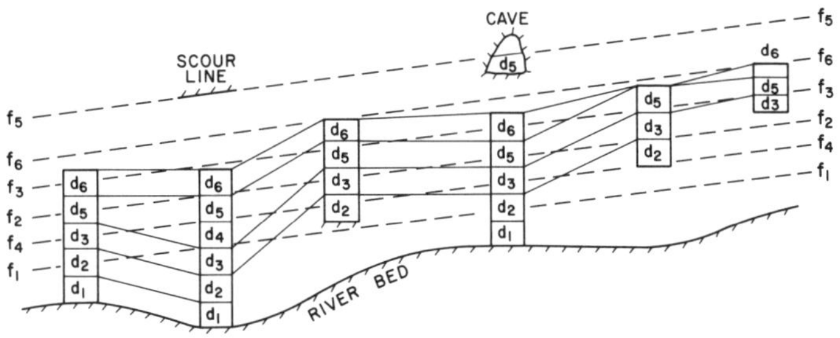

Figure 10. Idealized depiction of multiple paleoflood sites correlated along a study reach.

other, lower sites. Sites posing a very high threshold for a given flood will preserve the very highest deposits emplaced by that flood, thereby establishing its stage. For engineering studies the value of the increased accuracy achieved by such studies must be balanced by the increased expense of prolonging the field work. Further research is needed to define optimum levels of investment in such studies.

Although multi-site and stratigraphic correlation studies permit convergence to very accurate stage information, this stage data must be transformed to discharge values. Such transformation does not pose a problem in resistant boundary channels, since cross sections remain stable for the hydraulic flow modeling. However, in some applications, as in the Finke example, scour is possible. Because scour occurs at high stage, the maximum high-water marks may be associated with a smaller present-day cross section than prevailed at the time of the emplacement. Thus, the computed discharge value will be somewhat conservative, showing that a flood at least that large occurred.

Clearly there is a critical need to perform new research on the proper statistical treament for various kinds of paleoflood information. The above discussion has emphasized the nature of idealized paleoflood data in the hope of stimulating such work. A promising approach is the use of binomial-censor estimators (Stedinger and Cohn, 1986).

CONCLUSIONS

Our work has yielded a list of rules-of-thumb that we hope will aid other investigators who work with paleoflood hydrology of ungaged streams:

1. Single-site slack-water studies contain ambiguities concerning individual flood magnitudes, since paleostages may always have been higher than their emplaced deposits. However, important threshold data are contained at all slack-water sites.

2. Multi-site studies on a reach, especially where thresholds vary for slack-water stack accumulation, will yield much more accurate paleostage information for the responsible flow events than possible for single-site studies.

3. If a slack-water study reach has a potential for scour during floods, resulting discharge values will be conservative. However, excellent slack-water study reaches occur in areas of stable channel boundaries. The latter should be sought wherever possible.

4. Flow modeling by step-backwater analysis is the ideal procedure for transforming slack-water deposit heights to paleoflood discharges. The highest elevations achieved by the deposits or other high-water marks of a given flood are used to define the peak discharge of that flood.

5. Because of rules 1 and 3 most error in slack-water paleoflood studies tends toward systematic discharge underestimation for floods smaller than the maximum paleoflood. However, the latter may be very precisely constrained at a maximum value by high-level scour marks.

6. Errors in the dating of paleofloods can be reduced by stratigraphic geological analysis of their deposits and by modern ^{14}C geochronology. Floods in the period 1950-present can be dated essentially to the calendar year. Floods in the period 1650 to 1950 require supplemental dating by historical documentation, archaeology, dendrochronology, or other means. Floods in the period 10,000 to 350 years before present can generally be dated to high accuracy by conventional radiocarbon procedures or by tandem accelerator mass spectrometry.

7. A vertically-stacked sequence of flood slack-water deposits records a censored series of paleoflood stages, in which the censoring threshold (the elevation of each succeeding deposit) increases with time. The analytical problems in slack-water deposit paleoflood hydrology can be minimized by lateral stratigraphic tracing, inset relationships, or multi-site studies (rule 2).

8. Because of difficulties in evaluating assumptions incorporated into flood-frequency analysis, as developed for systematic data, such analysis may be inappropriate for paleoflood data. We emphasize the need to develop new statistical treatments, sensitive to the nature of paleoflood data, that make maximum use of that data in flood-frequency analysis.

ACKNOWLEDGEMENTS

This research was funded by the U.S. National Science Foundation (NSF) Division of Earth Sciences, Surficial Processes Program, and the NSF Division of International Programs, U.S.-Australia Cooperative Science Program, NSF Grant EAR-8300183. G. Allen, T. Jones, W. Mollah, D. Nelson, and T. Yocum assisted with the field work. We thank Donald R. Davis and Jery R. Stedinger for helpful discussions during early phases of our research.

REFERENCES

Baker, V.R. 1983. Paleoflood hydrologic techniques for the extension of stream flow records. Transportation Research Record, No. 922, pp. 18-23, National Research Council, Washington, D.C.

Baker, V.R. 1984. Flood sedimentation in bedrock fluvial systems. in: E.H. Koster and R.J. Steel (Editors), Sedimentology of Gravels and Conglomerates, pp. 87-98, Canadian Society of Petroleum Geologists Memoir, No. 10, Calgary, Alberta.

Baker, V.R., Kochel, R.C., Patton, P.C. and Pickup, G. 1983a. Palaeohydrologic analysis of Holocene flood slack-water sediments. in: J. Collinson and J. Lewin (Editors), Modern and Ancient Fluvial Systems, pp. 229-239, International Association of Sedimentologists Special Publication, No. 6.

Baker, V.R., Pickup, G. and Polach, H.A. 1983b. Desert palaeofloods in central Australia. Nature, Vol. 301, pp. 502-504.

Baker, V.R., Pickup, G. and Polach, H.A. 1985. Radiocarbon dating of flood events, Katherine Gorge, Northern Territory, Australia. Geology, Vol. 13, pp. 344-347.

Damon, P.E., Ferguson, C.W., Long, A. and Wallick, E.I. 1974. Dendrochronologic calibration of the radiocarbon time scale. American Antiquity, Vol. 39, pp. 350-365.

Ely, L.L. and Baker, V.R. 1985. Reconstructing paleoflood hydrology with slackwater deposits: Verde River, Arizona. Physical Geography, Vol. 6, pp. 103-126.

Faure, G. 1977. Principals of Isotope Geology, John Wiley, New York, 464 pp.

Feldman, A.D. 1981. HEC models for water resources system simulation, theory and experience. in: V.T. Chow (Editor), Advances in Hydroscience, Vol. 12, pp. 297-423, Academic Press, New York.

Hedges, R.E.M. and Gowlett, J.A.J. 1986. Radiocarbon dating by accelerator mass spectrometry. Scientific American, Vol. 254, pp. 100-107.

Hirsch, R.M. 1985. Probability plotting position formulas for flood records with historical information. EOS, Vol. 66, pp. 267-268.

Kochel, R.C. and Baker, V.R. 1982. Paleoflood hydrology. Science, Vol. 215, pp. 353-361.

Kochel, R.C., Baker, V.R. and Patton, P.C. 1982. Paleohydrology of southwestern Texas. Water Resources Research, Vol. 18, pp. 1165-1183.

Lane, W.L. 1985. Rare flood frequency estimation -- a case study of the Pecos River. EOS, Vol. 66, p. 267.

Stedinger, J.R. and Cohn, T.A. 1986. Flood frequency analysis with historical and paleoflood information. Water Resources Research, Vol. 22, pp. 785-793.

Stuiver, M. 1982. A high-precision calibration of the AD radiocarbon time scale. Radiocarbon, Vol. 24, pp. 1-26.

U.S. Water Resources Council 1982. Guidelines for Determining Flood Flow Frequency. U.S. Water Resources Council Bulletin, No. 17B, Washington, D.C., 28 pp.

Watson, I., Kingwell, R., Reader, B. and Hamrozi, A. 1970. Katherine town investigation of flooding from the Katherine River. Northern Territory Administration, Mines and Water Resources Branch, Darwin, Australia, 20 pp.

Yen, B.C. 1970. Risks in hydrologic design of engineering projects. Journal of the Hydraulics Division, Proceedings of the American Society of Civil Engineers, Vol. 96, No. HY4, pp. 959-966.

PALEOFLOOD HYDROLOGY AND DESIGN APPLICATIONS

Victor R. Baker
Lisa L. Ely
Jim E. O'Connor
John B. Partridge
Department of Geosciences
University of Arizona
Tucson, Arizona 85721

ABSTRACT. In China historical records of floods occurring over the past 2000 years are carefully interpreted in terms of magnitude and frequency. Such data play a major role in formulating design decisions for high-risk projects, including dams on the Chang Jiang (Yangtze River) that place a greater population at risk than that of the entire United States. In recent years it has become possible to reconstruct accurately paleoflood magnitudes in the U.S. over similar time scales as the Chinese historical floods. However, paleoflood procedures are generally not understood by the engineering design community in the United States, and a great resource of potential paleoflood information is being ignored. We present an example from central Arizona where construction cost estimates exceed $1 billion to modify an existing system of dams for new safety standards and increased storage capacity. Paleoflood studies on the two major rivers in the project area indicate that the largest floods to occur in the last 1000 to 2000 years significantly exceed flows recorded or observed in the last century. However, the paleoflood maxima are only 1/3 to 1/4 the design worst-case flood values (probable maximum floods) at the respective study sites. Such information can be of great value in the overall design decision, especially for major, high-risk projects. Where it can be obtained, paleoflood information should be incorporated into the decision process.

INTRODUCTION

In 1950 there were 300 hydrologic stations in China, and most of the record lengths were too short to provide useful design information (Luo Chengzheng, 1985). In order to generate information to design numerous major dam projects, the People's Republic of China embarked on a national survey of historical flood marks (Chen Chia-Chi et al., 1974; Hua Shi-Qian, 1985). In part this realization derived from hazardous under-design of several reservoir spillways based on flood-frequency analysis of inadequately short systematic records (Teng Wei-fen and Gu Chuan-zhi,

V. P. Singh (ed.), Regional Flood Frequency Analysis, 339–353.

1985). The philosophy of the historic flood survey was to collect as much documentary information as possible to test the representativeness of the systematic gage records and to reduce the potential error of extrapolation in frequency curves to extraordinary flood magnitudes.

Examples of Chinese historical flood documentation include studies of the Chang Jiang (Yangtze River) (Figure 1) where the major floods of 1153, 1227, 1520, 1560, 1788, 1796, 1860, and 1870 are well documented (Luo Chengzheng, 1985). Figure 2 shows one of 91 inscriptions and 306 flood marks that document the magnitude of the 1870 Chang Jiang flood, the largest in the last 1000 years. The importance of such data in China is emphasized by its use in defining the spillway design discharge for the Gezhouba Dam, a $2.4 billion project now nearing completion, and the Three Gorges Dam, a proposed $7.4 billion project (Shih Winshing, 1985). Failure of these dams would put a greater total population at risk than that of the entire United States.

The Huang He (Yellow River) has an even longer historical record, extending back to a recorded flood stage in 223 A.D. The Chinese have incorporated geological studies of flood sedimentation into design studies on this river in order to verify historical records (Shih Fucheng et al., 1985). They have rigorous procedures to evaluate the uncertainty of indirect discharge estimation for ancient flood evidence and to incorporate dated ancient floods into frequency analysis. They find that the largest events in the last several hundred to 2000 years exert a tremendous influence on frequency analysis (Shih Winshing, 1985).

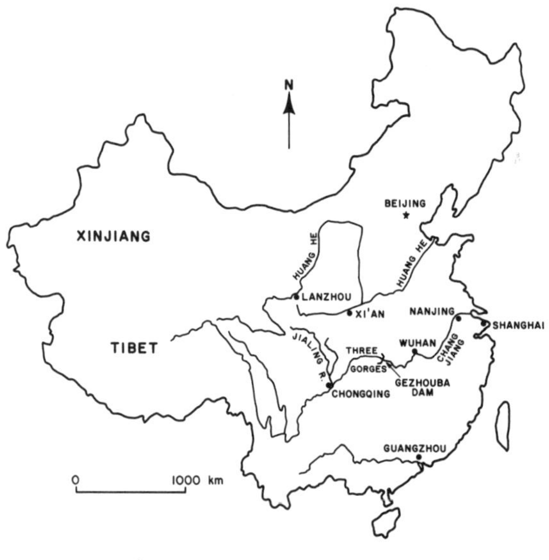

Figure 1. Map of China showing areas mentioned in the text.

Figure 2. The 1870 flood mark for the Jialing River at Beipei, near Chongquing, China.

Figure 3 shows two high-water marks for ancient floods on Boulder Creek, in south-central Utah. The marks consist of silt protected from erosion by a natural rock overhang. Both marks postdate petroglyphs carved between 1000 and 1275 A.D. The lower, younger silt line can be directly traced to a silt deposit that is radiocarbon dated at 110±90 years before present. The upper silt line was emplaced before the lower, but after the petroglyphs. It probably correlates to deposits dated at 180±50 radiocarbon years before present. Flow modeling of the hydraulics associated with the upper mark indicates a paleodischarge of 400 m^3/s (O'Connor et al., 1986).

The information described in the previous paragraph is an example of paleoflood hydrology. This techinque, developed to its present precision in the United States, is capable of generating information of floods over the past 10,000 years (Baker et al., 1979, 1983a). The information generated is analogous to that used in China as the most important basis of design for extraordinarily large floods. However, in the United States such information is rarely even considered for dam projects that involve great public expense and risk. Moreover, many of those projects inundate the very geological sites that would be useful in evaluating the safety of large reservoirs.

There are several reasons for the above paradox. First, the engineering design community is generally unaware that a revolution is occurring in the ability to precisely quantify the magnitude and frequency of paleofloods in certain, optimum geomorphic settings. Second, certain past paleohydrologic studies of floods were not sufficiently accurate for design purposes, and these studies have been judged by some engineers, unfamiliar with the field, to represent the full potential of

Figure 3. Silt lines (light streaks) on a sandstone panel of petroglyphs, Boulder Creek, Utah. The artwork was probably inscribed between 1000 A.D. and 1275 A.D. The upper silt line is associated with a paleodischarge of 400 m^3/s (O'Connor et al., 1986).

the method. Third, any new scientific approach, especially one involving multiple unfamiliar disciplines, disrupts established procedures that have evolved to most efficiently address design issues with conventional knowledge. An example of the latter is the inefficiency of the U.S. Water Resources Council procedure (Hydrology Subcommittee, 1982) at incorporating paleoflood data into flood-frequency analysis (Lane, 1985; Stedinger and Cohn, 1986).

This paper will briefly address these issues. However, a complete review of modern paleofood reconstruction procedures is beyond the scope of one paper. An in-depth review (Baker, in press) and several recent applications (Ely and Baker, 1985; O'Connor et al., 1986; Partridge and Baker, in press; Webb, 1985) illustrate current methodology. This field is undergoing rapid scientific advancement; it deserves the attention of the engineering design community.

PALEOFLOOD HYDROLOGY

Paleoflood hydrology is the study of past or ancient flow events which occurred prior to direct measurement by modern hydrologic procedures.

The definition of a paleoflood varies with the site of investigation. At a gage site there will be a systematic record with a length equal to the time period over which measurements were taken. There may also be a record of historic flood events which precede the gage record but for which stages were observed and recorded (Benson, 1950; Thomson et al., 1964). The extreme examples in China, noted above, involve up to 2000 years of historic data. At the opposite end of the scale, numerous ungaged sites have no systematic record, and may also have no historical flood record. Examples from northern Australia illustrate the use of paleoflood hydrology in cases where the technique is the only source of flood information (Baker et al., 1983a, 1983b, 1985, 1987).

Paleodischarge Estimation

A variety of procedures have been proposed for the reconstruction of paleoflood magnitudes. Past work has included the estimation of paleoflood depth employing relationships between bed particle sizes and shear stress (Knox, 1980), Shields parameter (Baker, 1974), and stream power (Costa, 1983). The error in such studies can be large (Maizels, 1983). Williams (1984) provides a comprehensive review of relevant formulae.

The ages of paleofloods have been estimated from the preservation of unique erosional and depositional features of a single large flood (Costa, 1978; Jarrett and Costa, 1982). While such studies provide useful information on the maximum floods that have occured in the geologic record of a given drainage basin, there may be significant error in evaluating flood magnitudes.

The procedure of paleoflood hydrology advocated for design purposes by this paper utilizes slack-water sedimentation sites (Baker et al., 1979, 1983a; Patton et al., 1979) and paleohigh-water marks (Ely and Baker, 1985; O'Connor et al., 1986; Webb, 1985). This procedure, when properly applied in the appropriate geomorphic settings, yields highly accurate reconstructions of multiple flood events.

The recommended approach for paleoflood discharge calculations is step-backwater analysis, as illustrated in several recent studies (Ely and Baker, 1985; O'Connor and Webb, in press; Partridge and Baker, in press). A significant advantage of the method is that the calculations can be run independently of the high-water indicator survey. Thus, problems in the geomorpholgy of paleoflood indicators can be separated from problems in specifying the hydraulics of the study reach. Recent advances in computer flow modeling (Feldman, 1981) have greatly facilitated the latter. Of prime importance to accurate flow modeling is a precise characterization of channel geometry. In stream channels with non-deformable boundaries (the preferred application) the survey cross sections should be chosen to account for the major energy losses that occur at channel expansions and constrictions.

Paleodischarge Estimation Error

Assuming that sufficient and appropriate cross sections are chosen to define the flow geometry of a reach (Davidian, 1984), error for slope-area calculations mainly occurs in the accuracy of measuring water-surface fall, Manning's n, expansion/contraction coefficients, and scour/fill relationships (Kirby, 1985). Major error problems occur for high-gradient streams (Jarrett, 1984), expanding reaches, and especially for alluvial streams with a high scour uncertainty.

Water-surface profiles produced by modern step-backwater flow-modeling procedures require input parameters of energy-loss coefficients, stage, and discharge. The initiating stage choice is of little consequence, since profiles generated at equal discharge will converge after a few computed steps. Moreover, several studies have demonstrated that step-backwater flow modeling of relatively deep, rare floods is remarkably insensitive to reasonable errors in estimating either Manning's n or the expansion/contraction coefficients (Dawdy and Motayed, 1979; O'Connor et al., 1986; Sauer et al., 1985). These observations apply to relatively narrow, deep flood channels with resistant boundaries. Alluvial reaches, subject to appreciable scour and filling, are generally not appropriate for paleoflood hydrologic studies.

Discharge estimation error in paleoflood hydrologic analysis can be reduced in several ways. The categories of error reduction include: (a) paleoflood cross-sectional stability, (b) paleoflood water-surface estimation, and (c) paleoflood flow coefficients. An important strategy is to separate the problems of categories (a) and (b), which depend on field conditions, from category (c), which depends on calculation procedure. This separation is not possible in the slope-area calculation procedure, since it requires that cross sections used in the analysis be located at sites where paleoflood highwater indicators are present.

Discharge estimation error in paleoflood studies is somewhat analogous to estimation error in historic flood studies. However, several aspects of paleoflood studies allow more precise controls on error than possible in historic flood studies or even some systematic flood studies. The main reason for this is the preservation of the best paleoflood information at exceptionally stable geological sites. Methods of reducing paleodischarge estimation error in the various categories are discussed by Baker (in press) and O'Connor and Webb (in press).

Design Applications

Paleohydrologic procedures have been advocated as sources of relevant data for risk analysis in the design of nuclear facilities, including power plants and waste containment sites (Foley et al., 1984). For the design of high-hazard dams, the National Research Council Committee on Safety Criteria for Dams (1985, p. 112) notes, ". . . that use of paleohydrologic methods could offer methods of assessing the magnitude of prehistoric great floods in some geographic areas. Further development of this approach should be considered." The latter committee (National Research Council, 1985, p. 235) goes on to state, "paleohydrologic techniques . . . have the best potential for illustrating what size floods can occur." A specific application is advocated in relation to the probable maximum flood (PMF) procedures used in the U.S. (in contrast to the Chinese approach described above). "When such techniques [ie., paleoflood hydrologic procedures] are applicable, they should be used to demonstrate that calculated PMFs are credible and are neither unreasonably large or small" (National Research Council, 1985, p. 235). These statements are not definitive, but they clearly show that paleoflood hydrologic data should no longer be ignored, especially where design concerns involve large uncertainties in risk determination for major, high-hazard projects.

CASE STUDY: CENTRAL ARIZONA

During 1984 a paleoflood hydrologic study was conducted on the Verde and Salt Rivers, central Arizona (Figure 4). The study was initiated by the Salt River Project, in relation to concerns about the design of major dams and storge reservoirs as mandated by the Safety of Dams Act. Initial field work determined that two study reaches, one on each river, displayed excellent qualities for achieving accurate paleoflood hydrologic determinations. Geologic evidence of paleoflood stages (slackwater sediments, scour lines, and driftwood accumulations) were well preserved at both reaches. Both reaches displayed exceptionally stable cross sections, amenable to flood discharge calculation by step-backwater flow modeling.

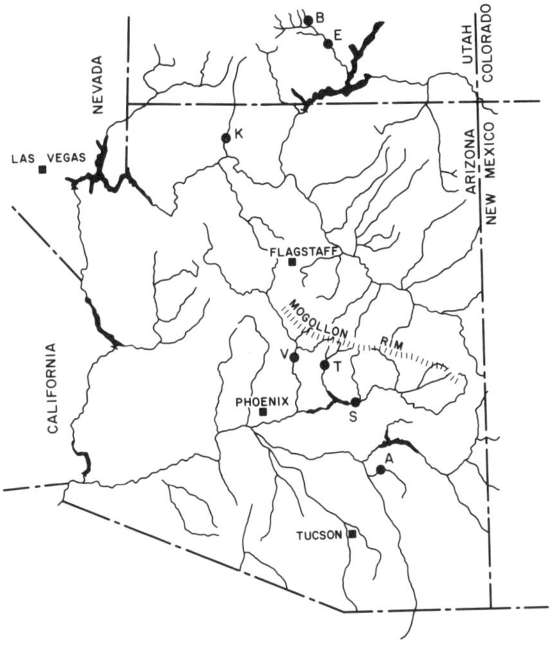

Figure 4. Location of the Salt (S) and Verde (V) study reaches. Other paleoflood studies have been completed at Tonto Creek (T), Aravaipa Creek (A), Kanab Creek (K), Boulder Creek (B), and the Escalante River (E).

Evidence of Paleoflood Magnitudes

The Salt River study reach is confined by granite and resistant conglomerate rock (Figure 5). Aerial photographs of both the Salt and Verde study reaches document essentially stable channels over the past 50 years. At the Salt study reach paleoflood slack-water sediment accumulations are most common immediately downstream of bedrock or colluvial protrusions into the channel. During extreme floods such channel margin irregularities act as flow obstructions, and suspended sand and silt is rapidly deposited in their lee.

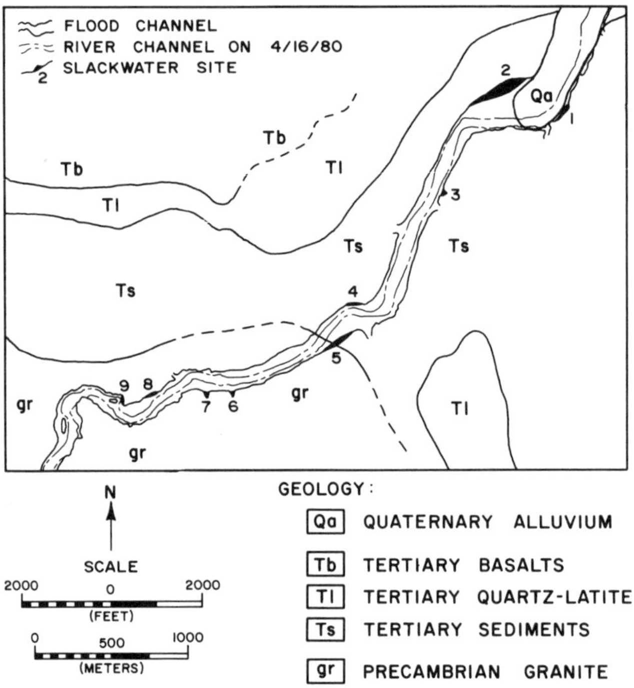

Figure 5. Sketch map of the Salt River study reach showing geologic units and numbered slack-water sedimentation sites. The flood channel is confined by bedrock.

In the Verde River study reach stable cross sections are achieved by local bedrock outcrops in the bed and banks, and by confining terrace walls of calcified Pleistocene river gravel. Paleoflood slack-water sediments occur at ineffective flow areas along the maximum flood channelways. These include (1) tributary mouths, (2) abrupt channel-expansions, (3) areas downstream of bedrock protrusions, and (4) areas downstream of protruding terrace gravel at river bends. In many settings the flood deposits comprise long benches which support dense groves of mesquite trees. Stratigraphic evidence documents 10 individual major flood events indicated by the sedimentation of these sties (Ely and Baker, 1985).

Maximum paleoflood stages in the study area are indicated by prominent flood scars that mark the soil-mantled valley sides. On granite outcrops this soil comprises a regolith of granite weathered to sand-sized particles, called grus. The grus was removed by the largest paleoflood and has not reformed below the flood scar in the time since the flood. In both study reaches the flood scars occur 10 to 12 m above low-flow levels in the river. On the Salt River the flood scar occurs about 2 m above the highest slack-water deposits.

Discharge Calculations

The HEC-2 water surface profile computer program (Feldman, 1981) was used to determine discharges for the paleoflood stages. Details of the model application, the hydraulic survey, etc. are described by Ely and Baker (1985) and by Partridge and Baker (in press).

At the Salt River study reach flow model accuracy could be evaluated by calibration of the modeled discharge to field evidence of the October, 1983 flood. Organic debris lines from this flood were carefully surveyed, and their elevations were used to relate the flow model to the 1700 m^3/s discharge that was determined for this flood at a nearby gage site. Because of the accuracy of the flow model in determining this gaged value, we are confident in its accuracy in evaluating the larger discharges corresponding to various paleofloods. Table 1 shows the discharge values determined for 8 major floods on the Salt River indicated by geologic evidence at the study reach.

Table 1. Paleodischarges for major floods on the Salt River, upstream of Roosevelt Lake, central Arizona (Partridge and Baker, in press).

Flood Unit Designation	Date of Flood or Age in Radiocarbon Years Before Present	Discharge[1] (m^3/s)
organic debris	1983	1700[2]
A	1980	2500
B	1952	2900
C	1941	2500
D	1891	>2300
E	200 yrs B.P.	3000
F	600 yrs B.P.	3200
flood scar	1000 to 2000	4100

[1] Determined by step-backwater flow modeling.
[2] Calibrated to nearby stream gage.

A check on our procedures for the Verde River study reach was achieved by a survey of maximum slack-water deposit heights and driftwood accumulations corresponding to the 1980 flood, the most recent major event prior to the survey. The flow model water surface that best corresponded to the 1980 paleostage indicators was generated at a discharge of 2700 m^3/s. Since the gaged value of the 1980 flood was 2680 m^3/s, it is clear that the flow model was very accurately depicting the hydraulics of the study reach.

Paleoflood Ages

Ages of paleofloods in the study areas were determined by a combination of relative and absolute dating described in detail by Ely and Baker (1985) and Partridge and Baker (in press). Stratigraphic analysis and correlation were used between multiple slack-water sedimentation sites as advocated by Baker et al. (1987). Absolute ages were determined by radiocarbon analyses. Ages to the calendar year were determined from 1950 to present by use of a calibration procedure described by Baker et al. (1985). Ages prior to 1950 were determined by correspondence to historic records, e.g. the great flood of 1891, and by standard radiocarbon analyses, which yield ages in radiocarbon years before present. Examples of the age determinations are given in Table 2.

Table 2. Ages of paleofloods on the Salt and Verde Rivers, central Arizona.

Flood Unit Designation	Radiocarbon Laboratory Number	Radiocarbon Dates[1]	Probable Dates[2]
A (Verde)	AA-928 AA-930	223±70 7±109	1891
Flood Scar (Verde)	AA-927	1010±95	1000 yrs B.P.
B (Salt)	TX-5124 TX-5120	<1950-1958 <1950-1958	1952
C (Salt)	TX-5121 TX-5123	<1950-1957 1956	1941
D (Salt)			1891
E (Salt)	TX-5116 TX-5115	90±280 300±150	200 yrs B.P.

[1] Reported in years before present (yrs B.P.) with ± error or as calendar date from curve of Baker et al. (1985).

[2] Derived from stratigraphic correlation between sites, comparison to observed flow record, and evaluation of possible contamination in radiocarbon samples.

The most critical age for design applications is that of the high-level, scoured granite outcrops. Approximately 15 to 20 cm of grus were stripped at the Salt River study site to expose the bare rock surface. Various studies of the rate of grus formation indicate that at least 1000 to 2000 years would be required to form such a soil layer (Partridge and Baker, in press). Since the flood scar cuts this soil, it is the largest flood to have occurred in at least 1000 to 2000 years.

A more accurate age estimate is possible for the Verde River study site. There, the flow model water-surface profile corresponding to the scour line can be traced to high-level slack-water deposits containing datable materials. The slack-water sediment, 12 m above low-water level in the river, is littered with diagnostic pottery shards. The pottery was traded into this region from the Anasazi and Sinagua cultures of northern Arizona around 950-1000 A.D. (A. Lindsay, personal communication, 1984). Small flakes of charcoal, probably cultural in origin, were collected 10-50 cm below the surface in the sandy slack-water sediments. A radiocarbon date of 1010±95 years B.P. (Laboratory Number AA-927) confirms that this maximum paleoflood occurred approximately 1000 years ago.

Design Implications

Results of paleoflood studies on the Salt and Verde Rivers indicate that the largest flood in the past 1000 to 2000 years for either river is significantly smaller than the theoretical worst-case flood. The latter was determined by probable maximum flood (PMF) procedures, as mandated for central Arizona by the Safety of Dams Act. Construction costs to implement water storage, flood control, and Safety of Dam's criteria for current central Arizona water projects total $1.1 billion in 1984 funds (Salt River Project, written communicaton, 1984). Of this, Safety of Dams changes in various existing structures, mainly to accommodate the PMF discharges, are estimated at $270 million (Salt River Project, written communicaton, 1984). Clearly, given the magnitude of this undertaking, it is absolutely essential to investigate all information sources that might affect the design, including paleoflood hydrology.

Table 3 compares theoretical PMF values for central Arizona sites with the results of this study. Note that the paleoflood maxima are significantly larger than flows in either the systematic or historic flood records, the former spanning 90 years for the Salt River. However, the paleoflood maxima are only 1/3 to 1/4 the PMF values at various study sites. This disparity raises a significant design question. Do we wish to make an immense investment of public money to protect for a theoretical worst case flood of unknown probability that is so rare that it is 3 to 4 times larger than the worst floods experienced in the last millenium? Because of greatly improved, accurate paleoflood hydrologic procedures it is now possible to address this issue at appropriate study sites.

Table 3. Flood peaks in central Arizona determined by several methods; discharges in m^3/s.

Location	Maximum Observed Flood Systematic	Maximum Observed Flood Historic	Maximum Paleoflood	Theoretical PMF
Salt River near Roosevelt	2900	2900	4600[2]	12,900
Verde River above Horseshoe Dam	2830	4250[1]	5400	21,000

[1]Estimated discharge of 1891 flood, considered to have relatively low accuracy (John Keane, Salt River Project, oral communication, 1984). A paleoflood estimate of this flood suggests a discharge of 3800 m^3/s (Ely and Baker, 1985).

[2]Flow augmented for contribution of a tributary in order to correlate the paleoflood site values (Table 1) to discharges recorded at a downstream gage site.

CONCLUSIONS

In China historical records of floods over the past 2000 years are interpreted in terms of magnitude and frequency and incorporated into the design of major dam projects. The historical flood data vary in accuracy and are not necessarily found at ideal sites for paleodischarge calculation. Nonetheless, the inclusion of those data in design, along with systematic records, is standard engineering practice for the People's Republic of China.

In the United States we generally lack long historical flood observations. However, we have local opportunities to generate excellent paleoflood records. The expense of such studies is minor in relation to planning costs for major high-risk projects such as nuclear power plants or large dams. At present these opportunities are mostly being ignored. Valuable paleoflood records are being destroyed by reservoir construction. In contrast, large public expenditures are made to salvage the archaeological information upstream of future impoundments, data which have no bearing on the safety of those impoundments.

In this paper we have only introduced the multidisciplinary complexity of paleoflood hydrologic studies. Such studies require expertise in geology and geomorphology as well as in hydrology and hydraulics. Lack of familiarity with concepts and terminology has hindered the adoption by engineers of paleoflood methodologies, but the potential contribution of paleoflood data should outweigh this artificial barrier. It is imperative that the design engineering community in the United Sates familiarize itself with modern advances in paleoflood hydrology.

At a minimum the physical evidence of large paleofloods should be considered to provide objective evidence of the likelihood and frequency of larger floods than can be documented in systmatic or historic flood records. Current U.S. thinking (National Research Council, 1985) prescribes the use of paleoflood data, when available, to test the credibility of calculated probable maximum flood values used in the design of high-hazard dams. Thus far there has not been an official U.S. advocation of probability-based design, incorporating, as in the Chinese procedure, information on extraordinary floods occurring during the past several thousand years. Perhaps this extreme position should be reconsidered. For critical projects the paleoflood data should at least be collected, appropriately weighted, and considered in the overall decision process leading to design.

ACKNOWLEDGEMENTS

Aspects of this study were supported by the United States Geological Survey through the States Water Resources Research Institute Program. Specific work on the Salt and Verde Rivers was funded by the Salt River Valley Water Users' Association. Contents of this publication do not necessarily reflect the views and policies of the United States Department of the Interior, nor does mention of trade names or commercial products constitute their endorsement by the United States government.

REFERENCES

Baker, V.R. 1974. Paleohydraulic interpretation of Quaternary alluvium near Golden, Colorado. Quaternary Research, Vol. 4, pp. 95-112.

Baker, V.R. In Press. Paleoflood hydrology and extraordinary flood events. Journal of Hydrology.

Baker, V.R., Kochel, R.C. and Patton, P.C. 1979. Long-term flood freqeuncy analysis using geological data. International Association of Hydrological Sciences Publication, No. 128, pp. 3-9.

Baker, V.R., Kochel, R.C., Patton, P.C. and Pickup, G. 1983a. Palaeohydrologic analysis of Holocene flood slack-water sediments. in: J. Collinson and J. Lewin (Editors), Modern and Ancient Fluvial Systems, pp. 229-239, International Association of Sedimentologists Special Publication, No. 6.

Baker, V.R., Pickup, G. and Polach, H.A. 1983b. Desert palaeofloods in central Australia. Nature, Vol. 301, pp. 502-504.

Baker, V.R., Pickup, G. and Polach, H.A. 1985. Radiocarbon dating of flood events, Katherine Gorge, Northern Territory, Australia. Geology, Vol. 13, pp. 344-347.

Baker, V.R., Pickup, G. and Webb, R.H. 1987. Paleoflood hydrologic analysis of ungaged sites, central and northern Australia. in: V.P. Singh and T.H. Illangasekare (Editors), International Symposium on Flood Frequency and Risk Analysis (This Volume), Baton Rouge, Louisiana.

Benson, M.A. 1950. Use of historical data in flood frequency analysis. Transactions of the American Geophysical Union, Vol. 31, pp. 419-424.

Chen Chia-Chi, Yeh Yung-Yi and Tan Wei-Yan 1974. The important role of historical flood data in the estimation of spillway design floods. Scientia Sinica, Vol. 18, no. 5, pp. 669-680.

Costa, J.E. 1978. Holocene stratigraphy in flood-frequency analysis. Water Resources Research, Vol. 14, pp. 626-632.

Costa, J.E. 1983. Paleohydraulic reconstruction of flash-flood peaks from boulder deposits in the Colorado Front Range. Geological Society of America Bulletin, Vol. 94, pp. 986-1004.

Davidian, J. 1984. Computation of water-surface profiles in open channels. Techniques of Water-Resources Investigations, Book 3, Ch. A15, 48 pp., U.S. Geological Survey.

Dawdy, D.R. and Motayed, A.K. 1979. Uncertainties in determination of flood profiles. in: Inputs for Risk Analysis in Water Resources, pp. 193-208, Water Resource Publications, Fort Collins, Colorado.

Ely, L.L. and Baker, V.R. 1985. Reconstructing paleoflood hydrology with slackwater deposits: Verde River, Arizona. Physical Geography, Vol. 6, pp. 103-126.

Feldman, A.D. 1981. HEC models for water resource system simulation: Theory and experience. in: V.T. Chow (Editor), Advances in Hydrosciences, pp. 297-423, Academic Press, New York.

Foley, M.G., Doesburg, J.M. and Zimmerman, D.A. 1984. Paleohydrologic techniques with environmental applications for siting hazardous waste facilities. in: E.H. Koster and R.J. Steel (Editors), Sedimentology of Gravels and Conglomerates, pp. 99-108, Canadian Society of Petroleum Geologists Memoir, No. 10, Calgary, Alberta.

Hua Shi-Qian 1985. A general survey of flood frequency analysis in China. Paper presented at U.S.-China Bilateral Symposium on the Analysis of Extraordinary Flood Events, Nanjing, China

Hydrology Subcommittee 1982. Guidelines for Determining Flood Flow Frequency. U.S. Water Resources Council Bulletin, No. 17B, Washington, D.C.

Jarrett, R.D. 1984. Hydrualics of high gradient streams. Journal of Hydraulics Division, American Society of Civil Engineers, Vol. 110, no. 11, pp. 1519-1539.

Jarrett, R.D. and Costa, J.E. 1982. Multidisciplinary approach to the flood hydrology of foothill streams in Colorado. in: Report on International Symposium on Hydrometeorology, pp. 560-565, American Water Resource Association, Denver, Colorado.

Kirby, W. 1985. Statistical error analysis of the slope-area method of indirect discharge determinaton. EOS, Vol. 66, p. 267.

Knox, J.C. 1980. Geomorphic evidence of frequent and extreme floods. in: Improved Hydrologic Forecasting: Why and How, pp. 220-238, American Society of Civil Engineers, New York, N.Y.

Lane, W.L. 1985. Rare flood frequency estimation -- a case study of the Pecos River. EOS, Vol. 66, p. 267.

Luo Chengzheng 1985. A survey of historical flood and its regionalization in China. Paper presented at U.S. China Bilateral Symposium on the Analsyis of Extraordinary Flood Events, Nanjing, China.

Maizels, J.K. 1983. Paleovelocity and paleodischarge determination for coarse gravel deposits. in: K.J. Gregory (Editor), Background to Palaeohydrology, pp. 101-139, John Wiley, New York.

National Research Council 1985. Safety of Dams; Flood and Earthquake Criteria, National Academy Press, Washington, D.C., 176 pp.

O'Connor, J.E. and Webb, R.H. In Press. Hydraulic modeling for paleoflood analysis. in: V.R. Baker, R.C. Kochel and P.C. Patton (Editors), Flood Geomorphology, John Wiley, New York.

O'Connor, J.E., Webb, R.H. and Baker, V.R. 1986. Paleohydrology of pool and riffle pattern development, Boulder Creek, Utah. Geological Society of America Bulletin, Vol. 97, pp. 410-420.

Partridge, J. and Baker, V.R. In Press. Paleoflood hydrology of the Salt River, Central Arizona. Earth Surface Processes and Landforms.

Patton, P.C., Baker, V.R. and Kochel, R.C. 1979. Slackwater deposits: a geomorphic technique for the interpretation of fluvial paleohydrology. in: D.P. Rhodes and G.P. Williams (Editors), Adjustments of the Fluvial Sytems, pp. 225-253, Kendall/Hunt Publishing Company, Dubuque, Iowa.

Sauer, V.B., Curtis, R.E., Santiago-Rivera, L. and Gonzalez, R. 1985. Quantifying flood discharges in mountainous tropical streams. in: F. Quinones and A.V. Sanchez (Editors), International Symposium on Tropical Hydrology and 2nd Caribbean Islands Water Resources Congress, pp. 104-108, American Water Resources Association, Bathesda, Maryland.

Shih Fucheng, Yi Yuanjun and Han Manhua 1985. Investigation and verification of extraordinarily large floods of the Yellow River. Paper presented at U.S.-China Bilateral Symposium on the Analysis of Extraordinary Flood Events, Nanjing, China.

Shih Winshing 1985. Application of historic flood in the design of Three Gorges Project. Paper presented at the U.S.-China Bilateral Symposium on the Analysis of Extraordinary Flood Events, Nanjing, China.

Stedinger, J.R. and Cohn, T.A. 1986. Flood frequency analysis with historical and paleoflood information. Water Resources Research, Vol. 22, pp. 785-793.

Teng Wei-fen and Gu Chuan-zhi 1985. China experience on estimation of design floods. Paper presented at the U.S.-China Bilateral Symposium on the Analysis of Extraordinary Flood Events, Nanjing, China.

Thomson, M.T., Gannon, W.B., Thomas, M.P. and Hayes, G.S. 1964. Historical floods in New England. U.S. Geological Survey Professional Paper, No. 1779-M, 105 pp.

Webb, R.H. 1985. Late Holocene floods on the Escalante River, south-central Utah. Ph.D. dissertation, University of Arizona, Tucson, Arizona, 204 pp.

Williams, G.P. 1984. Paleohydrologic equations for rivers. in: J.E. Costa and P.J. Fleisher (Editors), Developments and Applications of Geomorphology, pp. 343-367, Springer-Verlag, Berlin.

BOTANICAL EVIDENCE OF FLOODS AND PALEOFLOOD HISTORY

Cliff R. Hupp, Botanist
U.S. Geological Survey
A-413 Federal Building
Nashville, Tennessee 37203

ABSTRACT. High flood stages in conjunction with high water velocities and transported debris leave long-term datable deformations in the wood and habit of riparian and flood-plain vegetation. Tree-ring analysis subsequently may yield evidence of floods outside the period of record. Additionally, vegetation distribution patterns may delineate areas inundated by flows of distinct duration or recurrence interval.

Botanical evidence of floods along the downstream reaches of Passage Creek, Virginia, was recovered from increment cores and cross sections of flood-damaged trees. Flood years were checked against gage records and known floods on Passage Creek and on nearby Shenandoah River. Through regression of intermittent stage data from the Shenandoah River, the Passage Creek flood-stage record was extended from 1932 to 1870. Discharges of botanically determined floods were estimated using the current stage-discharge rating for Passage Creek and the results of the regression. Botanical evidence also showed that no major floods occurred between intermittent stage observations on the Shenandoah River. Six additional dates of flooding prior to 1870 were determined from botanical evidence and were assigned a minimum stage necessary to affect damage to bottomland trees, thus providing a history of flooding for 265 years. Additionally, patterns of species distribution were shown to delineate terraces (flood-return period greater than 3 years), floods plains (flooded on the average once every 2 years), channel shelves (10 to 25% flow duration), and depositional bars (about 40% flow duration).

INTRODUCTION

Paleofloods are floods that occurred before the time of continuous hydrologic record keeping or direct measurement; they may have occurred from less than a decade to millions of years ago (Costa, 1984). Floods have two long-term effects on bottomland woody vegetation. First, periodic floods of varying magnitude affect vegetation patterns by creating "new" areas such that suites of bottomland species with varying tolerances to flooding develop as bands parallel to the stream

V. P. Singh (ed.), Regional Flood Frequency Analysis, 355–369.

channel (Teversham and Slaymaker, 1976; Hupp and Osterkamp, 1985). Second, infrequent floods damage (usually not destroy) the bottomland plants such that their growth form reveals the effects of past floods either as outwardly evident stem deformation or as anomalous growth patterns in their serial tree-ring sequence (typically both). The purpose of this paper is to show that use of botanical evidence when combined with standard hydrogeomorphic techniques can extend flood records, reduce the margin of error in flood prediction particularly where gaging station records are short or intermittent, and allow for inference of the general flow regime.

In a landmark paper, Sigafoos (1964) showed that flood-plain trees yield evidence of past flooding from tree ages, trunk scars, and differences in properties of wood related to flooding. Dates of flooding determined from trees along the Potomac River correlated closely with those documented by gage and historical records (Sigafoos, 1964). A flood-frequency graph was produced by Harrison and Reid (1967) through analysis of tree scars in North Dakota. They analyzed the age and height of tree scars to determine the magnitude and frequency relation where streamflow records were only 25 years. Helley and LaMarche (1973) developed a 400-year record of large floods for parts of northern California from tree ages and radiocarbon dates of flood-entrained wood. Bottomland trees, with flood-related deformations in their stems or wood provided direct, on-site, evidence of past floods. Analysis of botanical evidence of floods usually is not a substitute for other methods for estimating flood frequency. However, when used with various hydrogeomorphic methods, botanical evidence can provide information suitable for incorporation into standard flood-frequency analyses.

Variation in bottomland vegetation patterns has previously been related to topographic variation by Wistendahl (1958), Sigafoos (1961), Hosner and Minckler (1963), Everitt (1968), Hupp (1982), Osterkamp and Hupp (1984), and Hupp and Osterkamp (1985). These authors suggest that differences in elevation above the stream channel are a major factor determining species distribution patterns. Small differences in elevation are directly related to substantial variation in hydrologic characteristics of bottomlands. Breaks in elevation on bottomlands are usually associated with different fluvial landforms and each of these landforms may have associated distinct suites of woody vegetation (Osterkamp and Hupp, 1984). Thus, bottomland vegetation patterns may be used to delineate hydrogeomorphic characteristics along a stream where direct hydrologic measurements are unavailable. Flood frequency, flow duration, and period of inundation are independent hydrologic factors that may be measured along stream reaches and have been related to vegetation patterns (Hack and Goodlett, 1960; Sigafoos, 1961; Bedinger, 1971; Teversham and Slaymaker, 1976; Hupp, 1982; 1983; Yanosky, 1982a; Hupp and Osterkamp, 1985).

RECONSTRUCTION OF BOTANICAL PALEOFLOOD HISTORY

The knowledge of flooding characteristics and magnitude-frequency information is needed by students of flood hydrology. Flood prediction for any given stream depends on knowledge of the magnitude and frequency of past floods. This information is obtained by analysis of records compiled at gaging stations and historical records of high stages. Ages of trees on bottomland landforms and of scars and sprouts from flood-damaged stems (Sigafoos, 1964; Harrison and Reid, 1967; Helley and LaMarche, 1973; Hupp, 1983) and differences in properties of wood anatomy related to flooding (Yanosky, 1983; 1984) are evidence of past floods. These data may be used to extend flooding history where records are short or lacking. Long flood records provided by botanical evidence of floods permit more accurate estimations than from short records; thus, procedures that extend flood records are potentially useful. Damage to trees during flooding commonly is severe but the proportion of trees killed may be low. High water velocities, turbulence, and large quantities of transported debris are characteristics of floods that are most damaging to bottomland vegetation (Sigafoos, 1964). However, even relatively gentle inundation can cause datable anomalies in intra-ring structure (Yanosky, 1983; 1984).

The mechanism of tree growth is well documented in botanical literature. Sigafoos (1964), Phipps (1967), and Yanosky (1982b) have described tree growth in relation to surface-water hydrology. The annual increment of tree growth has been the basis of many studies using tree rings in documentation of the magnitude and frequency of important hydrologic and geomorphic events (Sigafoos and Hendricks, 1961; Sigafoos, 1964; Harrison and Reid, 1967; LaMarche, 1968; Alestalo, 1971; Helley and LaMarche, 1973; Shroder, 1978; Yanosky, 1982a; 1982b; Hupp, 1983; 1984).

Trees grow vertically only from the tips of branches. Radial growth (growth in girth) is accomplished by a vascular cambium beneath the bark which forms a cone of new wood each year. This cone is called a tree ring when viewed in cross section. A count of tree rings from an increment core taken from the base of the stem yields the age of the stem so long as each year is represented by a single ring. The standard dendrochronological technique of cross dating (Stokes and Smiley, 1968; Cleaveland, 1980) is used to ensure that dating errors are not introduced by the possible occurrence of missing or multiple rings.

Types of Botanical Evidence of Floods

Corrasion scars. Scars on riparian trees and shrubs are the most conspicuous evidence of past flooding. Currently, the most reliable and accurate method of tree ring determined dates of flooding is the analysis of increment cores or cross sections through scars. Analysis of these samples yields the year of flooding. Because of differences in wood produced during the early and late parts of the growth season, it is often possible to determine the season when the scar was formed. Corrasion destroys the cambium (wood producing tissue) in the area of impact. Thus, growth ceases in the damaged tissues, and the event is recorded as undamaged tissue (callus) grows in annual increments around and over the scar (Fig. 1). Evidence of the scar may remain for

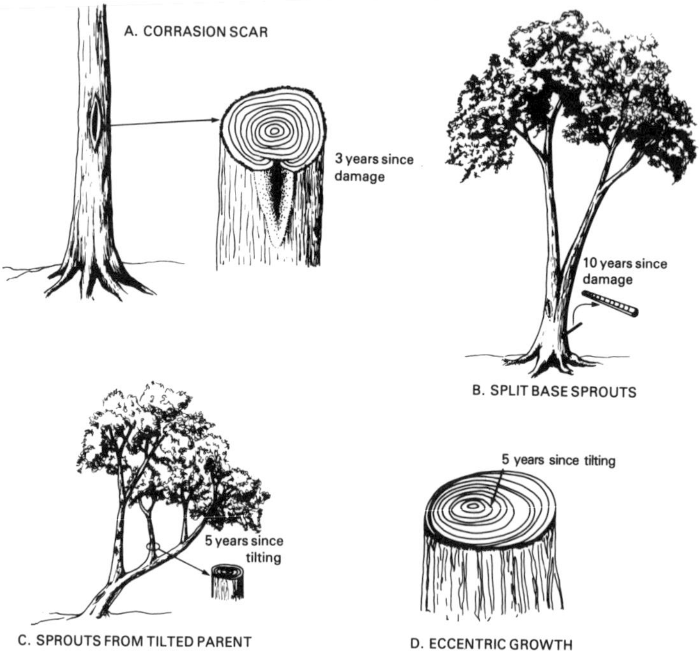

Figure 1. Types of botanical evidence of floods. Cores or cross-sections of affected trees are shown.

several years because of the presence of this callus tissue. After many years, outward evidence of the scar may disappear. The usefulness of scars as a dating technique diminishes with time, unless the tree is cut to reveal scars deep in the wood. Maximum scar heights on damaged trees can also indicate a minimum peak stage.

Adventitious sprouts. Sprouts from broken or inclined stems are easily recognized and can be used to determine the date of a flood by coring the base of the sprout. This type of evidence, after a few years, has the appearance of a split base (Fig. 1), or vertically growing sprouts from a stem that has been tilted in a downstram direction (Fig. 1). Flood training of riparian trees and shrubs is highly visible along streams subjected to periodic high-velocity floods. Use of the term sprout implies growth form or habit and not youth, as some sprouts may be hundreds of years old. The sprout may begin growth either the same year of flooding or during the growth season immediately after the flood. The age of these sprouts is usually within 1 year of the flood that caused the inclination or decapitation of the main stem. The accuracy obtained by coring sprouts is exceeded only by corrasion-scar analysis, which often has limited use in dating older events. The center ring occurs in the first year of growth and must be included in the increment core. Some trees may bear evidence of several floods.

Tree age. Flood-deposited sediments or flood-scoured areas provide sites where vegetation can become established. For example, flood-deposited longitudinal bars or point bars (Everitt, 1968; Hupp and Simon, 1986; Bryan and Hupp, 1984) may be rapidly colonized by certain woody species. The age of trees and shrubs growing on these "new" surfaces indicates a minimum time since initial deposition or scour. The population age structure can indicate whether the deposit is young enough to be dated dendrochronologically (Sigafoos and Hendricks, 1961; Hupp, 1984). When all the oldest trees on a deposit are within a few years of each other in age, it is reasonable to assume that the age of the oldest is close to the age of the deposit. Along actively migrating channels, tree ages can be used to estimate channel shifts through time. Of course, the age of some flood deposits exceeds the relatively limited life span of woody species. This technique has been used by Helley and LaMarche (1973) to extend the historic flood record for certain streams in northern California.

Ring anomalies. Floods may alter tree growth without leaving outward evidence as described above. This occurs as abrupt changes in ring width (Yanosky, 1982b), or as various alterations in intra-ring anatomy (Yanosky, 1983; 1984). Cores must be examined microscopically to detect and analyze anomalous ring patterns.

Eccentric ring patterns develop when a tree is tilted from the vertical. Slight inclinations can induce eccentric ring patterns without causing the formation of adventitious sprouts. Abrupt tilting results in subsequent rings that are wide on one side of the trunk while on the opposite side, the same rings are relatively narrow (Fig. 1). When this pattern follows concentric ring production, the year of the onset of eccentric growth is usually within 1 year of the event. This line of evidence is particularly useful in areas where conifers are the dominant species, because even severe tilting generally does not induce adventitious sprouts (Hupp, 1984).

Suppression and release tree-ring sequences appear as abrupt changes in ring width. This type of sequence can be caused by a number of events such as fire, disease, or windthrow. In general, suppression is the result of competition between the specimen tree and neighboring trees. Should these competitors be removed or damaged by a destructive flood, an undamaged tree may respond by producing an atypically wide ring or a series of wide rings. Conversely, severe crown damage may result in an extremely narrow ring or ring series (Yanosky, 1982b). Similarly, trees partially buried by flood-deposited sediments may undergo suppressed growth while new roots are growing into the new soil surface. These types of evidence can be used only when it has been clearly established that flood-growth responses are the most likely explanation.

Intra-ring abnormalities have been shown to be correlated with floods. Additionally, intra-ring evidence may occur even when the floods are of magnitudes below that necessary for substantial overt evidence (Yanosky, 1983). These abnormalities occur as a zone of enlarged vessels between the annual-ring boundaries. Thus, the position of these "flood rings" relative to ring width also permits more precise timing estimates of flooding, sometimes to within a few weeks (Yanosky, 1983). The estimated height of the tree in the year of the flood and its relative position on the bottomland can also be used to estimate the minimum peak stage.

Reconstruction of Passage Creek Flood History

The following is an example of an interdisciplinary approach to flood-frequency study. Botanical evidence has been combined with hydrologic and statistical methods to extend the flood record (62 years beyond gaging station records for Passage Creek in northwestern Virginia. Additionally, botanical evidence alone allowed for the determination of dates of major floods back to 1720.

Site description. Passage Creek drains the northern part of Massanutten Mountain in northwestern Virginia (Fig. 2). The basin has an area of 230 km^2 above the confluence of Passage Creek with North Fork Shenandoah River. Most of the basin is contained in a synclinal valley that is structurally controlled by ridges of Silurian Massanutten Sandstone. The stream flows through a gorge near the downstream end of the drainage basin. The sites selected for study of flood deformed trees consist of the gorge, adjacent parts of the bottomland above the gorge, and an alluvial fan below the gorge. These sites were selected because most of the forest in this area has not been disturbed by recent lumbering or agriculture and because evidence of floods in the vegetation is common in the gorge and adjacent reaches.

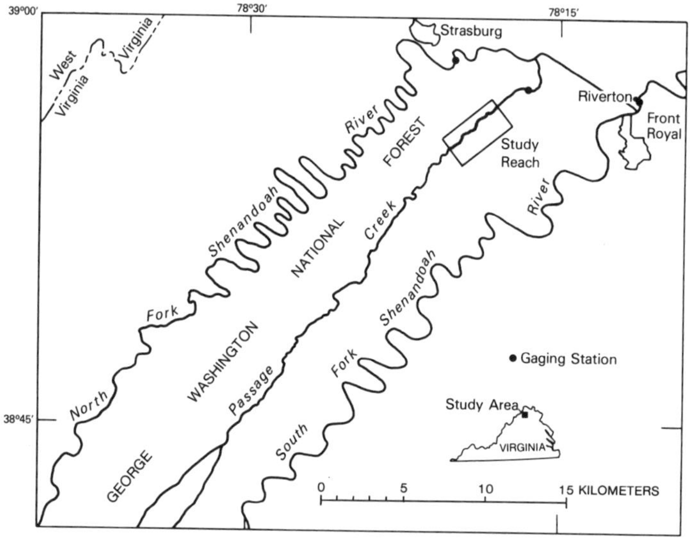

Figure 2. Location of study area in northwestern Virginia. Reach of intensive study along Passage Creek is indicated. Solid circles are gaging station sites.

In addition to the main channel, the flood plain in the study area is dissected locally with flood channels that carry water for flow events exceeding the 10 to 15% flow duration (Hupp, 1982). These higher flows produce a braided stream pattern; for extreme flows, the water surface is continuous across the valley bottom.

Streamflow in Passage Creek has been recorded continuously since 1932 at a site 2.1 kilometers (km) above its mouth, near Buckton, Va., and 3.5 km below the study area. According to surface-water records of the U.S. Geological Survey, (Water Resources Data for Virginia, 1981, published annually) the area of Passage Creek basin above the gage of 227 km^2. The average discharge for 45 years of record is 1.88 cubic meters per second (m^3/s), and maximum discharge for the period of record was 588 m^3/s on October 15, 1942. A flood associated with Hurricane Agnes (June 1972) left much evidence on the riparian vegetation within the gorge. Maximum discharge during this flood was 360 m^3/s, which has a recurrence interval of about 12.5 years.

Tree-ring flood dates and calibration. Over 200 trees and shrubs bearing evidence of flood damage were cored or sectioned to determine dates of flooding. These years were then tabulated chronologically (Table I). A flood year was incorporated into the list of possible dates when indicated by three or more different samples. This list was then compared to the annual flood series of the gaging station (since 1932) to determine coincidence of flood years. The period of record was used as a check, or calibration period, to determine the plausibility of earlier flood years indicated by tree-ring analysis.

Thirteen flood years, as indicated by tree and scar ages, occurred prior to the Passage Creek record. Supporting evidence of floods prior to 1932 was sought from gaging station records of nearby North Fork Shenandoah River near Strasbrug, Va., gaged continuously since 1926, and Shenandoah River at Riverton, Va., which has annual peak stages extending intermittently from 1870.

All dendrochronologically determined flood years, back to 1870, correspond with high peak flows listed in the gage, or stage records (U.S. Geological Survey, Water Data for Virginia, 1981, published annually) from Passage Creek (back to 1936) and from the Shenandoah River gages (back to 1870). Thus, a correlation exists between flood dates determined from the botanical evidence and gaging station records, as has been shown elsewhere (Sigafoos, 1964). However, at present, most botanical evidence cannot allow for precise inferences of the magnitude of a dated event.

Regression analysis. Peak stages from the Shenandoah River gage at Riverton, Va., were regressed against those of Passage Creek, from 1933 to 1986. Stage data were chosen for use in the regression analysis because the Riverton gage has only stage data. The rating curve at the Passage Creek gage has not changed substantially during the period of record; the site is under bedrock control. Thus, it is assumed that discharge at the gaging station has remained relatively time stable for the period of record. The regression results were used to estimate flood stages for Passage Creek for the period of record at the Riverton gage. The dates of floods indicated by the regression were checked

Table I--Flood dates from botanical evidence and type of botanical evidence on Passage Creek, Virginia.

Date	Type of Botanical Evidence		
	Corrasion scar	Inclined stem or stump sprout	Landform age
1972	X	X	X
1955	X	X	
1942	X	X	X
1936	X	X	X
1924		X	
1910	X	X	
1902		X	
1897		X	X
1889		X	X
1878		X	
1870		X	X
1866		X	
1858		X	
1852		X	X
1836			X
1825			X
1720			X

through botanical evidence to confirm whether they actually occurred on Passage Creek. The regression line for Passage Creek stage and the Shenandoah River at Riverton, Va., stage is shown in Figure 3. Shenandoah River at Riverton stage data accounted for 82.8% ($R^2 = 0.828$) of the variance of the Passage Creek record (Fig. 3). Results of this regression analysis were used to estimate stage for flood dates on Passage Creek indicated by botanical evidence from 1870 through 1932.

The combination of Passage Creek flood dates from botanical evidence with peak flows determined from the gage record and the Passage Creek-Shenandoah River regression analysis is illustrated in Figure 4. All years of flooding indicated by botanical evidence coincided with high peak flows estimated from the regression. The second-through-fifth highest stages of this historic record occurred during the period of record for the Passage Creek gaging station. It is impossible to estimate the relative ranking of the historic floods from the regression alone, because the Riverton record is intermittent and incomplete. However, the botanical evidence indicates that no higher floods occurred between those historic peaks shown on Figure 4 for the estimated Passage Creek record. The highest stage of the historic record is slightly higher than the 1942 peak. Analysis of Figure 4 shows that the minimum stage necessary to produce botanical evidence of a flood along Passage Creek is about 3.4 meters on the gage near Buckton, Va. This stage corresponds to a discharge of 155

m^3/s, determined from the 1981 stage-discharge rating for Passage Creek near Buckton, Va.

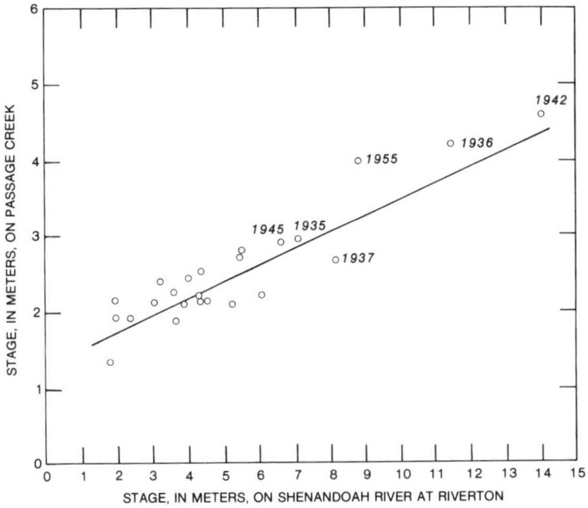

Figure 3. Regression results of data from Passage Creek and the Shenandoah River at Riverton. Dates of six highest peaks are shown.

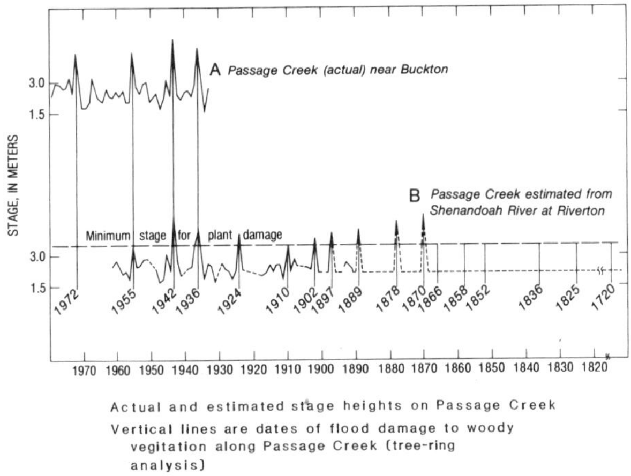

Figure 4. Actual and estimated stage heights on Passage Creek. Vertical lines are dates of floods indicated from botanical evidence. Dates prior to 1870 are assumed to have a minimum stage of 3.4 meters that is necessary for botanical evidence.

Botanical evidence of floods prior to 1870 are not supported by gage records. However, the correlation between flood years since 1870 and botanical flood evidence strongly suggests that flood dates prior to 1870 are valid. It is inferred that these floods had a minimum stage of 3.4 meters (155 m^3/s), based on a minimum necessary to affect damage to vegetation (Fig. 4). Furthermore, because trees with old evidence of floods have had to withstand disease, blowdown, and other floods, it seems reasonable to speculate that the floods prior to 1870 were quite large. Large floods usually leave more evidence than small floods.

Flood frequency. Discharges were computed using a stage-discharge relation for the peak stages of Passage Creek estimated from the regression analysis. These discharges were then used in a standard flood-frequency analysis (log-Pearson-Type III) as historic peaks. Only peaks which were assumed with confidence to be highest for the given year were used. The historically adjusted flood-frequency curve was then compared to the flood-frequency curve determined from gaging station records alone.

Flood-frequency analysis of peak-flow discharges is illustrated in Figure 5. Five historic peaks, determined through regression and tree-ring analysis, are included in one curve (Fig. 5), and the flood frequency curve without the addition of historical data is shown in the other (Fig. 5). The similarity of the two curves indicate a close match in flood frequency predicted between the systematic record and the record developed through botanical evidence of floods; this suggests that proxy flood information can be used with good reliability where the systematic records from gaging stations are incomplete.

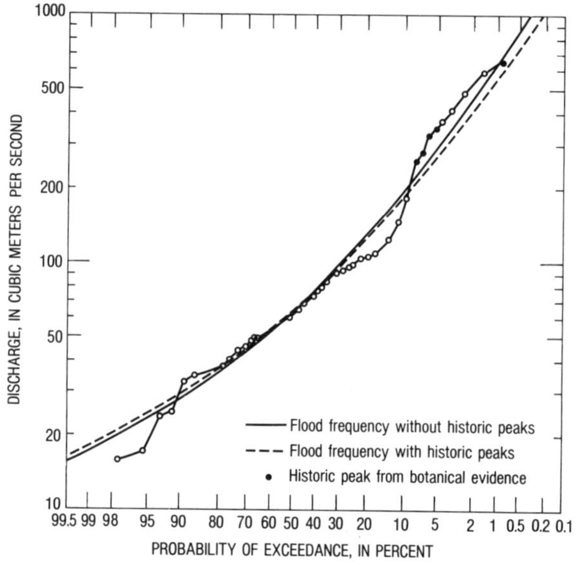

Figure 5. Flood frequency on Passage Creek. The unsmoothed line is actual annual peaks from gaging station records. Note the addition of botanical-historical data lowered the estimated discharge for low-frequency events (upper right).

Analysis of flood-damaged vegetation supplements other methods of flood-frequency estimation. The mechanism of plant growth provides for accurate dating of past floods through tree-ring analysis. This technique permits the extension of flooding history and provides physical evidence of past floods. Current methods in dendrochronologic analysis of flood-damaged vegetation are adequate to produce a historic record of flooding dates, but alone generally are not yet well suited for precise estimation of the magnitude of a given flood.

The botanical evidence of floods along Passage Creek, when combined with the regression results from the Shenandoah River, corraborates a 63-year extension of the timing and magnitude of major floods. These data provide the information necessary for incorporation into a standard flood-frequency analysis thereby extending the flood record from 1932 to 1870. From 1720 to 1870, botanical data provide strong evidence of six additional floods. Thus, botanical evidence alone permits a reconstruction of flood dates over a period of 265 years. The utility of this type of flood-record extension is limited to high-magnitude, infrequent flood events. Although the magnitude of infrequent floods is difficult to predict with short gage records, the recurrence interval of frequent flow events can usually be accurately determined from relatively short gage records. Intermittent stage or historical records of large floods along large streams extend substantially farther than the hydrologic record. This study illustrates the value of combining hydrologic data with tree-ring data to extend and verify the record of low-frequency floods. Reliable prediction of infrequent floods is critical where high flood stages can be destructive to human interests.

FLOODING AND VEGETATION PATTERNS

Many factors can influence the spatial distribution charateristics of plant species across bottomlands. Osterkamp and Hupp (1984) and Hupp and Osterkamp (1985) suggest that the characteristics of hydrogeomorphic processes associated with various fluvial landforms are most influential in affecting the distributional patterns of most bottomland species. The important fluvial landforms include depositional bar, active-channel shelf, flood plain, and terrace (Fig. 6). Each of these landforms can be defined hydrologically and support distinctly different vegetation types. Thus, it is possible, where the vegetation-hydrogeomorphology relations have been identified, to use easily mapped species distributions to delineate portions of the bottomland with distinctly different hydrologic characteristics. The following is a description of the plant ecological and hydrogeomorphic relations along northern Virginia high-energy streams (Osterkamp and Hupp, 1984; Hupp and Osterkamp, 1985).

Hydrogeomorphology

The landforms described herein are defined on hydrologic grounds. The independent parameters employed are flow duration (amount of time during which a level is reached or exceeded by streamflow) for surfaces below the flood plain, and flood frequency (recurrence interval of

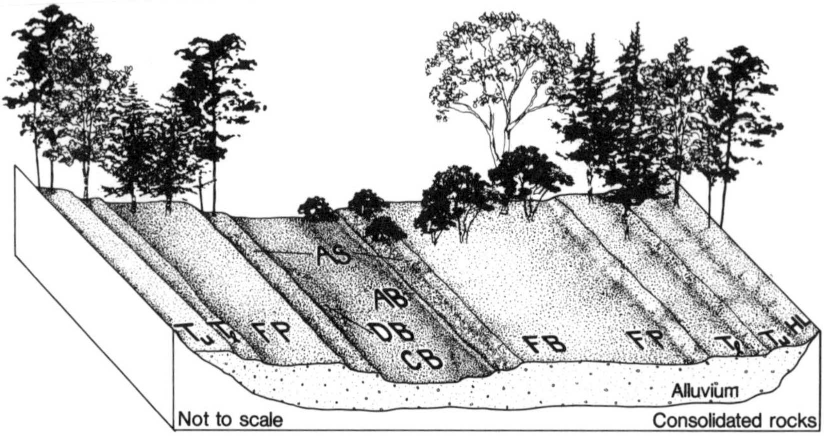

Figure 6. Block diagram showing fluvial landforms. From the lowest, the features are CB = channel bed, DB = depositional bar, AB = active-channel shelf bank, AS = active-channel shelf, FB = flood-plain bank, FP = flood plain, T = lower terrace, T = upper terrace, HL = hillslope. From Osterkamp and Hupp (1984).

flooding) for the flood plain and terraces. Fluvial geomorphic surfaces in northern Virginia have been related to flow frequencies using stage-discharge data from gaging stations (Osterkamp and Hupp, 1984).

The channel bed (CB) of a perennial stream (Fig. 6) is the surface that is wholly or partly covered by flows below mean discharge. Thus, at least part of the channel bed of a perennial stream is inundated at all times. Depositional bars (DB) occur in the active channel part of the valley section and are the lowest prominent feature higher than but within the channel bed (Fig. 6). The elevation of channel bars generally correspond to a stage slightly higher than the low-flow water stage, with a flow duration of about 40%. The active-channel shelf is a horizontal to gently sloping surface that normally extends a short distance between a relatively steep bank slope and the channelward limit of persistent woody vegetation (Fig. 6). The level of the active-channel shelf generally conforms to the stage of the average discharge; this level is typically reached between the 5 to 25% flow duration. Flood plains (FP) are a well-known (although often misinterpreted) reference level in the valley section. Flood plains are generally flat geomorphic surfaces (Fig. 6) that are inundated once ever 1 to 3 years (≈ 2-year recurrence interval). This surface and those lower in elevation are directly related to the present hydrologic activity of the stream. Terraces (T) generally represent incised former flood plains and occur at various levels above the modern flood plain. The likelihood of terrace inundation is always less than that of the flood plain, with recurrence intervals of 4 or more years.

These landforms are not mutually exclusive, although a given valley cross-section usually has one or more of the features lacking. The active-channel shelf and the flood plain appear to develop partially at the expense of one another. Gentle channel gradients tend to favor flood-plain development, whereas steep gradients tend to favor active-channel-shelf development (Hupp, 1986).

Plant Ecology

Regionally consistent and discrete relations between bottomland vegetation patterns and several fluvial landforms were found to exist in northern Virginia (Osterkamkp and Hupp, 1984; Hupp and Osterkamp, 1985). Apparently the plant distributions are at least in part controlled by inundation frequency and susceptibility of plants to damage by destructive flooding. Species composition, growth form, and frequency of inundation vary markedly from terrace levels down to the depositional bar. The effects of destructive flooding on vegetation increases from terrace levels toward the channel. The vegetation type and hydrologic characteristics of the defined landforms are summarized in Table II.

Table II--Summary of the relations between fluvial landforms, vegetation patterns, and hydrologic characteristics along northern Virginia streams.

Fluvial landforms	Vegetation type	Flow duration (percent of time)	Flood recurrence interval (years)
Depositional bar	largely absent	≈40	--
Active-channel shelf	riparian shrubs	5-25	--
Flood plain	flood-plain forest	--	1-3
Terrace	terrace assemblage	--	>3

Several species have been identified that can be used as key species whose presence is indicative of a particular landform. Depositional bars (DB) rarely have persistent woody vegetation, however they may be densely covered in water willow, *Justicia mariana*. The active-channel shelf supports a characteristic low-shrub thicket (riparian-shrub forest, Hupp and Osterkamp, 1985). Typical species of the active-channel shelf include: alder (*Alnus serrulata*), winterberry (*Ilex verticilata*), red willow (*Cornus amomum*), black willow (*Salix nigra*), and ninebark (*Physocarpus opulifolius*). The flood plain has a commonly diverse woody flora but supports one or more of the following indicator species: black walnut (*Juglans nigra*), American elm (*Ulmus americana*), bitternut (*Carya cordiformis*), silver maple (*Acer saccarhinum*), and hackberry (*Celtis occidentalis*). Terraces (T) have at least some individuals of upland species including: Chestnut oak (*Quercus prinus*), White oak (*Quercus alba*), mockernut (*Carya glabra*), Sassafras (*Sassafras albindum*), and Virginia pine (*Pinus virginiana*). Species such as sycamore (*Plantanus occidentalis*), boxelder (*Acer negundo*), ironwood (*Carpinus caroliniana*), and green ash (*Fraxinus pennsylvanica*), which generally are found on flood plains but occasionally present on active-channel shelves as well, are good indicators of bottomlands.

CONCLUSIONS

The precise role of periodic flood disturbance in affecting bottomland species distribution patterns is incompletely understood. Together, increased frequency of flooding and flood intensity on lower fluvial features relative to higher features represent a major factor influencing vegetation patterns by operating on species with variable tolerances to flood disturbance. In the absence of discharge data to relate surfaces to flow frequencies, plants provide a convenient method of identifying geomorphic levels and associated hydrologic characteristics with a relative level of confidence.

REFERENCES

Alestalo, J. 1971. 'Dendrochronological interpretation of geomorphic process.' *Fennia*, Vol. **105**, pp. 1-140.

Bryan, B. A. and Hupp, C. R. 1984. 'Dendrogeomorphic evidence of channel morphology changes in an East Tennessee coal area stream.' *Proc. Am. Geophysical Union*, Special Session on Paleoflood Hydrology, EOS Transactions, AGU, Vol. **65**, no. 45, p. 891.

Bedinger, M. S. 1971. 'Forest species as indicators of flooding in the Lower White Valley, Arkansas.' Pages 248-253 in *Geological Survey Research 1971*. U.S. Geological Survey Professional Paper 750-C.

Cleaveland, M. K. 1980. 'Dating tree rings in Eastern United States.' In: *Dendrology in the Eastern Deciduous Forest Biome, Blacksburg, Virginia*, VPISU, Pub. No. FWS-2-80, pp-124.

Costa, John E. 1984. 'Fluvial paleoflood hydrology.' *Transactions of the American Geophysical Union*, Vol. **65**, p. 890.

Everitt, B. L. 1968. 'Use of cottonwood in an investigation of the recent history of a flood plain.' *American Journal of Science*, Vol. **266**, pp. 417-439.

Hack, J. T. and Goodlett, J. C. 1960. *Geomorphology and Forest Ecology of a Mountain Region in the Central Appalachians*. U.S. Geological Survey Professional Paper 347.

Harrison, S.S. and Reid, J. R. 1967. 'A flood-frequency geoph based on tree-scar data.' *North Dakota Academy of Science*, Vol. **21**, pp. 23-33.

Helley, E. J. and LaMarche, V. C., Jr. 1973. *Historic Flood Information for Northern California Streams from Geological and Botanical Evidence*. U.S. Geological Survey Professional Paper 485-E, 16 pp.

Hosner, J. F. and Minckler, L. S. 1963. 'Bottomland hardwood forests of southern Illinois.' *Ecology*, Vol. **8**, pp. 29-41.

Hupp, C. R. 1982. 'Stream-grade variation and riparian-forect ecology along Passage Creek, Virginia.' *Bulletin of the Torrey Botanical Club*, Vol. **109**, pp. 488-499.

Hupp, C. R. 1983. 'Geo-botanical evidence of Late Quaternary mass wasting in block field areas of Virginia.' *Earth Surface Processes and Landforms*, Vol. **8**, pp. 439-450.

Hupp, C. R. 1984. 'Dendrogeomorphic evidence of debris flow frequency and magnitude at Mount Shasta, California.' *Environmental Geology*

and Water Sciences, Vol. 6, no. 2, pp. 121-128.
Hupp, C. R. 1986. 'The headward extent of fluvial landforms and associated vegetation on Massanutten Mountain, Virginia.' Earth Surface Processes and Landforms, Vol. 11, pp. 545-555.
Hupp, C. R. and Osterkamp, W. R. 1985. 'Bottomland vegetation distribution along Passage Creek, Virginia, in relation to fluvial landforms.' Ecology, Vol. 66, no. 3, pp. 670-681.
Hupp, C. R. and Simon, A. 1986. 'Vegetation and bank-slope development.' Proceedings of the 4th Federal Interagency Sediment Conference, Las Vegas, Nevada, 1986.
LaMarche, V. C., Jr., 1968. Rates of Slope Degradation as Determined from Botanical Evidence, White Mountains, California. U.S. Geological Survey Professional Paper 352-I, pp. 1341-1377.
Osterkamkp, W. R. and Hupp, C. R. 1984. 'Geomorphic and vegetative characteristics along three northern Virginia streams.' Geological Society of America Bulletin, Vol. 95, pp. 1093-1101.
Phipps, R. L. 1967. Annual Growth of Suppressed Chestnut Oak and Red Maple, a Basis for Hydrologic Inference. U.S. Geological Survey Professional Paper 485-C, 27 pp.
Shroder, J. F., Jr. 1978. 'Dendrogeomorphological analysis of mass movement of Table Cliffs Plateau, Utah.' Quaternary Research, Vol. 9, pp. 168-185.
Sigafoos, R. S. 1961. Vegetation in Relation to Flood Frequency near Washington, D.C. U.S. Geological Survey Professional Paper 424-C, pp. 248-249.
Sigafoos, R. S. 1964. Botanical Evidence of Floods and Flood-plain Deposition. U.S. Geological Survey Professional Paper 485-A, 35 pp.
Sigafoos, R. S. and Hendricks, E. L. 1961. Botanical Evidence of the Modern History Nisqually Glacier, Washington. U.S. Geological Professional Paper 387-A, 20 pp.
Stokes, M. A. and Smiley, T. L. 1968. An Introduction to Tree-ring Dating. University of Chicago Press, Chicago, 73 pp.
Teversham, J. M. and Slaymaker, J. 1976. 'Vegetation composition in relation to flood frequency in Lillooet River Valley, British Columbia.' Catena, Vol. 3, pp. 191-201.
Wistendahl, W. A. 1958. 'The flood plain of the Raritan River, New Jersey.' Ecological Monographs, Vol. 28, pp. 129-153.
Yanosky, T. M. 1982a. Effects of Flooding upon Woody Vegetation Along Parts of the Potomac River Flood Plain. U.S. Geological Survey Professional Paper 1206, 21 pp.
Yanosky, T. M. 1982b. 'Hydrologic inferences from ring widths of flood-damaged trees, Potomac River, Maryland.' Environmental Geology and Water Sciences, Vol. 4, pp. 43-52.
Yanosky, T. M. 1983. Evidence of Floods on the Potomac River from anatomical abnormalities in the wood of flood-plain trees. U.S. Geological Survey Professional Paper 1296, 42 pp.
Yanosky, T. M. 1984. 'Documentation of high summer flows on the Potomac River from the wood anatomy of ash trees.' Water Resources Bulletin, Vol. 20, pp. 241-250.

IMPLICATIONS OF FLUME EXPERIMENTS FOR THE INTERPRETATION OF SLACKWATER PALEOFLOOD SEDIMENTS

R. Craig Kochel and Dale F. Ritter
Department of Geology
Southern Illinois University
Carbondale, Illinois 62901

ABSTRACT

Experiments were conducted in a 12 m x 2.5 m flume to determine the properties of slackwater deposits produced during backflooding under different hydrologic and geomorphic conditions and to test some of the basic assumptions involved in using slackwater deposits to reconstruct paleoflood history. This was done by comparing observations made under controlled experimental conditions with those made in field studies along the Pecos River in west Texas. Twenty cement-lined tributaries were designed to enter a trunk channel at various angles and elevations above the main-channel floor. Four experimental runs were made, each having a different combination of flume slope and discharge. Thickness, grain size and maximum elevation of slackwater deposits with respect to peak main-channel stage were measured. The extent of backflooding and character of the deposits varied with junction angle between tributaries and the main channel, main-channel gradient, elevation of the tributary mouth, and tributary channel geometry. Properties of slackwater sediments deposited under various slope-discharge combinations were consistent with field observations. These flume studies support the validity of assumptions made when using the slackwater sediment method to reconstruct paleoflood history and provide a better understanding of what criteria should be used when searching for potential slackwater-sediment sites.

INTRODUCTION

Slackwater sediments are typically sand and silt-sized particles that accumulate during floods in protected areas where flow velocities are low. Analysis of slackwater sediment is becoming a widely used technique for determining the history of large paleoflood events (Patton et al., 1979; Baker et al., 1979; Kochel and Baker 1982; Kochel et al., and Baker 1986). In the appropriate geomorphic settings, slackwater or ungaged rivers. Flood frequency and magnitude records can, therefore, be extended for thousands of years.

The basis of the slackwater sediment method is that the highest elevation of sediment in backflooded depositional sites is essentially

V. P. Singh (ed.), Regional Flood Frequency Analysis, 371–390.

equivalent to the peak mainstream stage for the associated paleoflood. Stages are converted into paleoflood peak discharges using standard hydrological techniques, such as the slope-area method or step-backwater method. Paleoflood frequency can be estimated when the slackwater sediments are dated using radiocarbon analysis.

Field studies of slackwater sediments in tributary mouths along the Pecos River in west Texas show that sediment thickness and grain size vary between flood units at a site and between sites (Kochel 1980; Kochel et al., 1982; Patton and Dibble 1982). Between site variation is probably due to differing geometry and morphology of the tributaries in the mainstream. At-a-site variation most likely results from hydraulic differences between floods. If sedimentological variations can be explained in terms of flood hydrographs and/or depositional site geometry, valuable interpretations about ancient floods can be made. Regardless of its desireability, almost no concrete data exist which relate a particular sedimentologic property of slackwater deposits to a specific hydrologic or geometric control. We suspect, therefore, that complete confidence in the slackwater technique may be lacking until a more firm understanding about the genesis of the sedimentary characteristics is determined. With that goal in mind, this paper summarizes the results of preliminary flume experiments designed to study some of the controls of slackwater sediment thickness and grain size in backflooded tributaries along a stable bedrock channel. The results of the flume experiments will be compared to field observations made along the lower Pecos River of west Texas where the slackwater paleoflood method was first tested in detail (Kochel 1980; Kochel and Baker 1982; Kochel et al., 1982). The primary objective of our paper is not to quantify relationships, but to demonstrate trends which may lead us to understanding what thickness and textural variations in slackwater sediments are telling us about paleohydrologic events.

SLACKWATER PALEOFLOOD HYDROLOGY

Slackwater sediments are deposited rapidly from suspension in backflooded areas during large floods. The most favorable geomorphic settings for the accumulation and preservation of continuous slackwater-sediment sequences are: 1) in the mouths of tributaries; 2) in shallow caves along bedrock channel walls; 3) downstream from major bedrock and/or talus obstructions and in areas of dramatic channel widening, and 4) on high benches prone to overbank sediment accumulation. Slackwater sedimentation occurs in most river systems but is most pronounced and best suited for paleoflood studies in entrenched bedrock canyons like the lower Pecos River in west Texas (Kochel and Baker 1982; Baker and Kochel 1987). Slackwater sediment deposited at tributary mouths have provided the best data in recent studies (Kochel and Baker 1982; Kochel et al., 1982; Baker et al., 1983; Kochel and Baker 1987) and will be the focus of our discussion here.

Tributary mouth backflooding is accentuated during a large flood when the peak flow stage in the mainstream and in a tributary are not synchronous, or when parts of a watershed are unaffected by the storm that produced the trunk river flood. These situations are especially common in semi-arid regions where mainstream floods are often generated by intense convective rainfall over a small portion of the catchment while little flooding occurs in the downstream tributary drainages. As sediment-laden waters surge up the tributaries, slackwater sediments are deposited rapidly from suspension, leaving a record of the flood event. When these sediments accumulate repeatedly at the same location, their stratigraphy can be used to build a record of paleoflood magnitude and frequency.

Paleoflood Discharge

Reconstruction of paleoflood discharge requires an estimate of peak stage long after the flood has occurred. The slackwater approach uses the elevation of tributary mouth slackwater sediments to approximate peak flood stage. The use of any type of high-water indicator to reconstruct a past flood stage requires several qualifying assumptions. A critical assumption is that the mainstream channel cross-section has remained constant since the paleoflood event. That condition is best satisfied in bedrock channels and, therefore, these channel types are most conducive to paleoflood reconstruction (see Kochel and Baker 1982; Kochel et al., 1982; Baker et al., 1983; Kochel and Baker 1987).

The most important assumption in using slackwater deposits as paleostage indicators is that their elevations correspond to the maximum, peak-flood stages. In general, slackwater sediments at tributary mouths or in caves are conservative estimators of peak stage. Better estimates of paleoflood stage can be made by tracing slackwater flood units up the tributaries to determine the elevation where they pinch out. These pinch-out elevations more closely correlate with the peak flood stage for catastrophic historical floods along the Pecos River (Kochel 1980). In addition, the use of multiple sites having different base elevations above the mainstream greatly reduces the uncertainty of paleoflood stage estimates.

Once paleoflood stage has been estimated, the discharge can be derived using one of several hydraulic procedures, including the slope-area technique, or, for more complex bedrock channels, step-backwater methods. The application of these techniques to paleoflood slackwater studies has been thoroughly discussed in previous studies (for example, Kochel et al., 1982; Baker et al., 1983; Webb et al., 1987).

Paleoflood Frequency

After a relative sequence of paleoflood discharges has been established along a river, the absolute chronology of these floods can be determined because most flood sediments contain datable organic materials. The best material for radiocarbon dating is fine-grained organic detritus which normally is deposited on top of a given slackwater deposit during the waning stages of a flood (Kochel 1980; Kochel and Baker 1982; Baker et al., 1983). In addition, organic materials formed in-situ by pedogenesis of the flood sediments may yield respectable estimates of the time elapsed between floods. A complete discussion of the application of radiocarbon techniques and their limitations are given by Kochel (1980), Kochel and Baker (1982), Baker et al. (1983), and Baker et al. (1985).

FLUME SLACKWATER SEDIMENTATION EXPERIMENTS

In an effort to elaborate upon the relationships between slackwater sedimentology (thickness, geometry, and grain size) and flood hydrograph, gradient, and geometry and morphology of the tributaries and mainstream, we conducted experiments in an artificially molded bedrock channel within our flume. The main purpose of these experiments were: 1) to determine whether it is feasible to obtain slackwater sedimentation in a laboratory setting; 2) to study the controls on slackwater sediment thickness and grain size; and 3) to determine if any of the sedimentological variations observed in the field could be observed in the flume, where their genesis could be more easily studied than in a complex field situation. The flume model offers the flexibility to compare slackwater sediments in a wide variety of morphologic and hydraulic settings.

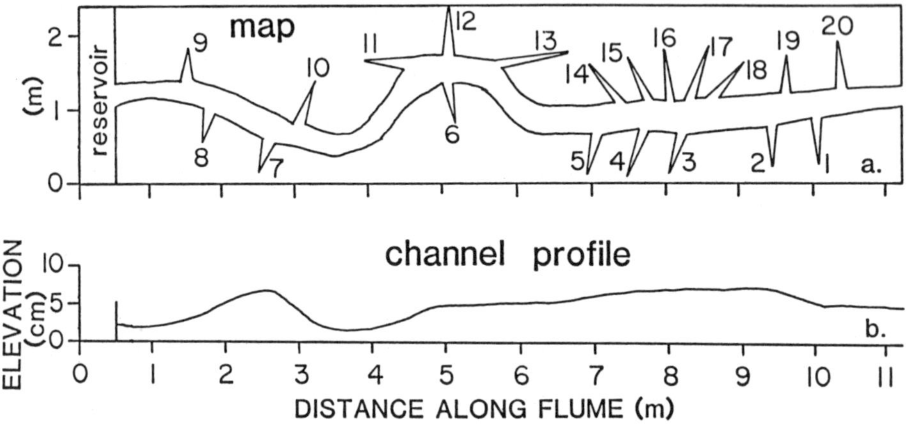

Figure 1. Schematic of the cement channel constructed in the flume. a) Map view of channel showing location of tributaries. b) Main-channel profile measured along the channel axis.

Design of Flume Experiments

A cement channel was constructed to simulate a non-erodible bedrock reach of a river. In accord with the approximate scaled-down channel dimensions of the Pecos River, a somewhat rectangular cross-section was constructed with an average width of 30 cm at the base and a depth of approximately 25 cm. Channel walls were nearly vertical, with slopes averaging between 75° to 85°. The plan view of the channel and twenty tributaries is shown in Figure 1. The channel extended over the entire length of the 12 m x 2.5 m flume. Tributaries were also lines with cement and designed to enter the mainstream at various angles and at different elevations above the main channel floor (Fig. 1). Lateral extent of the tributaries was restricted by the flume walls. This restriction ultimately resulted in problems with distal slackwater sedimentation during experiments done at high discharges because backwash from the tributary headwalls altered the depositional process.

Table I summarizes the basic controls of the four floods generated in which detailed measurements were made. The four runs were designed to compare slackwater sedimentation at low and high discharge variably combined with low and high mainstream slope. The range of slopes used in the flume was considerably higher than the range of slopes exhibited in the field, between 0.0015 - 0.0030 along the lower Pecos River (Kochel 1980).

TABLE I. SUMMARY OF FLUME EXPERIMENTS

RUN NUMBER	DISCHARGE (W)(D)(V) = Q (m)(m)(m/s)	SLOPE	DURATION OF RUN (hours)	LABEL
1	(.4)(.05)(.5) = .01	.020	1.15	low flow, low slope
2	(.49)(.1()(.6) = .03	.020	1.15	high flow, low slope
3	(.38)(.035)(.72) = .01	.030	1.15	low flow, high slope
4	(.45)(.09)(1.0) = .04	.030	1.15	high flow, high slope

The two slopes used in the runs described in Table I were achieved using the hydraulic jacks on the flume. Two oversteepened sections were molded into the channel bottom to accentuate the effects of mainstream gradient on slackwater processes near the upper and lower ends of the flume (Fig. 1b at tributaries 7, 10 and 2, 19).

Sediment was fed into the system by introducing approximately 10 kg of poorly-sorted sand, silt, and clay into a mixing reservoir at the upper end of the flume. Sediment injections occurred every 10 minutes throughout the duration of each 70 minute run. The recirculating flume continued to recycle the silt and clay fraction through the system. Suspended sediment concentration monitored midway through each run varied within 25% of the mean and very consistent sediment concentrations were obtained in runs of like discharges. Clearly, however, our method of sediment introduction did not insure a constant sediment load throughout a given run. Nonetheless, this does not negate our results since constant sediment concentration during the tenure of any natural flood is equally unlikely.

Table II summarizes the parameters measured during and after each of the four runs. Our measurements allowed for comparisons of sediment thickness, grain size, and bedforms for high and low discharges at high and low slopes.

Slackwater Sediment Distribution

Deposit Geometry. Backflooding occurred in all except one of the 20 tributaries during experiments run at low slope and in 2/3 of the sites during the high slope experiments. As expected, tributary slackwater deposits thinned rapidly away from the main channel, forming an up-tributary sedimentary wedge (Fig. 2). In the proximal part of the tributary (near the mouth), the distribution of sediment was markedly asymmetrical across the floor of the tributary. With few exceptions, slackwater sediments were thickest on the downstream sides of the tributary channels. This asymmetry gradually disappeared in the distal direction (up the tributary channel). Observations of flow motion using dye tracers showed that the depositional asymmetry resulted because backflood waters entered the tributary mouth along its downstream wall and exited along the upstream wall, thereby establishing a consistent eddy system at the tributary mouths. The depositional asymmetries were most pronounced where tributary junction angles were nearly 90°, and became less obvious where obtuse angles occurred, in particular where angles exceeded 120°.

TABLE II. FLUME MEASUREMENTS

PARAMETER	FREQUENCY	METHODS AND NOTES
Discharge	1/run	mid-run; based on cross-section measurements and surface velocity (checked via manometers)
Suspended Sediment	1/run	mid-run; based on a 300 ml sample collected in the upper 1/3 of the water column using standard sampler
Backflood depth	1/run	mid-run; measured at mouth of tributaries
Slackwater sediment thickness	1/run	post-run; measured at proximal, mid, and distal portions of all slackwater sites
Slackwater sediment grain size		post-run; samples at selected slackwater sites (see Table III)
Mainstream slope		pre-run; measured from flume tilt

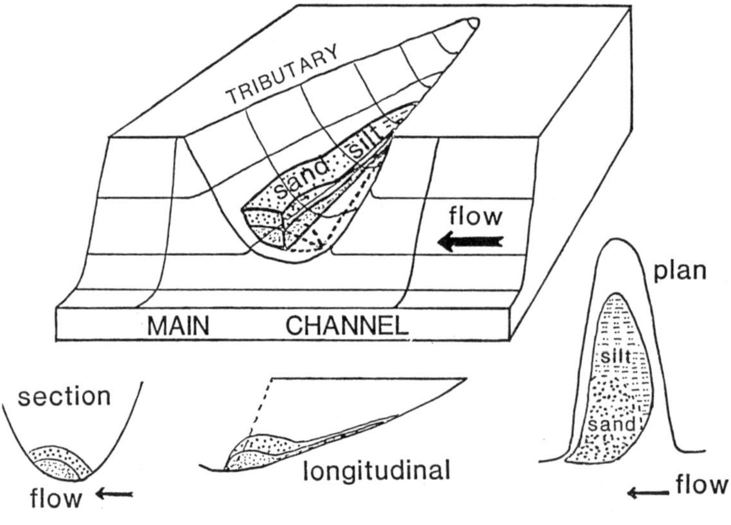

Figure 2. Schematic of a flume tributary mouth slackwater deposits. The sediments generally became finer distally and were concentrated along the downstream edge of the tributary canyons.

Correspondence of Flood Stage and Slackwater Sediment Elevation. The flume experiments provided an opportunity to test the important assumption that slackwater sediment elevation closely corresponds to peak flood stage. Figure 3 shows that during our low discharge runs, the correspondence between flood stage and slackwater sediment deposition at the distal part of the tributaries was good. Near the tributary mouths, deposit elevations grossly underestimated the stage of the flood. However, deposit elevation at the distal "pinch-out" margin closely corresponded to the flood-water surface elevation. The only exceptions were at tributaries 8 and 9. At site 8, the lack of correspondence can be attributed to oversaturated sediments sliding from their depositional position because they could not remain in place on the steep tributary gradient. At tributary 9, the lack of equivalence is probably due to interference by the tributary headwall. As explained earlier, tributary lengths were constrained by the flume width. As a result, backflooding in tributary 9 proceeded to the rear of the tributary and encountered the nearly vertical headwall. This prevented slackwater sediment from achieving its full distal extension and elevation. Thus, except in the cases controlled by our design, slackwater sediment elevations closely matched the stage of the peak flood in the main channel, thereby, lending support to the common assumption made in field applications of the slackwater technique. Similar close correspondence between the sediment and water elevations occurred for Run 3, low discharge at high slope (Fig. 5).

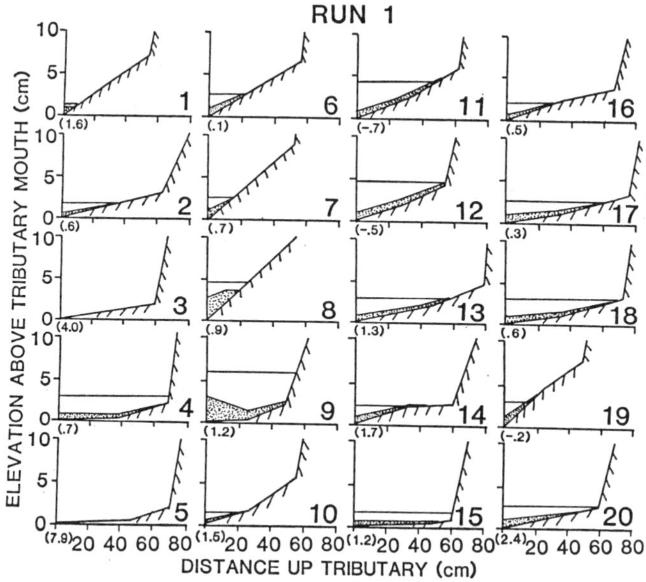

Figure 3. Results of Run 1, done at low flow and low slope. Cross-sections at each tributary show the depth of the backflooding, tributary profiles, extent of slackwater sediment deposition (dotted area), and elevation of tributary mouth with respect to the main channel (numbers in parentheses, in cm).

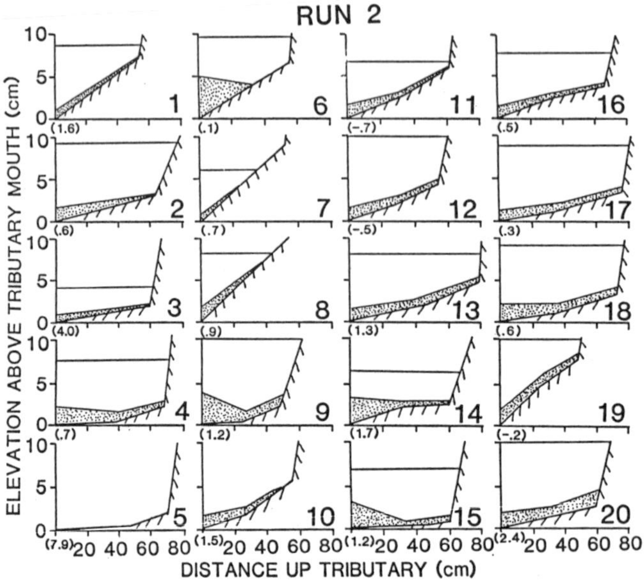

Figure 4. Results of Run 2, done at high flow and low slope. See Figure 3 for explanation.

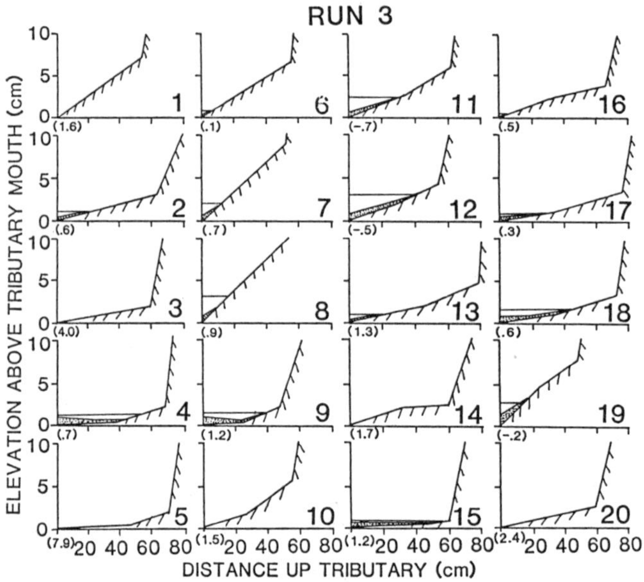

Figure 5. Results of Run 3, done at low flow and high slope. See Figure 3 for explanation.

Correspondence between peak flood stage and slackwater sediment elevation in the tributaries was less convincing in Runs 2 and 4 done at high discharge. During high discharge runs, slackwater sediments accumulated all the way to the nearly-vertical headwalls of the tributaries. We feel that the lack of equivalence to peak stage in these cases was primarily an artifact of design in that distal extension of deposition was prohibited by the vertical headwalls. In addition,

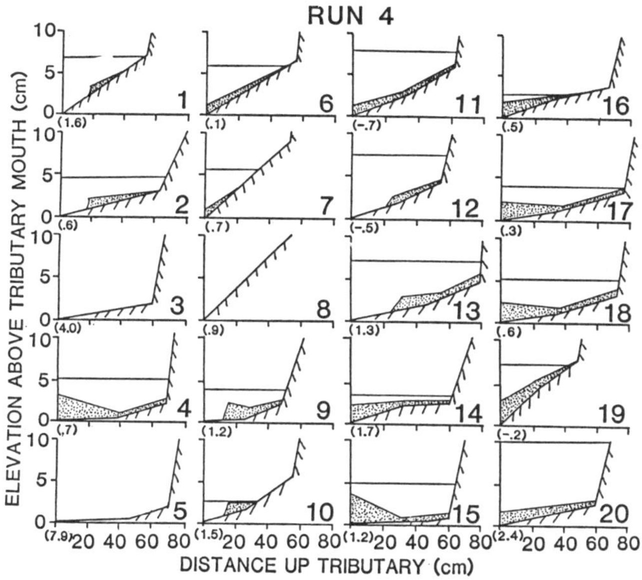

Figure 6. Results of Run 4, done at high flow and high slope. See Figure 3 for explanation.

backwash became a factor because the tributaries were inundated well above the level of the floors at their distal ends. Support for this belief is the relatively thick deposits of slackwater sediment in the distal ends of many of the tributaries. In fact, a reversal of the normal distal thinning trend can be observed at several sites (Fig. 4, sites 15, 20). Figure 7 shows an example of the deep flooding of distal tributary regions that occurred during this run.

Figure 7. Photo showing extensive backflooding of tributaries that occurred during high flow. Flow depths at the tributary headwalls typically exceeded 3 cm. View is downstream showing tributaries 13, 6, 11, and 12, clock-wise from the top. Scale is 10 cm.

Controls of Slackwater Sediment Thickness

Reconstruction of paleoflood histories over extended periods requires the presence of continuous stratigraphic sequences of slackwater sediments. These sequences are found in the field where slackwater sedimentation occurs during major floods and where a depositional site is shielded from erosion by subsequent mainstream and tributary flows. Field studies (Kochel and Baker 1982; Baker et al., 1983) suggest that important factors affecting the potential for the formation and preservation of slackwater deposits include: 1) the geometry of tributary junction angles; 2) the tributary gradient; 3) the mainstream gradient; 4) morphometry of the tributary basin; 5) base elevation of the tributary above the mainstream; and 6) the location of the accumulation site with respect to the mainstream and tributary morphology and flow characteristics. The thickness of individual slackwater sedimentation units and the cumulative pile of slackwater sediments varies significantly along a river like the Pecos due to the above factors.

Variations in cumulative or single-flood slackwater sediment thickness between depositional sites is probably due to differences in: 1) cross-sectional geometry of the mainstream and tributary channels; 2) geometry of junction angles; 3) mainstream and tributary gradients; and 4) base elevation of the tributary above the mainstream. Our flume experiments were not designed to test for the effects of varying mainstream and tributary cross-sections. However, cross-sectional area certainly affects the stage of peak floods for a given discharge and affects the velocity of tributary backflooding. In general, we expect that slackwater sediments become increasingly thick as tributary canyon widths decrease and as slackwater stage increases.

Effect of Tributary Junction Angle. Backflooding must occur efficiently for a significant thickness of slackwater sediment to accumulate in tributary valleys. Field observations along the Pecos and Devils Rivers in west Texas (Kochel 1980) indicate that the extent of backflooding, and hence, the thickness of slackwater sedimentation during a flood is affected by the geometry of the mainstream and tributary, i.e., the junction angle. A suite of eight flume tributaries (Fig. 1 tributaries 11-17), were designed specifically to determine the effect of junction angle on slackwater sedimentation.

Maximum backflood sedimentation occurred when tributary junction angles were between 50° and 130° to the mainstream (see Fig. 4, tributaries 4, 6, 9, 11, 12, and 14-18, for example). Mainstream floodwaters tend to bypass tributaries when junction angles became less than 45°. In contrast, where tributaries joined the mainstream at angles greater than 130°, mainstream water flowed up the tributary channels at high velocity. Significant slackwater sediment may be deposited during these surges up the obtuse-angled tributaries, but the total accumulation is seldom very thick because pre-existing slackwater sediments are eroded during these floods. Therefore, continuous stratigraphies are rare in obtuse-angled sites, but the deposits can be useful for documenting the most recent slackwater floods. Field investigations of a site along the Pecos River with a similar geometry documented complete flushing of the site by the most recent major flood on the Pecos River in 1974 (Kochel 1980), while neighboring tributaries entering the Pecos River at acute angles contained slackwater sediment sequences reflecting 4 to 6 floods that occurred over the last 2,000 years.

Horizontal lamination is the dominant sedimentary structure observed in slackwater sediments in the field. Bedforms are rarely reported, probably because the surfaces of the fine-grained sequences are rapidly reworked by aeolian processes and bioturbation. We do know, however, that bimodal cross-bedding of various scales is preserved with slackwater sediments (Kochel 1980, Kochel et al., 1982; Baker et al., 1983). Most cross-beds indicate paleoflow directions up the tributary canyon away from the mainstream. Cross-beds indicating down-canyon flow are generally smaller than up-canyon flow indicators and are most common in the upper part of slackwater sedimentation units.

In our flume studies, rippled surfaces were very common on the slackwater sediments (Fig. 8), particularly in Runs 2 and 4 done at high discharge. Wavelength and height of ripples varied directly with discharge and particularly velocity. Therefore, wavelength of the ripples, called ripple chord, can be used as an index of the relative velocity of tributary backflooding in the flume. Table III shows the ripple chords from Run 2 (high discharge, low slope). In most cases, the ripple chord decreased away from the mainstream. This would be expected as backflooding velocities must diminish rapidly up the tributary channels where slackwater sediment is being deposited quickly from suspension.

Figure 8. Ripples formed on slackwater sediment surfaces in the flume experiments. These bedforms were common high flow runs. Scale is 10 cm. The main channel is to the right.

Ripple chord also varied between sites during the same flood (Table III). Smaller ripple chord, indicating lower backflooding velocity, occurred in tributaries with particularly acute junction angles. Therefore, ripple chord variations documented the effect that junction angle had upon backflooding efficiency. Considerable variation in thickness of slackwater sedimentation during a given flood event occurred in tributaries with variable junction angles. Figure 9 summarizes these relationships and shows similar trends from the Pecos River.

Effect of Mainstream Gradient. Slackwater sediments were usually thin in flume tributaries where the mainstream gradient was oversteepened (Figs. 3-6, tributaries 7, 10), even during high discharges. In these reaches, backflooding into tributaries was limited because of the continuity requirements for discharge. At constant discharge (i.e., during a given run), the effect of an increased slope is to increase velocity while simultaneously decreasing the depth. Therefore, mainstream flow quickly passed the potential tributary mouth slackwater sites, leaving them dry or invading them with a much-reduced stage compared to other lower-gradient reaches of the mainstream. The same phenomenon occurred when at equal discharge, the entire flume was set at a higher slope (compare Runs 1 and 3 or Runs 2 and 4). For the same discharge, slackwater sediments were much thinner in tributary mouths when the flume (mainstream) slope was higher. During high slope experiments, backflooding failed to occur at many sites that were previously inundated by the same discharge conditions during the low slope.

TABLE III. RIPPLE CHORDS ON SLACKWATER SEDIMENT SURFACES

TRIBUTARY	RIPPLE CHORD (mm)			RUN & CONDITIONS
	Proximal	Mid	Distal	
2	--	9	4	Run 1
9	--	18	11	Low Discharge
12	18	--	10	Low Slope
13	23	--	--	
18	--	10	--	
20	10	6	--	
3	--	25	--	
6	--	25	20	Run 2
9	--	23	18	
10	--	15	15	High Discharge
11	40	27	17	
12	--	20	18	Low Slope
13	40	29	20	
14	--	10	7	
15	--	10	9	
16	45	28	12	
17	60	25	20	
18	80	30	10	

A similar relationship between slackwater sediment thickness and mainstream gradient can be seen in the field data from the Pecos River (Fig. 10). Therefore, tributaries joining a high gradient reach of a trunk river would probably not preserve a continuous record of slackwater sedimentation, and would be unlikely targets for paleohydrologic investigations.

Effect of Tributary Morphology and Morphometry. There is little doubt that it is better to analyze the effect of tributary valley morphology on slackwater deposition in the natural field situation, primarily because there is no way to ascertain tributary response to precipitation in a flume. Nonetheless, our flume studies do support some of the general perceptions about the relationship between tributary morphology and slackwater sedimentation. For example, tributary gradient probably affects the preservation potential of slackwater sediments in several ways. First, our flume observations indicate that it is exceedingly

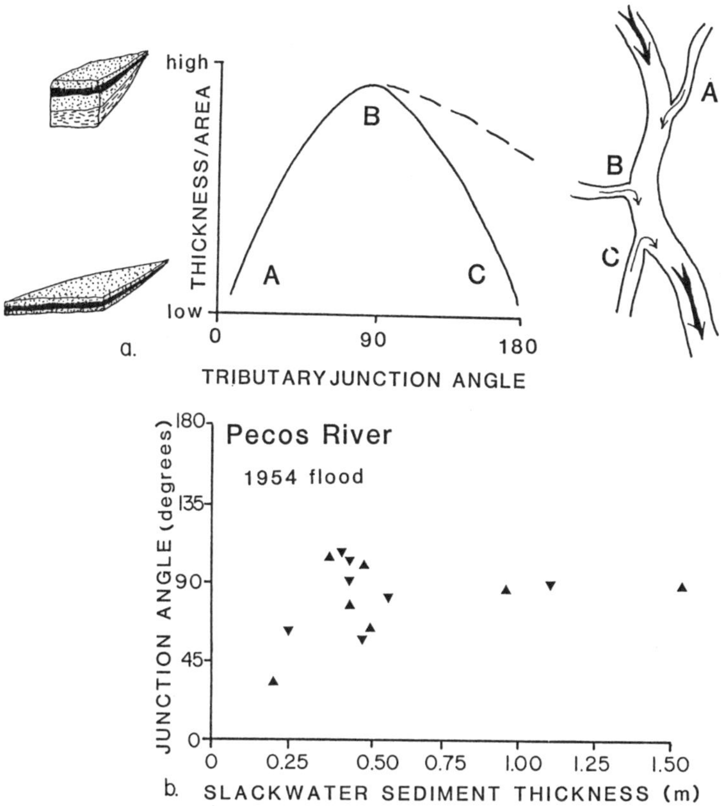

Figure 9. Effect of junction angle on slackwater sedimentation. a) Schematic showing effect of junction angle. b) Field relationships from the Pecos River slackwater sediments showing increased slackwater sediment thickness at tributary sites whose junction angles are near 90°.

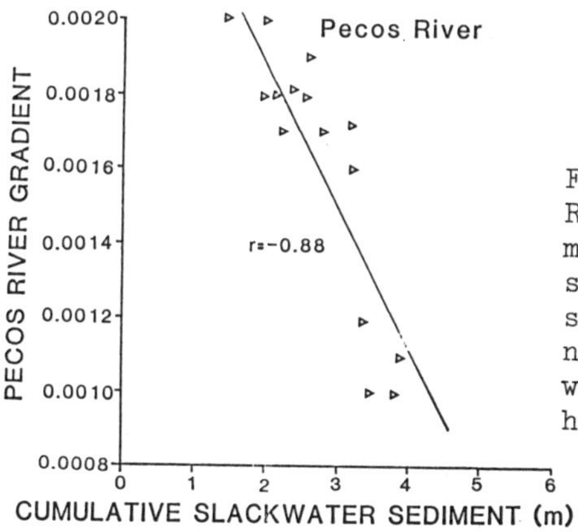

Figure 10. Field relationships from the Pecos River illustrating the effect of mainstream gradient on slackwater sediment thickness in tributary sites. Cumulative sediment thicknesses are significantly thinner where mainstream gradient is higher.

difficult to transport significant amounts of slackwater sediment up high-gradient tributaries (see for example Fig. 3, tributaries 7, 8). Second, backflood flows returning to the mainstream on steep tributary slopes are more likely to wash potential slackwater sediment back into the main channel before it is deposited. In natural tributary basins, gradient is also a factor. A steep slope in such tributaries is prohibitive to the preservation of continuous slackwater sediment sequences because high tributary flow velocities are likely to erode the sequence between episodes of slackwater sedimentation.

The morphometry of natural tributary basins probably exerts little control on the potential for slackwater sediment accumulation in their mouths. However, it may have significant influence on the long-term preservation of continuous stratigraphic sequences. Tributary basins with flashy flood hydrographs are likely to flush slackwater sediment accumulations from their mouths during periodic tributary floods of local nature, i.e., intense morphometry likely to produce large peak floods includes high relief, high drainage density, high gradient, and high first-order stream frequency. These characteristics, combined with the sparse vegetation, steep slopes, thin soils, and catastrophic rainfall regime of west Texas, provide conditions likely to result in frequent flushing of slackwater sequences. Neighboring tributary mouth slackwater sequences from the lower Pecos River sometimes show dramatically different stratigraphies. For example, one canyon contained a slackwater sequence recording several floods extending over 1100 years. The neighboring tributary contained several units deposited in the last 40 years (Kochel et al., 1982). Older slackwater sediments had been flushed from the second site within the last 40 years. This underscores the necessity for using multiple sites in any investigation of paleoflood history of a river.

In most cases, our field studies show that the thickness of slackwater sediments from a given flood increase with paleoflood stage or discharge (Fig. 11). However, the rate of thickness increase is not constant because it is probably influenced by the local cross-sectional area of the mainstream and tributary. Figure 11 shows that thickness becomes greater as channels become narrower.

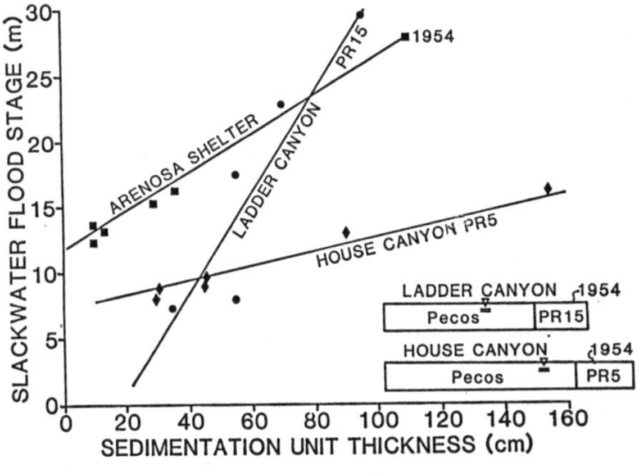

Figure 11.
Relationship between slackwater sedimentation unit thickness and associated paleoflood stage for three sites along the lower Pecos River. Inset sketches show the relative cross-sections of the mainstream and tributary for two sites. Arenosa is a cave site. The slope of the lines appears to be affected by the geometry of the flooded cross-sections.

Effect of Tributary Base Level. Along most rivers, tributaries are graded smoothly to their local mainstream base level. This is not always the case for bedrock channels. Tributaries sometimes enter the main channels as hanging valleys or in a reach of oversteepened gradient along their lower few hundred meters. Therefore, the level of tributary valleys above the mainstream channel may become an important factor in controlling the thickness of slackwater sedimentation during mainstream floods. Assuming that a lower sediment concentration exists in the upper part of a flow column, one would expect deposits to be thinner where tributary channels stand higher above the mainstream bed. Such a relationship is apparent between sites of various base level elevation in the field for cumulative slackwater sediment thickness and for thickness of single-flood units (Fig. 12). Similar trends can be seen when comparing the thickness of slackwater sediments during a

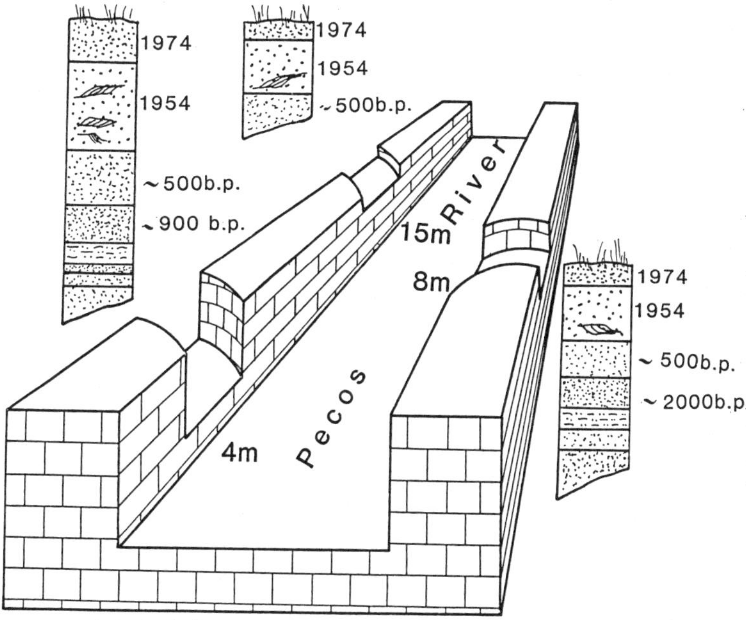

Figure 12. Schematic showing how thickness of slackwater is affected by elevation of the tributary above the mainstream along the Pecos River. Deposits from a given flood become thinner with increasing elevation above the mainstream.

given run in the flume experiments. In Run 2, for example, 2-3 cm of sediment accumulated in tributaries 14-18 whose base elevations were less than 1 cm above the mainstream, while less than 0.5 cm of sediment was deposited in tributary 3 where the base elevation was 5 cm above the mainstream (Fig. 4).

The recognition of slackwater sediments in tributary mouth sites standing at various elevations above their mainstream base level can be extremely useful in bracketing the actual peak paleoflood stage. If the range of elevation is large enough, and if enough sites are available for study, the flood stage is assumed to be the level between where deposition occurs at a lower site but fails to occur at a slightly higher site.

Controls of Grain Size of Slackwater Sediments

Field Observations. Grain size of slackwater sediment appears to be more complexly controlled than sediment thickness. Mean slackwater sediment size is partially a function of the sediment heredity, i.e., it depends upon the lithology and geomorphic processes upstream from the depositional site. The size and volume of sediment carried by a river during a flood is likely to vary with the lithology of the basin and the climate of the region. Grain size of known slackwater sediment varies from gravel in the Pleistocene Channeled Scablands deposits (Baker 1973) to silt in Mississippi River slackwater deposits of southern Illinois. In contrast, some deposits like those of the Scablands floods in eastern Washington, have the size of slackwater sediment controlled by competence, because backflood velocities are unable to transport the largest particles suspended in the mainstream flow.

Slackwater sediments along the Pecos River range from coarse silt to coarse sand. Between-flood variability in grain size in the Pecos River deposits is typically minor. Most differences are probably accounted for by: 1) distance of the depositional site from the mainstream; 2) changes in competence of backflooding resulting from diverse tributary channel cross-section and geometry; 3) elevation of older slackwater sediments in the tributary mouth; and 4) variations in the character of the flood hydrograph. On the Pecos River, we found a trend of increasing grain size with increasing paleoflood discharge (Kochel and Baker 1987), however, the textural range was slight, only from very fine sand (3.8Ø) to fine sand (2.9Ø) (Fig. 13).

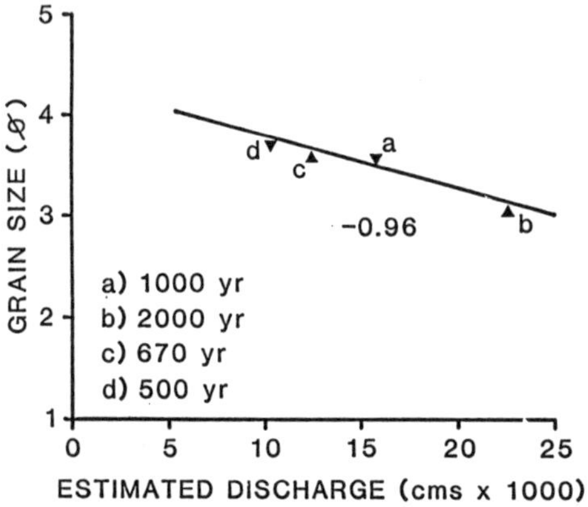

Figure 13. Relationship of paleoflood discharge and grain size of slackwater sediments along the Pecos River. Data is averaged from 15 sites.

Flume Observations. Inspection of the sediment samples extracted from the proximal and distal portions of several slackwater sites in the flume after each of the four runs showed that slackwater sediments became finer with increasing distance from the main channel, but changes were slight (Fig. 14). This is because slackwater sediments are dominantly deposited from suspension; hence, large lateral variations in mean grain size and sorting should not be expected. Figure 15 shows similar trends in the field data along the Pecos River. Relationships between discharge and grain size were inconclusive due to the limitations of our experimental model.

Figure 14.
Photos showing typical slackwater sediments in the proximal sand wedges and ripple trains in more distal silts. Scale is 10 cm. Tributaries 14-16 show sites having junction angles near 90°.

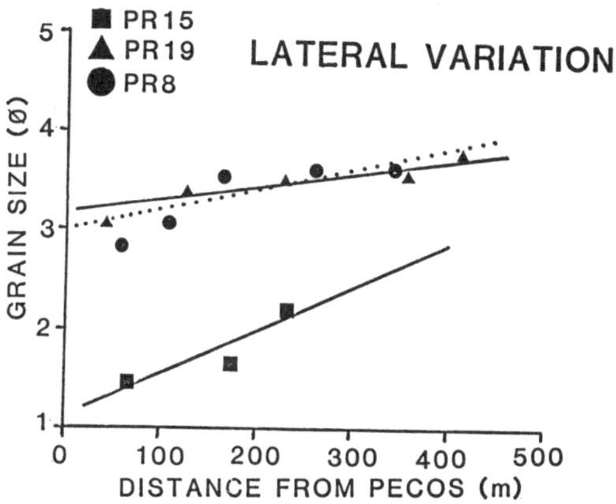

Figure 15.
Distal fining of sediments observed in slackwater sites along the Pecos River.

DISCUSSION

Backflooding Processes

Backflooding occurred during all floods simulated in the flume. The degree of backflooding varied greatly between tributary sites depending upon their junction angle and level above the mainstream, and their location with respect to mainstream channel morphology. Backflooding was most efficient in tributaries with junction angles greater than 50°. Bypassing occurred at tributary sites having more acute junction angles. As expected, slackwater sedimentation was most extensive in the sites where backflooding was best developed. Continuous slackwater sediment sequences were best preserved at junction angles within 40° of normal. More acute junctions prohibited significant slackwater sediment accumulation. Tributaries with junction angles in excess of 130° received slackwater deposits, but had their pre-existing slackwater sediments eroded by high backflooding velocities.

Mainstream gradient played an important role in determining the extent of backflooding and slackwater sedimentation in tributary mouth sites. As mainstream gradient increased at constant discharge, mainstream flow velocity increased as flow depth decreased. Mainstream flow depth is much more important than mainstream velocity in determining whether slackwater sites will be inundated during a flood of a given discharge. Thus, as the stage of tributary mouth backflooding was reduced by raising the mainstream gradient, the thickness and extent of slackwater deposits for a given discharge was reduced in tributary mouth sites. The reduction in backflooding caused by the increased gradient occurred at all tributary sites, regardless of their junction angle geometry. Many sites that were backflooded at low slope during a particular discharge remained dry when mainstream gradient was increased under the same discharge.

Grain size variations between floods of different discharge appeared to be nominal in these experiments, a phenomenon also observed in field studies. This suggests that the range of sizes transported in suspension did not change significantly in our experimental channel over the range of discharges produced. High discharge was only about three times that of the low discharge. However, there was a distinct increase in the volume of coarse grained sediment deposited in slackwater sites during the high discharge runs. This was particularly well demonstrated in the proximal portions of the tributary channels. Grain size decreased rapidly with distance from the main channel as expected and as observed in the field.

The two runs made at high flow yielded additional observations of note. First, during the run with high slope and high flow, considerable scour of the proximal region of tributary mouth slackwater sediments occurred in many sites (Fig. 6, tributaries 2, 9, 10, 12, 13). This occurred because intense turbulent eddies developed where mainstream flows backed into the tributaries and collided with the downstream tributary walls. Little effect was noticed farther upstream in the tributaries.

Second, backflooding waters were not as calm as might have been expected. Instead, a significant seiche-like pulsating flow up and down the tributaries dominated the backflooding process. Some pulsation was observed during low flow, but it was much more evident at high discharges. It is unclear whether this pulsatory flow pattern is normal in backflooding environments, or whether it was an artifact brought on by the flume design because tributary headwalls produced an unnatural backwash. On the other hand, field observations of bimodal cross-bedding in Pecos River slackwater sediments deposited by a single flood might be explained by this phenomenon. In addition one eyewitness of the 1954 flood on the Pecos River described similar rocking motions on a much large scale in a tributary mouth site on his ranch (Ted White, personal communication 1979).

Slackwater Deposits as Paleoflood Stage Indicators

As stated earlier, one of the fundamental assumptions in the slackwater technique is that deposit elevation corresponds closely with the peak flood stage of the mainstream. Our flume studies showed this to be a reasonable assumption. In almost every case where the headwall of the tributary did not impede upvalley flow of backflood water, peak mainstream stage and elevation of the slackwater deposit were the same, within acceptable error limits of less than 10%. This range of error is certainly reasonable when applied to paleoflood estimations for large floods.

Criteria for Field Selection of Slackwater Sites

One of the most significant results of our flume studies and their comparison with field sedimentology may be in the refinement of criteria used to select potential sites for slackwater sediment studies. Previous studies have discussed these criteria (Kochel et al., 1982; Baker et al., 1983; Webb et al., 1987). Most considerations relate to the assumptions necessary to insure nominal changes in channel cross-section; a prerequisite for successful application of slope-area procedures used to estimate flood discharges. These criteria are summarized in Table IV. Table IV also summarizes the additional insights gained from our study which should increase the chances for success in paleoflood studies by making the best site selection along the river in question.

TABLE IV. CRITERIA FOR SUCCESSFUL SITE SELECTION

Criteria Likely to Have Stable Channel Cross-Sections

1. Bedrock channel reaches
2. Reaches where scour and fill during floods is likely to be minimal

Criteria Likely to Yield Thick and Continuous Slackwater Sequence

1. Tributaries with junction angles between 50 - 130
2. Tributaries along low gradient reaches of the mainstream
3. Tributaries along fairly straight reaches of the mainstream
4. Low gradient tributaries
5. Tributaries with basins having morphometries indicative of low flash flood potential.

CONCLUSIONS

Paleoflood studies using slackwater sediment analyses have records of large magnitude flooding over several thousands of years. Records of this length are required to adequately estimate recurrence intervals of major floods and to properly assess the risk associated with events of this kind. Paleoflood studies are especially important for rivers that have highly skewed annual maximum flood distributions such as those described in semi-arid regions (for example, Kochel and Baker 1982; Kochel et al., 1982; Baker et al., 1983; Ely and Baker 1985; Baker et al., 1985; Kochel and Baker 1987).

Our preliminary flume studies have provided an opportunity to observe backflooding processes involved in slackwater sediment deposition. Our results support some of the assumptions inherent in the slackwater method. The flume studies have also increased our understanding of the nature of processes leading to slackwater deposition and have provided valuable insights into the controls on thickness and grain size in these deposits. Finally, analysis of the flume results in combination with field data from the Pecos River permit us to refine our thoughts on developing criteria useful in choosing field sites most likely to lead to successful slackwater paleoflood investigations.

ACKNOWLEDGEMENTS

Research support was provided by the Department of Geology, Southern Illinois University. We thank the following for their assistance in the construction and operation of the flume experiments: Jay piper, David Simmons, Greg Riley, and Bret Robinson. We thank Betty Atwood for word processing. Andy Mason and DeAnn Kirk drafted some figures.

REFERENCES

Baker, V.R. 1973. Paleohydrology and sedimentology of Lake Missoula flooding in eastern Washington. Geological Society of America Special Paper 144, 79 p.

Baker, V.R. and Kochel, R.C. 1987. "Flood sedimentation in bedrock rivers". in: V.R. Baker, R.C. Kochel, and P.C. Patton (Editors), Flood Geomorphology, John Wiley, New York, in press.

Baker, V.R., Kochel, R.C. and Patton, P.C. 1979. "Long-term flood frequency analysis using geological data". International Association Hydrological Science Publication 128, p. 3-9.

Baker, V.R., Pickup, G. and Polach, H.A. 1985. "Radiocarbon dating of flood events, Katherine ?Gorge, Northern Territory, Australia". Geology, vol. 13, p. 344-347.

Ely, L.L. and Baker, V.R. 1985. "Reconstructing paleoflood hydrology with slackwater deposits, Verde River, Arizona". Physical Geography, vol. 5, p. 103-126.

Kochel, R.C. 1980. Interpretation of flood paleohydrology using slackwater deposits, lower Pecos and Devils Rivers, southwest Texas. Ph.D. dissertation, Austin, University of Texas, 360 p.

Kochel, R.C. and Baker, V.R. 1987. "Paleoflood hydrology". Science, vol. 215, p. 353-361.

Kochel, R.C. and Baker, V.R. 1987. "Paleoflood analysis using slackwater deposits". in: Baker, V.R., Kochel, R.C. and Patton, P.C. (Editors), Flood Geomorphology, John Wiley, New York, in press.

Patton, P.C., Baker, V.R., and Kochel, R.C., 1979. "Slackwater deposits: A geomorphic technique for the interpretation of fluvial paleohydrology". in Rhodes, O.D., and Williams, G.P., eds., Adjustments of the Fluvial System, Dubuque, Kendall-Hunt, p. 225-253.

Webb, R.H., O'Connor, J.E. and Baker, V.R. 1987. "Paleohydrologic reconstruction of flood frequency on the Escalante River, south-central Utah". in: Baker, V.R., Kochel, R.C. and Patton, P.C. (Editors), Flood Geomorphology, John Wiley, New York, in press.

SUBJECT INDEX

AUTHOR INDEX